The Microbial Models
of Molecular Biology

The Microbial Models of Molecular Biology

From Genes to Genomes

ROWLAND H. DAVIS

OXFORD

UNIVERSITY PRESS

2003

OXFORD

UNIVERSITY PRESS

Oxford   New York
Auckland Bangkok Buenos Aires Cape Town Chennai
Dar es Salaam Delhi Hong Kong Istanbul Karachi Kolkata
Kuala Lumpur Madrid Melbourne Mexico City Mumbai Nairobi
São Paulo Shanghai Taipei Tokyo Toronto

Copyright © 2003 by Oxford University Press, Inc.

Published by Oxford University Press, Inc.
198 Madison Avenue, New York, New York 10016

www.oup.com

Oxford is a registered trademark of Oxford University Press

Library of Congress Cataloging-in-Publication Data
Davis, Rowland H.
The microbial models of molecular biology : from genes to genomes/
by Rowland H. Davis.
p. ; cm.
Includes bibliographical references and index.
ISBN 0-19-515436-3
1. Molecular biology—History.   2. Molecular genetics—History.   3.
Microbial genetics—History.   4. Molecular microbiology—History.
[DNLM: 1. Genetics, Microbial—history.   2. Models, Biological.   3.
Molecular Biology—history.   4. History of Medicine, 20th Cent.
QH 506 D263m 2003]   I. Title.
QH506 .D395 2003
572.8—dc21   2002152360

The quotation in chapter 10 on pages 125–126 is from pages 270–271 of *The Statue Within* by François Jacob. English translation copyright © 1988 by Basic Books, Inc. Reprinted by permission of Basic Books, a member of Perseus Books, L.L.C.

The quotation in chapter 12 on page 163 is reprinted by permission of the publisher from page 6 of *Of Flies, Mice, and Men: On the Revolution in Modern Biology* by François Jacob, translated by Giselle Weiss. Cambridge, Mass.: Harvard University Press, Copyright © 1998 by President and Fellows of Harvard College.

9 8 7 6 5 4 3 2 1

Printed in the United States of America
on acid-free paper

To Margot

Preface

The Microbial Models of Molecular Biology describes the role of microorganisms in an unusual period in the history of biology. That period, roughly 1925 to the present, saw the rise of genetics, biochemistry, and molecular biology as they contributed to our understanding of some of life's most fundamental attributes. The use of microbes was vital to the success of the enterprise, but no recent general treatment of the period has focused on the way microorganisms were chosen, their particular roles, and how they passed their prime. The time was special, and we are unlikely to see anything like it in the future.

This book explores the introduction of microbial model organisms in the historical context in which they were chosen. Their use allowed geneticists to pursue problems formulated earlier with the fruit fly, humans, corn, and the mouse. The coherence of microbial and molecular genetics becomes clear as we see what model organisms contributed and why some of them were abandoned in favor of others. As the examples accumulate, we discern a general distinction between a model system—an organism or experimental arrangement used for a specific purpose—and a *model organism*. I define the latter as a species that undergoes genetic domestication and intense genetic analysis at all levels of organization. A model organism eventually becomes known so thoroughly through genetic dissection that biologists take it as the key representative of the family, order, or phylum of which it is a member—in some cases even of life itself. While the fruit fly, mouse, and corn were model organisms before the introduction of model microbes, their universally significant contribution had been the elaboration of the chromosome theory of inheritance. The microbes, from the early 1940s onward and particularly after the discovery of the structure of DNA in 1953, opened the door to more fundamental understandings of many features of living systems, and we are only a short way on this quest even now.

The 20th-century research I describe demonstrates the power of a reductionist program in biology. Model microorganisms, in their simplicity, permitted the formulation of quite exact questions about basic features of gene structure and function. But insensibly, molecular biologists drifted toward highly typological thinking, in which highly selected stocks of fungi, bacteria, bacterial viruses, protozoa, and algae were taken as adequate representatives of the living world. The undoubted success of the reductionist approach gave confidence that life processes were basically simple. A faith prevailed that emergent properties and complex interactions could be resolved into straightforward diagrams, without the transcendent, irreducible complexity of life that most biologists had previously emphasized.

The Age of Model Organisms, as I call it, has come to a close with the rise of genomics. With genomic sequencing, the study of the diversity of organisms and the complexity of their development and function has become far more accessible to the modern biologist. Even the first findings in this new era have changed the attitudes of molecular biologists. No longer is a geneticist confined to crossing males and females belonging to the same species to acquire information about an organism; one may compare genes of distantly related species directly. No longer is a functional biologist confined to studying the role of individual genes; he or she may study the expression of all genes at once during embryonic development or after transfer of the organism to a new environment. The several model organisms now in use—many of the microbes I describe, as well as the mouse, the fruit fly *Drosophila*, the weed *Arabidopsis*, the worm *Caenorhabditis elegans*, and, yes, *Homo sapiens*—will continue their important roles as models. But their roles will be more as points of reference than as Platonic representatives. I argue that microbial models were the last to offer (and only briefly) findings of universal relevance in biology. Nevertheless, they have positioned us for our 21st-century, post-genomic explorations of biological variation and complexity. These were the very preoccupations of physiologists, embryologists, and evolutionary biologists that were shouldered aside by reductionist programs at the beginning of the 20th century.

In the first chapters of the book, I describe the origin of genetics and the elaboration of the chromosome theory of heredity by Thomas Hunt Morgan with his model organism, the fruit fly. When this program had reached its technical limits in exploring the structure and function of the gene, Morgan's associates and successors chose simpler systems from among the fungi—*Neurospora*, *Aspergillus*, and the yeast *Saccharomyces*—to explore these matters further. These eukaryotic microorganisms (carrying their genetic endowment in true nuclei) could reveal the role of genes in cell function more easily than the complex, multicellular organisms. The brief reign of fungi was followed by the introduction of prokaryotic microbes (bacteria and bacterial viruses) to explore their novel genetic systems that nevertheless promised answers to many questions posed by the study of more complex forms.

The discovery of DNA in 1953 was a fulcrum of sorts between the inferential, exploratory genetic work of the first half of the century and the understandings emerging since. The second part of the book shows how this discovery legitimized bacteria as models for living things at the level of DNA and its role in inheritance and cell function. The confidence in the rationales of bacterial molecular biology were extended to the study of yeast, which returned to the scene, invigorated by molecular approaches, to become a commanding model of eukaryotic microbes and, in many ways, of all multicellular organisms. Along the way, I discuss the model eukaryotic microbes *Paramecium*, *Chlamydomonas*, and yeast, as they were used to explore the role of the cytoplasm in inheritance. The last chapter of this part of the book summarizes the manner in which *Neurospora* became reestablished as a model organism, but for the more restricted group of filamentous fungi, important to us not only in basic science, but also in agriculture, medicine, and the food and fermentation industries.

The final three chapters concern the role of biochemistry in the development of the field of molecular biology, the advent of genomic analysis, and the current and likely future place of model organisms. In the final chapter I argue that the use of microbes has been a bridge, sturdy in its use, but soon to become a monument, between the dimly perceived mechanics of living things as seen in the late 1800s and the growing awareness of how complex life really is as we enter the 21st century. Owing to the contributions of the model microbes of the 20th century, biologists are prepared to tackle biology with tools unimaginable even 20 years ago.

Many informed readers will feel that some species that qualify for model organism status have been left out. Among these are the prokaryotes *Bacillus subtilis*, *Caulobacter*, *Haemophilus influenzae*, *Azotobacter*, *Streptomyces*, *Mycobacterium*, and the phages Mu, ΦX174. Qβ, and f2. Eukaryotic forms include *Physarum*, *Dictyostelium*, *Schizosaccharomyces pombe*, *Podospora anserina*, *Ustilago maydis*, *Coprinus cinereus*, and *Schizophyllum commune*. I respect the view that I have neglected some important organisms. I have held the same view toward many histories of the era that have neglected the fungi and biochemical genetics. However, my choices represent to my mind those that best carry a continuous narrative of discovery in the 20th century. Even in the discussions of organisms included, I have omitted many highly significant findings and their discoverers, omissions I regret. I have made choices that make this book brief but informative for scientists in other fields as well as for those already familiar with microbial and molecular biology. I have included brief appendixes on elementary genetics, molecular biology, and the rationales of genomics to satisfy the needs and curiosity of those unfamiliar with these subjects.

Although I am a biochemical and molecular biologist, I have read much of the historiographic commentary about 20th-century molecular biology. I mention these writings as I go along, but I claim no special authority based on training in historical methods or in the use of interviews and archival sources. The book focuses on organisms as much as on the scientists who studied them. Because scientists have been overly interviewed, and because their views and memories, like mine, are fallible, I have concentrated on the published scientific record as adumbrated and corrected by other commentators. In the process, I have blown on the embers of my own past to revive my sense of biology in the mid-20th century. As implied above, I have indulged a special affection for the fungi and for biochemistry, often forgotten by historians who have focused so narrowly on bacteria, bacteriophages, and DNA. I have sought thereby to bring a fresh and broader perspective to this period than one finds in available histories.

Two sorts of acknowledgment are appropriate in a book that covers ground the author has experienced as well as one he has studied in retrospect. The first thanks go to the people who shaped my outlook on biology, both in breadth and orientation. My undergraduate and graduate education in biology at Harvard University (1950–58) brought me into contact with three important teachers. Paul Levine, my graduate advisor, despite his using *Drosophila* and *Chlamydomonas* as experimental organisms, nevertheless fostered my earliest work on *Neurospora*. John (Red) Raper, an equally important graduate advisor, greatly deepened my knowledge of fungi, especially the Basidiomycetes. Ernst Mayr, a member, like Paul and Red, of my dissertation committee, gave me from the start a feel for the modern synthesis of genetics and evolution. An early course in graduate school, taught by the then-visiting Boris Ephrussi, was crucial in shaping my fascination with microorganisms. Even as undergraduates, Nicholas Gillham and I became friends; he would have a distinguished career in the study of organelle inheritance in *Chlamydomonas*. Many of the scientists I discuss, including Watson, Pontecorvo, Hershey, and Demerec, came to or passed through Harvard during my years there and contributed to my current, still vivid sense of those times. I did my postdoctoral work with Herschel Mitchell and, more informally, with Norman Horowitz and Sterling Emerson at Caltech between 1958 and 1960, sharing a laboratory for a year with Franklin Harold. Herschel gave me great autonomy as I began work in the biochemical genetics of *Neurospora* that I continued for the rest of my research career. While at Caltech, and during several of the earlier summers I spent at Woods Hole, many other persons I discuss in the book, including George Beadle, Alfred Sturtevant, Edward Lewis, Max Delbrück, Robert Edgar, Harris Bernstein, Mary Mitchell, Jacques Monod, Matthew Meselson, Ruth Sager, and Franz Moewus, were present or passed by as I watched.

After joining the faculty of the University of Michigan in 1960, an informal journal club formed, with Myron Levine and Robert Greenberg, devoted to keeping up with the emerging field of molecular biology, in which the university was conspicuously behind the times. That group was joined in the following years by Charles Radding, Hamilton Smith, David Botstein, and a number of other gifted scientists from whom I learned a great deal outside of my particular work in biochemical genetics. In several periods of service on National Science Foundation and National Institutes of Health grant-review panels, I had the fortune to discuss both the history and the future of the new genetics with many leaders of the field. I need not list them here: they, too, are represented in the following pages. At the University of California, Irvine, after 1975 I continued to learn more biochemistry and molecular biology as my work led me further and further in those directions. The special periods as a sabbatical visitor to the laboratories of David Stadler (1968) and Charles Yanofsky (1985) deserve mention; I got to know Herschel Roman during the first, and it was during the latter that I became certified as a molecular biologist.

A second sort of acknowledgment is more direct. I deeply thank reviewers of parts or all of the manuscript as it evolved. Oxford University Press prevailed upon Dr. Franklin Stahl, Dr. Jan Sapp, and a third, still anonymous reviewer to review the entire first draft of the book. Their comments, together with those of reviewers of the early outline, have been invaluable in shaping the final version of the book. Their efforts were solicited by my editor at Oxford, Kirk Jensen, whom I also thank warmly. Other reviewers kindly consented to my own requests to review individual chapters. These include Robert Metzenberg and Norman Horowitz (chapters 3 and 6); N. Ronald Morris (chapter 4 and 6); John Pringle (chapters 5 and 15); Joshua Lederberg (chapters 7 and 9); Stanley Maloy (chapters 8, 9, 11, and 12); Roy Britten (chapter 11); John Preer and Sally Allen (chapter 13); Nicholas Gillham and Anthony Griffiths (chapter 14); and Walter Fitch (chapter 18). The errors of fact, emphasis, or understanding that remain are entirely my own.

I especially thank Karla Pace, the copy editor, and Lisa Stallings of the production staff of Oxford University Press.

Unattributed epigraphs are my own.

Finally, I have greatly enjoyed writing this book, so I can hardly thank those close to me for their forbearance as I struggled with the prose. Instead, I thank my wife Margot in particular for her encouragement to write as well as I could, and for her tolerance of my preoccupation with seeing this work through to a happy end.

Contents

The Microbial Models
of Molecular Biology

1

Model Systems, Model Organisms

For all their ubiquity and familiarity, living organisms are truly strange objects.

—F. M. Harold

Making Sense of Creation

As long as we humans have sought our place in the universe, with all its regularity, sustenance, and danger, living organisms have fascinated us. Until recently we saw the earth as the center of creation, under an obliging sun. And on this earth living organisms swarmed, with special qualities and a spectacular diversity that we increasingly appreciate. During the millennium just ended, biologists explored the most intimate details of living organisms with increasing discipline. We have learned more about biology in the last 100 years than in all previous history. But in the course of the 20th century, we have lost much of our prehistoric awe and respect for our biological companions.

Natural philosophers and biologists of the past sought to make sense of the many types of plants and creatures that, if not made for humans, were at least created by imaginative gods. As the Renaissance yielded enlightenment, the classification of organisms and a study of their structures embodied a new attempt to subdue the mystery of life. The effort was satisfying in confirming the systematic intentions of the designer. But as the last millennium progressed, the urge simply to recognize and sort plants and animals gave way to an ambition to understand their origins, their functions, and their diversity. Detached from constraints of authority by the examples of Galileo and Newton, the early explorers of the living domain began to ask empirical questions buried for centuries, latent in the human imagination.

Our subject here is the choice and use of organisms in answering these questions. The organisms now living on earth are so diverse that one must ask how biologists made such choices. Let us look at the organisms that brought biology to the beginning of the 20th century, when particular models began to guide experimental research and for a time limited our appreciation of organismic diversity. In doing so, we may see how very recent our understanding of living things really is, and why we saw such an acceleration of biological research in the 20th century.[1]

Life processes, even the definition of life, preoccupied many natural philosophers of the Renaissance. Pope said that the proper study of mankind is man, and indeed, biological interests had focused early on him (yes, him) from antiquity. Ancient speculations were passed on for centuries in unchallenged texts. It was only in the 16th century that Andreas Vesalius (1514–64) deconstructed human bodies for our first accurately recorded anatomy. But even then, natural philosophers realized that humans would not easily serve as objects of analysis and biological experimentation. Having some taste for generalization, they looked to other organisms for insights into the human condition. Many historians mark the proposal of William Harvey (1578–1657) that the heart was a pump and that blood must circulate as a milestone in the scientific understanding of biological processes. Harvey's curiosity arose upon recognizing valves in human and animal veins that enforced flow in one direction. Marcello Malpighi (1628–94) was the first to announce the existence of capillaries in the tail of a tadpole, vessels that connected the arterial and venous systems, after the introduction of the microscope by Antonie van Leeuwenhoek (1632–1723). In his long career Leeuwenhoek would use a multitude of organisms and other material for his microscopic investigations. He showed, among other things, that the semen of all animals contained sperm cells, allowing later investigators to assign to sperm a role in fertilization. In fact, Harvey had worked on the origin of the embryo, and showed, in the deer on the grounds of Charles I's palace, that embryos were not as well formed as the fawn that would later be born. This became evidence against the preformationist theories of the origin of animals and supported what became an epigenetic view of development. Curiously, Harvey never saw a mammalian ovum, and of course he did not know how it was "activated" by semen. Nevertheless, Harvey could infer that the egg, obvious in other animals, was the origin of the organism, as he intoned in his famous phrase, *omne vivum ex ovo*.

The pattern of using experimental organisms continued, gathering strength. Johannes van Helmont (1579–1644) studied the growth of a willow tree in weighed amounts of soil after watering it with known amounts of water. He concluded, logically but in error, that water was the source of the organic substance of the tree. Francisco Redi (1626–97) acquired evidence against spontaneous generation by showing that maggots would not appear in putrefying meat if gauze impeded access by flies. René de Réaumur (1683–1757) used hawks to show that food held in a small, incompressible cage before being swallowed was nevertheless digested. This contradicted the prevalent theory that the stomach simply ground up the food. Lazzaro Spallanzani (1729–99)

used a vast repertory of animals, including bats and sponges, in his varied researches on physiology, spontaneous generation, and sexual reproduction.
Spallanzani was an early exponent of animal and plant models for these researches, choosing optimal material within his ken for each investigation.

In the early 19th century, German romanticism emerged with the idea of
the unity of plant and animal life. The peculiar origin and organic functions of
living things seemed incompatible philosophically with the mechanical views
of René Descartes (1596–1650). Johann Goethe (1749–1832) popularized the
idea that life continually evolved in a dynamic and mysterious way. But owing
to the underlying unity posited, many of life's patterns and processes could be
understood by the study of a single individual or species. This notion gained
support from generalizations based on wide knowledge of many organisms.
The cell theory, which held that all organisms were cells or cell aggregates,
was one of the most important of these generalizations. Teodor Schwann (1810–
82) proposed this formally in 1839 on the basis of his and Matthias Schleiden's
(1804–81) earlier investigations on many plant and animal cells and tissues.
The theory recognized that cell types as disparate as sperm, the mammalian
ovum (discovered only in 1827 in dogs), and protozoa all shared fundamental
structural attributes. The idea of the origin of cells by cell division was left to
Rudolf Virchow (1821–1902), working with pathological human tissues; his
dictum *omnis cellula e cellula* was pronounced in 1855.

The dialectic use of particular organisms for individual investigations and
the exploration of diverse organisms that justified generalizations about biological phenomena set the stage for the greatest generalization of all: the theory
of evolution by natural selection. In the middle of the 19th century, Charles
Darwin (1809–82) proposed how living things diversified and in the process
partially demystified the living world and the place of humans in it. The taste
for generalization and unification in biology led naturally to the fusion, in
biologists' minds, of heredity, embryology, and evolution, all special properties of organisms as they perpetuated their kind in nature.

As biologists focused on embryology, a wide variety of organisms was used.
These included the chicken, in which germ layers were first named; the frog,
which Wilhelm Roux (1850–1924) used to study the intrinsic and extrinsic
factors governing embryogenesis; and the sea urchin, used by Theodor Boveri
(1862–1915), Hans Driesch (1867–1941), and Hans Spemann (1869–1941) for
seminal studies of early development. These organisms were both available to
the investigators and clearly suitable for investigations that they initiated. Their
model status is certified by their use to the present day and by the application
of the principles formulated with them to many other animals.

Why were the organisms in the early days of experimental biology chosen?[2]
In fact, biologists rarely chose them as deliberately as we might think. The
natural world, close at hand, provided everyday organisms such as rats, chick-
ens, frogs, dogs, and fish that sparked the natural interests of curious observers. Thus common organisms found themselves from the outset in the limelight
of experiment. As life processes became domesticated in laboratories, a need
to optimize observations led to choices of more suitable material. Certainly

during the development of physiology and biochemistry in the late 19th and early 20th centuries, changing from one organism to another to surmount technical limitations presented few problems. In fact, the desire to generalize observations impelled biologists to use a variety of organisms in many experimental regimes. This tended initially to mute the search for a model. Most biologists did not perceive the need, even if such a need existed, to standardize stocks of their experimental organisms. The laboratory rat was caught or bought from suppliers having little interest in or record of pedigrees.[3] Marine biologists collected sea urchins and other organisms from the shore for work in embryology and physiology. The inability to repeat experiments in the literature could be ascribed as often to differences in the material as to different experimental conditions, incompetence, delusion, or fraud, all of which had their place in the polemics of the time. But the choices and changes of organism gradually improved our sense of the workings of living things, usually focusing on one attribute at a time. I refer to organisms widely used for particular research projects as *model systems*.

Descriptive work on plants and animals continued in the academy and at the amateur level. The turn of the century, during which experimental biology flourished, also saw vast accumulations of organisms in museums and the emergence of phylogenetic trees, following Darwin's example. (Many taxonomists of plants and animals, owing to the similarity of the classical Linnaean and the more modern Darwinian phylogenies, failed for some time to respond adequately to the implications of variation and evolution.) Bird-watching, botanizing, microscopic investigations of pond water, and mushroom forays occupied professional and amateur biologists alike. But generalizations based on observation and experiment, broad as they were, could not be applied as widely as one would think. Plants, for example, remained remote from the front lines of embryology and thus did not immediately challenge ideas about the sanctity of a sequestered set of cells—the "germ line"—emerging from studies of animal development. Fungal biology and evolution remained distant from the genetics of higher diploid organisms. Bacteria could not be studied adequately even in the early 20th century: their very classification as true living things was in doubt. Many of these areas would be illuminated only by genetic research in the later 1900s.

Model Systems for Genetics

Mendel

In 1865, a few years after the appearance of Darwin's *Origin of Species* (1859), Gregor Mendel (1822–84; fig. 1.1) proposed a quite novel conception of heredity.[4] Its significance was not recognized until the first years of the 20th century. Only at that time were the phenomena of heredity adequately detached from the study of embryology and evolution for scientists to define questions of inheritance as Mendel had. Standing alone, genetics then began to play a major role in modern biology.

Fig. 1.1. Gregor Mendel (*left*) and Thomas Hunt Morgan (*right*). Courtesy of the Archives, California Institute of Technology.

Mendel, an Augustinian monk and eventually abbot of a monastery in Brno (once known as Brünn, now in the Czech Republic), had an interest in science, had taught mathematics shortly after taking the orders of his church, and studied physics thereafter. The origin of his interest in heredity arose from his study of plants. He chose to work on heredity in the garden pea because it was easy to grow, its progeny were numerous, and a number of agriculturally desirable, inbred lines with distinct hereditary differences were available to him in the region surrounding Brno. Mendel's findings, published in 1866, were an anomaly for their time, owing to the novel mode of analysis of a fundamental question. We have increasing difficulty at this late date understanding how the most basic rules of inheritance could have escaped discovery until 1865. Breeding of agriculturally important animals and plants might have revealed the rules of inheritance of visible traits to anyone that carried pedigrees beyond one or two generations. Yet the discovery awaited someone who might perceive the formal implication of progeny ratios and the use of an organism with sufficiently large progenies that the regularity of those ratios could be appreciated. Much has been written about Mendel, what he brought to his task, whether he "cheated" in his counting of peas, and just how much he could infer about the contributions of parents to offspring.[5] I do not summarize these peripheral matters here, but simply describe Mendel's basic findings as an introduction to what will follow (see also appendix 1).

Mendel had at hand highly inbred, and thus genetically pure, varieties of peas. They differed in traits such as the color of the peas (yellow or green), height, and texture (smooth or wrinkled). The inbred character of the lines he started with stemmed from the fact that the flowers of pea plants have both female and male elements (carpels, with ovules; anthers, bearing pollen), and self-fertilization occurs even before the flower opens. Mendel cross-bred different varieties by opening flowers before they matured, removing the anthers, and fertilizing the stigma (the tip of the carpel) with pollen from another variety of peas.

In a cross between pure-breeding, parental plants having green and yellow peas, he found that the peas developing from the artificially fertilized flowers were all yellow. On first blush, this suggested that the green trait had been lost. Mendel, however, grew the peas from this first generation, which he called the F_1, into mature plants, and allowed them to self-fertilize in the normal fashion. The peas that then formed on the F_1 plants, called the F_2 progeny were a mixture of yellow and green. Mendel's novel contribution was to make sense not only of the reappearance, in undiluted form, of the green trait, but of the ratio of green to yellow peas. Regarding the disappearance and reappearance of the green trait, Mendel concluded that its reappearance showed that the character had not been lost, but only masked in the F_1. (It would be some time, perhaps 40 years, before this matter was clarified by distinguishing the "determinant," the gene, from the trait itself.) By systematic counting, Mendel showed repeatedly that such traits, masked in the F_1, appeared in only one-quarter of the F_2 progeny. They *segregated*, as he said, in this second generation.

We now interpret this phenomenon as follows (fig. 1.2): the green determinant is *recessive*, susceptible to being masked by the *dominant*, yellow determinant. Second, each plant has two determinants for each character (in the language of genetics, the plants are *diploid*). However, each parent makes ga-

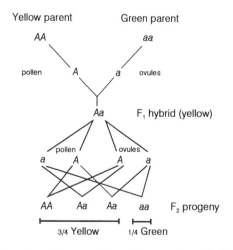

Fig. 1.2.   Cross of pure-breeding pea plants. These strains, if self-fertilized, will bear uniformly yellow or green peas generation after generation. In cross-fertilization, however, their gametes (pollen and ovules) combine to form yellow peas on the female parent; these peas are the beginning of the F_1 hybrid generation. Adult F_1 plants self-fertilize, and bear both yellow and green peas (the F_2 generation) in a 3:1 ratio. Self fertilization of the F_1 plants is accomplished as shown by random combinations of the two types of pollen with the two types of ovules. In this cross, the yellow determinant (A) is dominant over the green determinant (a) such that Aa diploid plants are yellow.

metes, pollen or ovules, which carry only one of the two determinants. The gametes are thus *haploid*, and when pollen and ovule fuse, the resulting diploid grows into a ripe pea, and this may be grown to a mature plant. Thus the yellow (*AA*) and green (*aa*) parents yield yellow *Aa* F_1 progeny in which *A* masks the expression of *a*. When the hybrid F_1 plants self-fertilize, the formation of haploid gametes yields equal numbers of *A* and *a* ovules and equal numbers of *A* and *a* pollen. Random combinations of *A* and *a* gametes of both types yield the ratio of one-quarter *AA*, one-half *Aa*, and one-quarter *aa* zygotes. (Notice that *Aa* progeny form in two ways, *A* ovule + *a* pollen and *a* ovule + *A* pollen, accounting for twice as many *Aa* as either *AA* or *aa*.) Only the last category, one-quarter of the total, will be green peas; the rest will be yellow. We speak of the phenomenon of 3:1 segregation as Mendel's First Law. Two other terms, invented much later, describe the genetic constitution of the plants: *homozygous* plants (*AA* and *aa*) are pure for a given trait, whereas the mixed, or hybrid, plants (*Aa*) are *heterozygous*.

Mendel demonstrated that the parents contributed equally to the progeny: the results of fertilizing *A* ovules with *a* pollen were no different than fertilizing *a* ovules with *A* pollen. It would be almost 40 years until Thomas Hunt Morgan (1866–1945) would detect a systematic exception to this rule in the discovery of sex linkage in fruit flies.

Mendel then showed that two recessive traits in the same cross segregated independently of one another. Thus a random distribution of two pairs of alternative traits such as yellow/green and smooth/wrinkled, each segregating in a three-quarters to one-quarter fashion, would yield progeny in the following ratio:

¾ yellow × ¾ smooth = 9/16 yellow, smooth
¾ yellow × ¼ wrinkled = 3/16 yellow, wrinkled
¼ green × ¾ smooth = 3/16 green, smooth
¼ green × ¼ wrinkled = 1/16 green, wrinkled

We refer to this outcome as *independent assortment*, or Mendel's Second Law, which Mendel demonstrated empirically in several crosses. The reader should appreciate the abstract way in which the idea of randomness is formulated so that it yields statistically precise predictions and explanations. This was an important part of Mendel's contribution to the biology of a later time, well after his death in 1884.

In the last five years of his scientific investigations, Mendel attempted to generalize his findings. On the somewhat uninformed advice of the famous botanist Karl von Nägeli (1817–91), Mendel sought to demonstrate the phenomenon of regular segregation in the hawk-weed, *Hieracium*. The choice was grossly inappropriate in the event, owing to the parthenogenetic nature (bypassing gamete formation and fusion) of many of the parents and all of the hybrid plants they produced. Mendel's failure to replicate his results in another organism and the inattention to his widely distributed publication of 1866 for some years is one of the most curious chapters in the history of biology. The story of the birth and death of the pea as a model system, used by Mendel, is one of the shortest on record in genetics. Its moral is that if fundamental findings using a model system are not or cannot be generalized, interest will quickly lapse. One could say that the garden pea never became a working experimental model for anything but itself, *sui generis*.

de Vries

In 1900, plant scientists independently discovered Mendelian phenomena in other plants, only to find that Mendel had anticipated them by 35 years. Whatever disappointment they might have felt, they may have been comforted by the authority of history. With their recognition of Mendel's achievement, Hugo de Vries (1848–1935), Carl Correns (1864–1933), and Erich von Tschermak (1871–1962) did what Mendel could not: they generalized Mendel's findings to a variety of plants, and, in fact, Tschermak confirmed them in peas.[6] From 1900 to 1909 other investigators called attention to or discovered many examples of hereditary variation in humans, plants, and higher animals that could easily be interpreted in Mendelian terms.

De Vries offered the most substantial confirmation of Mendelian principles in plants, starting in 1900. However, well before that, he had worked extensively on the mechanism of biological variation, anxious to test ideas of inheritance in light of Darwin's hypothesis of the origin of species. He had chosen the plant *Oenothera lamarckiana* as a model system for this work because it displayed unusual variability. Darwin had proposed, as a working hypothesis, that organisms transmitted their properties to the next generation in the form of "gemmules," particles that originated in the tissues of the adult and were passed on collectively to the next generation by the gametes. This of course allowed for the possibility of the inheritance of acquired characteristics, although Darwin subordinated this mechanism to the idea of natural selection of individuals from variable populations. De Vries developed a similar view of inheritance, under the name of *pangenesis*, in which both the number and the character of "pangenes" collected in the gametes would explain the subtle quantitative and the more obvious qualitative variations within natural populations. He even saw in the more discontinuous variation of *O. lamarckiana* a possible model for the origin of a new species. He coined the term "mutation" and proposed a theory of evolutionary change by way of these "saltations."

Unfortunately for de Vries, he had chosen a most atypical organism for his studies, and generalizations based on its behavior could not survive. Much of the discontinuous variation in *O. lamarckiana* arose from rare chromosomal rearrangements and changes in chromosome number, not from mutation of individual genes. De Vries's work thus failed to advance the general field. Although he usefully introduced the concept of mutation, the word now reflects entirely different phenomena from the ones he studied. This is another short story, like that of *Hieracium* in Mendel's hands, of an inappropriately chosen organism. Both plants were studied in the 1930s in more and more obscure detail, more as oddities than as models.

Morgan

The first decades of the 20th century saw the beginnings of genetics as we know it. During this time, geneticists became aware of the parallels between the behavior of chromosomes and genes in meiosis (the cellular process by which a diploid cell makes haploid gametes) and discovered linkage, crossing over, and sex linkage (see appendix 1).[7] Many of these seminal discoveries emerged between 1909 and 1915 from the laboratory of Thomas Hunt Morgan (fig. 1.1), made with a model system used to this day, the fruit fly *Drosophila melanogaster*. Morgan initially studied embryology, a subject closely linked to heredity and evolution. Morgan was aware of the cytological findings about fertilization and the importance of egg and sperm nuclei that emerged between 1875 and 1890. These discoveries established the continuity of nuclear and chromosomal contents between generations. They connected the contributions of parents to the already well-established constancy of nuclear contents during the proliferation and differentiation of cells of the animal embryo.[8] These findings attached themselves intellectually at the time to the Darwinian concept of evolution, in which the life cycle repeated itself over the generations, with variation subjected to natural selection. A central question began to vex biologists: how could adult cells of such different character emerge from the fertilized egg? Several investigators had proposed the nucleus as the seat of hereditary information passed between generations, but this conflicted with the obvious variation among cells that retained a constant nuclear content as they arose during embryogenesis.

Morgan initially wanted to study experimental manipulations of embryonic development with a highly empirical attitude, free of the speculative tendencies then common in the field.[9] His work on regeneration of earthworms and planarians led him to argue against the ideas then prevalent. The ability of isolated cells after the first few divisions of fertilized animal eggs and of small parts of worms to regenerate whole organisms led him to two conclusions. First, the information for all the parts of an organism was intrinsic to each cell or part. In others' minds, this strengthened the case for the known similarity of nuclear contents throughout development, but Morgan could not accept this in the face of differentiation. Second, the idea of a special germ line, segregated early in embryogenesis from somatic cells that went on to differentiate, should be abandoned. He felt that perhaps the position of cells in the embryo or regenerating worm might be the key to differentiation. This work demonstrates the reductionist, hypothesis-testing habits that characterized Morgan's later work.

Morgan, who read widely, became interested in evolution and embraced de Vries's notion of speciation occurring in a stepwise fashion through drastic "mutations" as seen in *Oenothera*. He criticized Darwin's incrementalist hypothesis about the origin of species and the somewhat Lamarckian ideas that underlay his hypothesis of heredity.[10] Morgan therefore embarked on a study of changes that might be detected after many generations of inbreeding, finally using, after 1908, the fruit fly *D. melanogaster*. Morgan knew from others that fruit flies could be bred easily, quickly, and in large numbers. He therefore suggested to a student that he use *D. melanogaster* for an experiment in evolution: would flies become eyeless if they were raised for many generations in the dark?[11] They did not. Morgan then realized that the organism might satisfy his own needs for experiments in which he applied selection for the intensity of pigment on a part of the thorax of the fly. In particular, he thought such variants would appear in a stepwise rather than in a gradual fashion. He was disappointed that, although selection was successful, changes soon ceased as generations went on. The scale of his experiments (using a small, easily reared fly to generate very large populations) probably assured success in an unexpected direction. Beginning in 1910, mutants of a sort familiar to modern geneticists began to appear; by the end of the year, he had more than 12 variants, including the sex-linked white-eye mutant about which he published in that year.[12]

We see that Morgan did not choose *Drosophila* as a model system for genetics, but for a failed study of evolution. His, and our, ideas of heredity owe much to his exploitation of observations at variance with his expectations. In developing his empirical focus, Morgan then decided that he should dissociate the study of heredity intellectually from evolution and the complicating aspects of an organism's life cycle. By concentrating on the variants that finally arose in his *Drosophila* cultures, he and a team of young students began an intense study of the inheritance of the mutant traits. In just six years, his group collected extensive evidence describing the "particulate" nature of the gene (uninfluenced by an alternative form in heterozygotes, like Mendel's green and yellow determinants), gene linkage, mutation, sex linkage, and the relationship between genes and chromosomes. His studies, summarized in the 1915 book, *The Mechanism of Mendelian Heredity*, and a later, more mature version in 1926, *The Theory of the Gene*, are described briefly in the next chapter.[13] A strict empiricist, Morgan insisted that the gene be treated as a theoretical entity. The gene was to be defined by operational features of breeding experiments without speculating about its structure or its mode of action. At first, Morgan was reluctant even to accept the notion that genes resided on chromosomes.[14] Even though his books depict chromosomes as bearers of genes, he did not insist on the relationship in his first book, preferring to say simply, "it would be folly to close one's eyes to so patent a relation."[15] In this way, he protected the evidence about the inheritance of genes from the dispute then current about their location.[16]

Drosophila, owing to Morgan's extraordinary success in demonstrating so many fundamental principles of heredity, became and remains a popular model for many other investigations. The fields include mutation, population genetics, linkage studies, cytogenetics, development, behavior, and biological rhythms. As the use of the organism in these fields grew in the 1920s, 1930s, and early 1940s, so did the number of mutants and the body of technical lore about the organism. Gene nomenclature, cultivation methods, and mutant screening methods were standardized. A *Drosophila* stock center was established; mutant indexes were updated frequently, a *Drosophila* newsletter appeared, and workers met regularly at *Drosophila* Information Conferences. These developments illustrate a process by which a model system for genetics evolved into a model organism for biological research.

The Emergence of Model Organisms

What is a *model organism*? I use this term to designate organisms whose biology is understood at many different levels—organisms that can be used as a representative or basis of comparison for a large taxon or kingdom of life, or even for life itself. An integrated understanding of the biology of such organisms starts with genetics. Among the mutants investigators isolate, some mutations might suggest interesting new lines of research, entirely unanticipated at the outset. The interest arises in the specificity of the mutations and the possibility that more mutations might allow a fine dissection of a particular phenotypic trait. By the time such organisms become models, investigators have a growing body of mutants, chromosome maps, and diverse research programs. But in the process, the community becomes trapped: all genetic approaches are constrained by intraspecific crosses. Knowledge grows thereafter through findings at many levels and through the connectivity of information.[17] Surrounding the organisms are communities of investigators that meet regularly, share information, observe common experimental conditions, use standardized genetic stocks, and, ultimately, use common terms of discourse.

The members of these communities, trapped in bonds of affection for their experimental organism and for one another, change to another organism with considerable risk and reluctance. Soon every new finding has a rich context in which it may be interpreted. Findings from other members of the taxon can immediately be interpreted further according to similar knowledge about the model organism. Findings from another organism that have no counterpart in the model organism may immediately capture the interest of investigators for that very reason. A model organism thus becomes a "communal intellectual resource."[18]

With the investigations on model organisms of the 20th century came a passion on the part of leading biologists for reduction and universality. I use the term *reductionism* loosely and colloquially to indicate the resolution of complex problems into logical, experimentally accessible elements.[19] In practice, a reductionist habit of mind leads scientists to use the simplest or most easily used systems that retain properties to be investigated, to avoid appeal to vague, emergent properties of a system, and to stress the generality of findings.

Model Organisms: A 20th-Century Phenomenon

The thesis of this book will emerge from an appreciation of the entry of microbes into genetics and the revolution that followed. I want to show how a few microorganisms came, purposefully or casually, to be used for work in molecular biology in the mid-20th century, the "century of the gene," in Evelyn Fox Keller's phrase.[20] Second, I want to show how each organism became vulnerable to its own limits as a model, or how fashion dictated that particular model organisms be demoted to serve as representatives of smaller taxa. Third, I argue that the highly focused use of a few model organisms in biology has been a long, transitional phase of biological research. That phase consisted of an initial period of speculation about the functions of organisms and the basis of heredity, followed by a period of coherent molecular representation after the certification of DNA as the genetic material. We are now at a point between the study of the complexity and diversity of organisms, widely appreciated in the 19th century, and a new approach to these attributes emerging from genomic analysis in the 21st century.

The accomplishments of molecular biology have occasioned misgivings in observers, both within and outside the field. The misgivings arise from a sense that the appreciation of life has been displaced from organisms to DNA; that complexity is ignored in favor of strict "upward causation" emanating from DNA; and that the language of life sciences has been transformed from the concrete and expressive to formal engineering analogies.[21] These misgivings are as justified as the regret we feel at an irreversible loss of innocence. Biology is a big science now, much less accessible than in the past to amateurs, and much less widely accessible even to professionals. What we knew 50 years ago is giving way to "systems biology," the modeling of the workings and integration of elements at every level of organization of living things, from metabolism, cell physiology, development, and organismal physiology to population biology and ecology. The enterprise depends on tough-minded ambition, sophisticated instrumentation, transnational and transdisciplinary efforts, high-throughput data gathering, and global database management. This collective, almost industrial effort has led many working biologists to forget the awe with which they might once have beheld, in a wood, on a lake, or in a humble classroom, how intricate a design a blind universe had imparted to life.

2

Morgan's Progeny

It is the "organism as a whole"; it is a "property of the system as such"; it is "organization." These words, like those of Goldsmith's country parson, are "of learned length and thundering sound." Once more, in the plain speech of everyday life, their meaning is: We do not know.

—E. B. Wilson

Morgan and His Findings: Keep It Simple

Morgan, like most embryologists, initially considered heredity, development, and evolution inseparable, all aspects of the continuity of life. This mindset hampered clear thinking about each subject until well into the 1930s. Many biologists in 1900–10 had become convinced that the diversification of cells in development and the basic plan of embryogenesis lay in postulated but unseen cytoplasmic patterns or determinants. This idea was a default position, resting on the well-documented cytological evidence that chromosomes, the main content of nuclei, remained constant during the divisions and differentiation of cell types during embryogenesis.[1] At the time, Mendelian genetics was in its deferred infancy, with no compelling evidence of what genes did, where they were, or what they were made of. In fact, many argued that genes (the word was coined only in 1909) might be a small and special class of hereditary factors that came into play only at the end of embryonic development.

This problem became sharper even before Morgan's studies began. As biologists became aware of the generality of the rules of Mendelian genetics, Sutton and Boveri independently recognized a striking correlation between the behavior of chromosomes in sexual reproduction and the behavior of genetic traits in crosses. The Sutton-Boveri hypothesis, formulated clearly in 1903, stated a formal correlation that could not be ignored.[2] Cytological study of the formation of gametes, both female and male and in both plants and animals, showed that each organism normally had two of every type of chromosome. The members of the pair are now called *homologs*. During the process of meiosis that leads to gamete formation, and at no other time, the homologous chromosomes divide into two *chromatids* and then pair with one another (fig. A1.1, appendix 1). The four-stranded bundles would then, through two cell divisions, be distributed to four gametes, the first division separating the homologs, the second separating the members of chromatid pairs. The net result was that each gamete would be haploid, having only one copy of the particular chromosome, whereas the diploid parent has two. Mendel had, of course, stated, as discussed in the last chapter, the same general rule for his traits in peas, each parent having two copies, each gamete containing only one. Thus chromosomes seemed the logical location of genes. With this question in the air in 1910, Morgan began work with *Drosophila*, but for different reasons.

The life cycle of *D. melanogaster* is simple, and the generation time (egg to egg) is about 14 days. Fertilized eggs, laid by inseminated females, undergo embryogenesis and hatch in about 24 hours. The larvae that emerge feed on yeast and other nutritive materials. In the laboratory, the flies are raised in pint milk bottles or in small vials stoppered with cotton plugs. The food provided in the early days was a paste of bananas and yeast, which supports rapid growth of larvae. The larvae molt twice as they grow, and after about 72 hours, they crawl up the side of the bottles and form a hard pupal case, originating with the larval skin, around themselves. During metamorphosis of the pupa, small groups of cells (disks, or *Anlagen*) present in the larva develop into the various organs of the adult: wings, eyes, legs, mouthparts, genitalia, and so forth. After about 100 hours, flies emerge from the pupal cases and, spreading their wings, reveal themselves as familiar fruit flies. Males and females are easily distinguished by secondary sexual characteristics (fig. 2.1). The females remain sterile for about 18 hours. This allows the geneticist to narcotize the flies with carbon dioxide or ether and to separate females from males if the females are to be used in controlled matings thereafter. *D. melanogaster*, like peas, are diploid, having two sets of the four chromosomes characteristic of the species. As we have seen above, males and females each pass one set of chromosomes, via haploid eggs or sperm, to their progeny.

Morgan, having found his first mutants in 1910, attracted students who had previously become fascinated by the new science of heredity, evolution, and allied subjects in biology. Of these students, Alfred Sturtevant (1891–1970), Herman J. Muller (1890–1967), and Calvin Bridges (1889–1938) made the most significant contributions. In Morgan's "fly room" at Columbia University, work progressed rapidly. The discoveries of sex linkage, crossing over, and the idea of chromosome mapping (see appendix 1) emerged from concentrated discussion and argument, with Morgan presiding, listening, and gradually building a unified picture of gene transmission—formal genetics as we now know it. During this time, Morgan defined the gene in strictly operational terms, that is, in terms of breeding experiments. To him, it was a theoretical factor that effected transmission of traits to the next generation. He was rigorous enough to

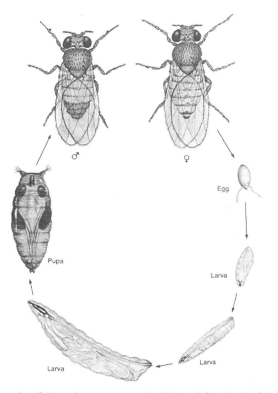

Fig. 2.1.   Life cycle of *D. melanogaster* male (*left*) and female (*right*). The male and female adults (*top*) are easily distinguished by the dark tip of the abdomen and the sex combs (small dark bumps in this figure) on the forelegs of the male. The larva hatches from the fertilized egg and proceeds through three *instars*, as shown, molting between each one. The last instar forms itself into a pupa, which then metamorphoses into an adult. © 1988, Carolina Biological Supply Company. Reprinted with permission.

delay adherence to the view that the chromosomes bore the genes, despite the Sutton-Boveri proposal.[3] His ability to ignore the embryological counter-argument as well, and to maintain a nonspeculative approach to the question, served him well. Critics would have to criticize his findings about gene transmission, rather than any inferences he might make about the genes' location. And the findings were so thorough, self-consistent, and compatible with findings in other organisms (Mendel's peas, among them) that they could not easily be faulted.

Morgan's main findings deserve a brief description as a prelude to the work of his intellectual descendants. Curiously, several of Morgan's clear-cut mutations did not conform precisely to Mendel's laws, and for good reasons. The mutations included white eye (*w*), yellow body (*y*), and rudimentary wings (*r*), all of which were *sex-linked*; that is, carried on the X chromosome (see appendix 1). Previous cytological investigations had shown that sex in insects was correlated with a chromosomal difference, the female having two X chromosomes, and the male having one X chromosome and, in *D. melanogaster* and most other cases, a morphologically different Y chromosome. The X and Y chromosomes, despite their different size and gene content, behave as homologous chromosomes, pair in meiosis, and segregate as strict alternatives in the two meiotic divisions. For this reason, the eggs (all with X chromosomes) may be fertilized with an X-bearing sperm, yielding a female zygote, or with a Y-bearing sperm, yielding a male zygote. Recessive mutations that occur on the X chromosome will be expressed when they are present in the male, because the Y chromosome of males does not carry a second copy of the gene in question. This explains why recessive, sex-linked mutations were among the first to be found in Morgan's laboratory.

I must pause here to introduce an important genetic term: *allele*. Alleles are simply alternate forms of genes. Alleles are related by mutation (e.g., the wild-type w^+ and its mutant form w are alleles), and therefore lie in the same position on homologous chromosomes. This was one of the concepts so well developed by Morgan's group. And because homologous chromosomes are separated during the meiotic process, alleles *segregate* as a matter of course into different gametes. Gametes can carry only one copy of a given gene, consistent with Mendel's conclusions about traits.

Morgan's initial crosses with the white-eye gene, while not following the Mendelian pattern, could be reconciled with Mendel's conception only by recognizing its correlation with the inheritance of the anomalous sex chromosomes (fig. A1.3, appendix 1). But even the correlation of sex-linked inheritance and sex chromosome behavior did not induce Morgan to embrace what he later propounded as the chromosome theory of inheritance. No one could say definitively that sex chromosomes were the cause of sexual differentiation; indeed, they might rather be a result, a secondary sexual characteristic.[4] Morgan would shortly accept the causal connections among sex chromosomes, sex determination, and sex linkage, but he remained publicly skeptical of the chromosome theory in general for some time thereafter. Meanwhile, Morgan's genetic work after 1911 demonstrated the orderly transmission of many genes that were not sex linked, in keeping with observations in other organisms such as corn, mice, and humans.

Sex-linked genes again became important in Morgan's laboratory in the discovery of genetic linkage. The sex-linked mutations w and y established most clearly the principle that two genes might be carried by a single chromosome, and not assort independently.[5] More important, the association of the genes might be broken occasionally, leading to recombinant progeny. Thus a mating of a homozygous white-eyed, yellow-bodied female (*wy/wy*) and a normal (w^+y^+) male would yield wy/w^+y^+ females and wy males among the F_1 progeny, the latter having only one copy of each gene, together with a Y chromosome. If these flies were interbred, the F_2 progeny were predominantly red-eyed, gray or white-eyed, yellow flies, with only about 1.5% red-eyed, yellow and white-eyed, gray flies. This finding (figs. A1.3, A1.4, appendix 1) demonstrated that some genes tended to stay together (to be linked) during the formation of gametes rather than being randomly assorted. It simultaneously showed that linkage need not be complete: a mechanism that broke up the linkages must also prevail. These observations began to answer the question of how there could be so many more gene combinations than would be allowed by the number of whole chromosomes. Mendel's results with multiple-gene crosses had displayed only the phenomenon of independent assortment. But cytological observations by others in Morgan's time had hinted at exchange of chromosomal parts during meiosis. Morgan suggested that the degree of linkage (that is, the tendency of genes to remain together) might be a measure of how close they were on the chromosome. Sturtevant used data involving *y*, *w*, and four other mutations on the X chromosome to make the first chromosome map based on recombination frequencies.[6]

Plainly, the sex linkage of many genes of *Drosophila* had an unanticipated advantage, perhaps a vital one in the discovery of linkage and recombination. This is because recessive mutations of genes on non-sex chromosomes, called *autosomes*, cannot reveal themselves in heterozygous flies due to the usual dominance of the wild-type copy. However, as Morgan's group accumulated autosomal mutations that displayed orthodox Mendelian behavior (segregation and independent assortment), linkage studies were extended to other chromosomes. As more autosomal mutations accumulated, they could be grouped into *linkage groups* that corresponded in number to the four types of chromosomes in *D. melanogaster*. The first summary of this work was published in 1915.[7]

After about 1920, Morgan no longer preoccupied himself with experiments, which he left more and more to his able group. He became a spokesman for the view (that of the "Morgan School") that nuclear genes underlay both heredity and evolutionary change. As we will see, this view opposed that of biologists who favored a substantial contribution of the cytoplasm in both. In 1928, now a highly regarded scientist, Morgan accepted an invitation to establish a Biology Division at the California Institute of Technology—Caltech, as I refer to it hereafter. He brought his group, which continued its work as he resumed his interest in embryology and administered an institution with increasingly diverse, modern investigators. Morgan was awarded the Nobel Prize in Physiology and Medicine in 1933 for his work in genetics.[8]

Morgan's book of 1934, *Embryology and Genetics*, did not succeed in integrating the two worlds, nor would this happen until the latter half of the 20th century.[9] His accomplishment was to enliven interest in the possibility that the contributions of the nuclear genes to embryonic development might someday be understood. Embryologists could no longer ignore the findings of formal genetics, which had turned up almost no hereditary variants that could not be explained in Mendelian terms. The mystery remained of how nuclei of a developing embryo, containing the same chromosomal endowment, could direct developmental diversification without some program of mutational change. The possibility that the cytoplasm directed differentiation persisted, although few clear hypotheses could be formulated without biochemical information about the cytoplasm itself—information that was yet to come.

Ephrussi and Beadle: What Did Genes Do?

As the study of genetics, referred to as the chromosome theory, became an accepted body of knowledge, the collection of mutants in *Drosophila* and many other organisms yielded some unusual derangements of development. Such mutants presented an opportunity to solve questions of development in a genetically domesticated organism. One such opportunity originated in the work of Sturtevant, who in 1920 discovered abnormal, genetically mosaic flies. The most striking mosaics were called *gynandromorphs*, a term signifying a combination of two sexes. Gynandromorphs are derived from female (XX) zygotes that lose an X chromosome during early mitotic cell divisions after fertilization. Some cells remain XX (female), while the aberrant cells retain only one X chromosome, and are therefore genetically male. (In *Drosophila*, the Y chromosome does not impart maleness. It carries only genes required for sperm formation. It is therefore not required for normal development, but only for fertility.) The tissues derived from the two sorts of cells may in many cases be identified as male or female. In the extreme, the two entire sides of the fly have recognizably different sexual characteristics. Moreover, if the XX zygote is heterozygous for a recessive gene such as forked bristles (f^+/f), the male tissues would often (about half the time, the remaining X carries the mutation f) reveal the mutant phenotype. On this basis, Sturtevant defined f as an *autonomous* mutation, in which the cell phenotype is determined by the genetic constitution of the cell itself, and mosaic flies showed patches of mutant and normal tissues. Other mutations however, like the vermilion (v) eye-color variant, are *nonautonomous*. Here, the mutant eye cells carrying the v mutation on the retained X are rendered phenotypically normal by the presence of normal cells (XX, having the heterozygous genotype v^+/v) in the same fly, even cells on the other side of the fly. Morgan and Bridges' original report suggested that genetic characteristics were autonomous; Sturtevant demonstrated that, in contrast, vermilion (v) was nonautomomous.[10] Mosaic flies implying loss of other chromosomes during development extended this type of analysis to autosomal genes.

Nonautonomous mutations suggested that diffusible substances allowed cells to communicate during development. In fact, such substances might be isolated and characterized chemically. This hope had been entertained previously by embryologists who proposed that specific chemical signals, originating in foci that induced morphological patterns during development (e.g., the "organizer" and the neural crest), might be transmitted by cell contact or by signals that traveled short distances in the developing embryo.

Boris Ephrussi (1901–1979), an established experimental embryologist, wanted to use *Drosophila* in an experimental program in which he could trace the action of genes in development. He had successfully cultured mammalian cells outside the organism and was impressed that they retained their basic phenotypes. By reasoning based on gynandromorphs, one might say that these cells had autonomous phenotypes; that is, they retained their appearance despite the different environment of a culture medium. In some experiments, however, Ephrussi explanted cells from mouse embryos with lethal genotypes—

embryos that would die before term. He found that the explanted tissue would survive in culture, multiply, and even differentiate. This indicated either that the lethality was a nonautonomous trait, imposed by surrounding tissues, or that the environment of the culture dish overcame the intrinsic lethality. Relating these observations further to autonomy and nonautonomy begs the question. However, Ephrussi's more general interest lay in the possibility that cell differentiation could be analyzed by exposing differentiating and differentiated cells to different environmental conditions.

Because Ephrussi could not study these problems further in mice, and because a wealth of different mutations was available in *Drosophila*, he came to Caltech in 1934 on fellowship to learn the requisite techniques from the Morgan group, specifically from Sturtevant.[11] As noted above, Sturtevant had shown that a substance, later named the v^+ substance (not necessarily the pigment itself), could diffuse within a gynandromorph and "repair" the eye-color defect of the deficient tissues. The observation suggested that a developmental hormone could be isolated and characterized, and perhaps this would yield a clue to the fundamental action of a gene. In Ephrussi's quest to familiarize himself with *Drosophila*, we see the impact of a model organism in studies outside the scope of Morgan's original work.

George Beadle (1903–89) came to Caltech in 1931 as a postdoctoral fellow after finishing his doctoral work in corn genetics. After some further work on the cytogenetics of corn, he became interested in *Drosophila*, concentrating on crossing over and other aspects of formal genetics. This program was not a mere extension of Morgan's program. By that time, it appeared that studies of recombination might yield some information about the structure of genes. This interest complemented that of Ephrussi, who arrived in 1934 with the hope that a more chemical approach to the nature of the gene might be appropriate. Together, they decided to pursue a genetic and developmental analysis of eye-color mutants.[12]

Ephrussi and Beadle initially wanted to see whether cells of a mutant would develop normally if they were transplanted into a normal fly. This required that the clusters of larval cells (larval disks, or *Anlagen*) that were destined to become organs of the adult during metamorphosis be transplanted to the body cavity of genetically different larvae. This difficult technique was perfected and used to determine whether a normal (v^+) host could impart normal color to an eye originating from an eye disk taken from a vermilion (v mutant) larva. The experiment showed that the transplanted eye was indeed normal. The investigators then hoped that the v^+ substance could be isolated and identified. This program was pursued by Beadle and Ephrussi at Caltech, in Ephrussi's laboratory in France, and by Beadle as he moved from Paris to Harvard, then to Stanford University during the next few years. Ephrussi and Beadle tested many eye-color mutants, but found only two (v and cinnabar, cn) that were nonautonomous. Therefore, the program did not develop as one of general application. Although the basic experiments worked, and the order of v^+ and cn^+ in a linear sequence of action could be discerned, the project foundered on the difficulty of identifying the v^+ substance in the small amounts present in fruit flies. In fact this required some time, even with the help of a biochemist, Edward L. Tatum (1909–75). The v^+ substance was finally isolated (as a substance with activity in imparting color) from a contaminating bacterium and identified as kynurenin.[13] Before Beadle and Tatum published their results, a group working under the German biochemist Adolf Butenandt (1903–95) had identified the substance through an entirely different experimental strategy.

The Beadle-Ephrussi work proceeded with directness and simplicity. Like Morgan, they formulated a problem without trying to reconcile all ideas about genes and their action. Ephrussi could ignore competing ideas about embryogenesis as he and Beadle devised their transplantation experiments. Beadle could ignore his prior recombinational approach to the gene at the same time. Both, of course, were interested in the structure and action of the gene, and Ephrussi maintained a comprehensive awareness of embryology and the question of the role of cytoplasm in differentiation. Their experiments, in contrast, were highly focused on a simple question: could the final product of a gene be identified, and would this give them any clue about how the gene produced it? Beadle and Ephrussi were not as close to the formulation of the one-gene, one-enzyme theory as textbooks would have us think.[14] This is indicated by their use of the word "hormone" when speaking about the v^+ substance—a chemical produced by one tissue that affected the activity of another within the body. They entertained the hypothesis that the diffusible products of a gene might be catalysts (enzymes?) that produced a product, or precursors of a product, but they did not commit themselves to a particular view. Ephrussi, in favoring the idea that the substances were hormones, somewhat obscured the idea in many biologists' minds, that of a more direct gene–enzyme link.

In the late 1930s, the time of Ephrussi and Beadle's work, the chemical nature of proteins and enzymes was poorly known, and knowledge of the nature of the genetic material was some years away. Although Ephrussi had chosen Drosophila, a model system for genetics at the time, as a feasible tool for his purpose, there was trouble ahead even in 1938 regarding its continued use in the program. Work on cytology, formal genetics, population genetics, and mutation was proceeding apace, but as a subject for embryology and chemical analysis, Drosophila had to face the hurdle of its small size, its highly differentiated nature, its exoskeleton, a tough chorion around the embryo, and the possibilities of contamination during microsurgery. The time was ripe for the choice of a still more favorable organism, one that would be chosen not for the pursuit of embryological questions, but instead for the pursuit of biochemical ones.

As Beadle and Ephrussi's work unfolded so naturally from the efforts of Morgan and Sturtevant, it brought together genetics and embryology, a long-standing hope of Morgan. The experiments Beadle and Ephrussi performed involved no breeding experiments in which the progeny were analyzed statistically. Instead, the experiments used embryonic transplant techniques, devised for the purpose of identifying diffusible substances required in normal development.[15] The work thus became one of the leading precursors of the field we know as biochemical genetics, using mutations to dissect the course of metabolism. It broke out of the severe strictures of Morgan's approach, where speculation about gene action was inappropriate until testable propositions could be formulated.

Several other groups worked toward the same ends with similar ideas at about the same time. At the John Innes Horticultural Institution in England, a group including Muriel Wheldale (1880–1932) and Rose Scott-Moncrieff (b. 1903) studied the inheritance of flower colors, and by 1940 they had related many mutant genes affecting color to particular chemical conversions.[16] However, Scott-Moncrieff (unlike some who cited her work) did not speculate further about how the genes might exert their effects. This may explain why those who cite the later work of Beadle and Tatum, described in chapter 3, ignore the John Innes work. But a more powerful reason lies in the attention given to the discovery, made in the first decade of the 1900s by Archibald Garrod (1857–1936) and William Bateson (1861–1936), of the genetic basis of alkaptonuria, a metabolic disorder of humans. Garrod, a physician, discovered that rare patients could not metabolize the aromatic amino acids phenylalanine and tyrosine. A peculiar compound, called homogentisic acid, therefore accumulated in the blood and was passed in the urine, causing it to blacken on exposure to air. To Garrod, this implied that one of the several steps in the pathway converting the aromatic amino acids to their normal, final product was blocked. The familial occurrence of this disorder and its pattern of inheritance suggested to Bateson that a single recessive mutant gene underlay the deficiency. Garrod summarized this work in a book, *Inborn Errors of Metabolism*, published in 1909, which included other cases of inherited, metabolic deficiencies, including albinism.[17] Garrod's main interest was not in genetics, however, but in the medical implications of his work; it fell to others to promote his contribution to the beginnings of biochemical genetics.[18] Nor was Bateson's enthusiasm focused on the relation of genes and enzymes, but for the support Garrod's findings gave to his notion that evolution might proceed mainly through loss of genes.

Beadle, in commenting on his later findings with the mold *Neurospora* that led to the one-gene, one enzyme principle (see chapter 3), cited Garrod's work as the beginning of the one-gene, one-enzyme theory. The historian Jan Sapp feels that Beadle, in addition to acknowledging this past work, was investing Garrod as a neglected founder of biochemical genetics, comparable to those in 1900 who credited Mendel as the father of genetics.[19] (Beadle also later credited the John Innes group with some influence on his work with Ephrussi.)[20] By the time of Beadle and Ephrussi's work, and certainly by the time of Beadle and Tatum's *Neurospora* work, the idea that enzymes and genes were connected in some fashion was hardly novel. Indeed, J. B. S. Haldane (1892–1964) had explicitly stated this relationship in 1920, and he gave Garrod credit for anticipating the one-gene, one-enzyme hypothesis in 1942.[21] But Beadle and Tatum had the grander goal of generalizing the notion that single genes endowed single enzymes with their specificity, a point that drew some resistance to their earliest work (see chapter 3). Sapp suggests with some force that Beadle, in adopting Garrod as a father of the field, tried to deflect some of this criticism by pointing to long neglected cases of one-mutation, one-enzyme relationships. Beadle thereby created a myth that served his interests.[22]

Another group working in the 1930s was based in Germany and included Richard Kuhn (1900–67) and Franz Moewus (1908–59). Moewus worked on sex hormones (now called pheromones) of the one-celled alga *Chlamydomonas*. Sex hormones evidently primed these cells for mating. Moewus developed the genetics of *Chlamydomonas* sufficiently to study the inheritance of mutations affecting the chemical conversions required in the synthesis of pheromones. Moewus's results were widely known in Europe. Moewus's work became known in the United States after World War II, in part through the advocacy of Tracy Sonneborn (1905–81), a major figure in *Paramecium* genetics (see chapter 13). Sapp describes Sonneborn's repeated efforts to persuade geneticists that Moewus had anticipated the Beadle-Tatum work by several years.[23] Moewus did indeed have ideas, well stated, that genes and biochemical reactions are related in a simple way. I believe, like Morange, that Moewus's efforts with an array of excreted, related substances more closely resemble Beadle and Ephrussi's work than the later, more general one-gene, one-enzyme formulation of Beadle and Tatum.[24] (It is of interest that the substances of interest in *Chlamydomonas* and the v^+ substance of *Drosophila* were both referred to as hormones—an indication of the influence of language and precedent in this new area of investigation.) Unfortunately, the major genetic and chemical data on which Moewus's conclusions rest are now considered to be untrustworthy or even fraudulent. They are therefore not credited as part of the mainstream of discovery. I return to this matter in chapter 14.

Beadle, frustrated in 1939 or 1940 by the difficulty of achieving a goal that had seemed within reach, took a radical step. He used his findings to formulate a new problem, biochemical rather than embryological, and pursued it in an entirely different organism, the fungus *Neurospora*. We will see in chapter 3 that the transition from *Drosophila* to *Neurospora* represents one of the most conscious and important choices of an experimental organism in the history of biology.

H. J. Muller: Unchained Speculation

The Morgan group adopted, initially as a part-time participant, a student of high intelligence, Herman J. Muller.[25] Muller joined the fly room informally as a friend of his undergraduate acquaintances, Sturtevant and Bridges, already associated with Morgan. All three had shared an intense interest in heredity and other new discoveries in biology. Even at that time, Muller displayed an uncanny, imaginative way of assimilating and interpreting experimental information. When he joined the group as a part-timer, not yet a candidate for a degree, he witnessed the work on sex-linked inheritance, crossing over, and chromosome mapping developing before his eyes. He would systematically take interpretations to the logical extremes permitted by current knowledge, in contrast to Morgan, who rigorously restricted speculations beyond the bare facts, particularly if they did not suggest experiments. Muller's different style, however uncomfortable it may have been at times, benefited the laboratory. Morgan, recognizing Muller's value to the group, allowed him to join his lab as a doctoral student. The work in the years 1911–15 progressed rapidly, with one discovery coming on the heels of the one before and discussions going on all the time. Muller's contributions were rarely acknowledged in the early, published papers that made such a major impact, but he is one of the four coauthors of the 1915 summary of Morgan's group's findings.[26] Soon thereafter, the clash of styles would lead to an estrangement of Morgan, Sturtevant, and Bridges on one side, and Muller on the other, with battles for priority for certain discoveries persisting to the end of their days.

Muller, a complex man, was driven by passions for understanding genetics, for promoting social justice, and for applying genetics for the betterment of humankind. He would become one of the most incisive theorists of the gene in the period 1920–50. In 1927, he discovered the mutagenic effects of X-rays, for which, with other work emanating from it, he was awarded the Nobel Prize in 1946. Throughout his professional life, spent at various times at Columbia University, Rice University, the University of Texas, Moscow, the University of Edinburgh, Amherst College, and Indiana University, he would lecture and write about the nature of the gene and the significance of genetics to society. In the former effort, he speculated constantly not about the rules of inheritance, but about what the gene might be and what it might do. His interest in the Darwinian mode of evolution led him to work on "modifier genes," genes that affected the expression of others having major phenotypic effects. In doing so, he bridged the early gap between those who saw Mendelian mutations as highly discontinuous and those who saw genetic differences in natural populations and agricultural varieties in terms of blending, or continuous variation. Long before Morgan would do so with certainty, Muller had concluded that the chromosomes contained or bore the genes.

To Muller, the gene was a fascinating, unknown, particulate factor that can duplicate faithfully and mutate to various forms. The ability to isolate *multiple alleles*, mutants of the same gene with different degrees of impairment, greatly impressed Muller. He took special note of the dual capability of genes to direct complex functions within the cell and to direct their own replication. Anyone with a physicochemical cast of mind would expect on first principles that if a complicated structure were damaged by something as undiscriminating as X-rays, it should become impaired in most of its other capabilities. Yet, as Muller emphasized, genes could replicate as well in their mutated form as in their normal, wild-type form. Muller's work on mutation led him to speculate on the physical nature of the changes induced by X-rays and those arising as spontaneous mutants. Before the discovery of the role and structure of DNA, he was already speculating on mutations as deletions (*amorphs*), changes in the strength of their expression (*hypo-* and *hypermorphs*), or changes in specificity (*neomorphs*). He pressed the issue of how genes were arranged on the chromosome, what, if anything, might lie between them, and what structure (perhaps linear and divisible) they might have. His experimental designs were brilliant and rigorous, and despite his holistic attitudes the experiments were highly focused. This ability had been one of Muller's major contributions to Morgan's laboratory. He had great ambitions and was from the start curious, aggressive, and insatiable in his wish to unify genetic knowledge. We will see more of Muller's theoretical contributions to gene structure in chapter 4.

Returning to Morgan, one is struck by his tolerance of intellects in his laboratory quite unlike his own. He dedicated his efforts to simple, important issues. He stuck by the facts of experiments and did not allow himself the liberty of dreaming up models that could not be tested. He initially refused to allow the disputed chromosome theory to cloud his main points about the behavior of the gene. He severely criticized those who attributed unproven or unprovable properties to the gene or to the cytoplasm. He was an able polemicist because he read widely and thereby maintained an awareness of the experimental basis of many theories in biology. His broad awareness made it easy for him to discuss anything with his younger associates. They used their imaginations much more freely in interpreting the findings made in the fly room and often convinced the skeptical Morgan, against his predisposition, of the worth of doing experiments based on such speculations. Thus it was Sturtevant who made the first chromosome map, relating the different probabilities of recombination to the possible linear arrangement of the genes on the chromosome.[27] It was Bridges who finally convinced Morgan of the chromosomal basis of gene behavior through his studies of abnormal chromosome segregation, a point about which Morgan remained skeptical even in the face of his own earlier studies of sex-linked genes.[28] Thus Morgan depended crucially on their collective imagination as he led his group, on a tight rein, to the harmony of evidence they assembled.

Muller was invited to the USSR after difficult years in Austin, Texas, and to Berlin for a year just as the Nazis came to power in 1933. He was offered much more in resources as director of the Institute of Genetics, soon to move to Moscow, than he might have hoped for in America. America was in the midst of the Great Depression, and Muller found it difficult to return for both personal and political reasons. In Moscow he fared well until the increasing tyranny of Stalin's regime imposed political constraints on his laboratory. Matters came to a head with the ascendancy of Trofim D. Lysenko (1898–1976), whose fraudulent, Lamarckian work Muller felt compelled to criticize in 1935 at the Lenin All-Union Academy of Agricultural Sciences.[29] The political climate quickly became worse and, in fact, dangerous. Muller left the USSR in 1936 by way of a short stint in a medical unit of the Republicans fighting in the Spanish Civil War. After a brief stay with Ephrussi in Paris, he found his way to a visiting position in Edinburgh, Scotland, where he spent the next three years working in the Institute of Animal Genetics. He would shortly be joined by an Italian expatriate, Guido Pontecorvo (1907–99), who would propagate and extend many of Muller's ideas by use of another model organism, *Aspergillus nidulans*. We will meet Pontecorvo formally in chapter 4.

3

Neurospora

A theory has to be reasonable, but a fact doesn't.
— G. W. Beadle

The Fungi: An Unusual Lifestyle

The true fungi include the Ascomycetes, comprising common unicellular yeasts and many filamentous plant saprophytes and pathogens, and Basidiomycetes, comprising rusts, smuts, mushrooms, and puffballs. They are non-photosynthetic and must find their energy sources in organic materials in the environment. (The slime molds and the water molds, having some resemblance to true fungi, are now thought to be more closely related to amoeboid unicells and certain algae, respectively.) Several yeasts have been used for centuries for making bread and alcoholic beverages. With the studies of Pasteur that sought to improve French industries dependent on yeast, the study of modern biochemistry advanced greatly, specifically in the areas of respiration and fermentation. In agriculture, fungi are conspicuous as plant pathogens, and many efforts in crop improvement have been directed to breeding crop strains resistant to fungi. And, of course, the collection and cultivation of edible mushrooms has had a long history. These practical interests in fungi stimulated the development of mycology as a science, and this field became a strong component of agricultural stations, botanical institutes, and university departments in Europe and North America by the early 20th century.

Biologists found the filamentous forms of the Ascomycetes and Basidiomycetes intrinsically interesting because of their unusual lifestyles. The huge number of fungal species, adapted to a variety of habitats, rivals that of insects. Unlike the unicellular yeasts, to which we will return, filamentous forms have hyphae as the basic cellular unit. *Hyphae* are tubular cells with thick cell walls.

27

The cells have many nuclei in a common cytoplasm, a condition called *coeno-cytic*. Growth proceeds by extension of hyphal tips, and their branching generates a colony called a *mycelium* that radiates from the point of inoculation. The vegetative stage usually includes the formation of asexual spores on aerial hyphae. These spores, called *conidia* in many Ascomycetes, offer the geneticist considerable facility in studies of mutation and, if the spores are uninucleate, the easy establishment of pure lines. The sexual phase of these fungi differentiates them taxonomically into a huge number of genera.

Fungi require an organic carbon source from living or dead plant or animal material for growth. For experimental work, many species are grown on media such as cornmeal agar or potato dextrose agar. These media are cheap and easy to prepare. However, many fungi grow on extremely simple mixtures of salts and a simple sugar, with few, if any, additional organic nutrients such as vitamins or amino acids. Such synthetic media offer geneticists considerable control in experimental work. Finally, microbial stocks, including fungi, may be stored indefinitely in an ordinary deep freeze or refrigerator at trivial cost. Therefore, large numbers of mutant strains, unthinkable to those working with *Drosophila* or mice, are available to the fungal geneticist.

The mycelial vegetative form of fungi is the form generally found in nature—whether as a bread mold or the substance of a mushroom. Growth at hyphal tips requires a constant flow of the cytoplasm in the direction of growth. Fungal nuclei are generally haploid, which makes the genetic outcome of matings easy to interpret (see appendix 1). The haploid condition is of particular benefit to geneticists because strains carrying a recessive mutation express the mutant character without the masking effect of a dominant, normal gene, seen in diploids such as *Drosophila*. However, this masking problem nevertheless arises in the common condition known as heterokaryosis. *Heterokaryons* are mycelia with two or more genetically different nuclei coexisting in the same cytoplasm, which is ordinarily multinucleate. In such cases, the presence of a recessive mutation in one nucleus is masked by the normal allele carried in another nucleus of the same cell. This can be a useful attribute, allowing geneticists to test mutations for their dominance relationships and for the interaction of mutations carried by different nuclei. Asexual reproduction yields spores that are frequently or always uninucleate, depending on the species. In sexual reproduction, meiosis almost always yields sexual spores (ascospores or basidiospores) that are genetically pure (*homokaryotic*), each derived from a single meiotic product. With both sexual and asexual spores, pure haploid stocks of mutants can be established with ease.

The haploid phase of fungi is extended, with the diploid phase lasting only long enough to provide a nucleus that undergoes meiosis. This is in contrast to animals such as *Drosophila*, in which the haploid phase is represented only by gametes. Among fungi with a known sexual stage, the events of the life cycle are genetically the same as in diploids (appendix 1). In both cases, haploid nuclei fuse to form a diploid nucleus. In *homothallic* fungi, the fusion may occur between identical nuclei of a single mycelium. In *heterothallic* fungi, strains differ in mating type, and only nuclei of different mating type can fuse, a restriction

that promotes outbreeding. The diploid nuclei resulting from fertilization, usually ensconced in a fruiting body, then undergo meiosis in special cells (fig. A1.2,
appendix 1). The four products of meiosis are haploid, and they are usually retained as a group called a *tetrad*, even if they divide once more by simple mitosis. In the Ascomycetes, tetrads form as *asci*, and the ascospores contained in the
asci germinate and grow into haploid mycelia, completing the life cycle. In this
and the following two chapters, we will explore the filamentous *Neurospora* and
Aspergillus and the unicellular Ascomycete *Saccharomyces cerevisiae*.

Before fungi were used specifically for the study of genetics, several areas
of fungal biology had developed along genetic lines. In studies of plant pathogens, the "dual phenomenon" had been recognized as a recurrent variation of
certain species. The term refers to the appearance of strictly mycelial strains
on the one hand and conidiating (sporulating) variants on the other hand during subculture of a single mycelium. The first variant is adapted to invasive
growth in plant tissues, while the second appears more readily on superficial
parts of the host, from which the conidia can disperse freely. The ability of
single strains to yield both variants reflects the heterokaryotic nature of the
unstable strain. In the multicellular filaments, genetically different nuclei corresponding to two types of growth habit coexist and segregate during growth
and spore formation into the two homokaryotic types.

Another area of research that became well developed in the 1930s concerned
the genetic determination of the mating-type systems of fungi, including water
molds, Ascomycetes, and Basidiomycetes. One finds heterothallic species in
all of these classes of fungi, but the mating-type systems differ. In heterothallic water molds and Ascomycetes, there are only two mating types, and these
behave as strict alternatives, determined by two stable alleles at single genetic
loci. In many higher Basidiomycetes such as the common mushroom, two loci
often govern mating type, yielding four mating types among the progeny of
any cross. This pattern is known as tetrapolar sexuality. The interest in this group
lies in the fact that each of the two loci has many alternative forms (multiple
alleles), so that large collections of a basidiomycetous species would include
hundreds of mating types. Mating of Basidiomycetes does not involve specific
male and female gametic cells or structures. Governed by the mating type system, fusion of vegetative mycelia takes place in the first step of a mating, and
diploid nuclei are found only in specialized cells of the fruiting body, where
the nuclei of different mating type fuse before meiosis. Genetic research with
Basidiomycetes in the 20th century therefore concentrated more on this unusual mating-type system and its associated developmental attributes. Due to
the unusual nature of their mating-type systems and lifestyle, Basidiomycetes
did not become model systems for attributes shared widely among living things.

Discovery of *Neurospora*

Cornelius L. Shear (1865–1956) and Bernard O. Dodge (1872–1960) discovered and named the genus *Neurospora* in 1927.[1] Dodge (fig. 3.1) had found

Fig. 3.1.   Bernard O. Dodge (*left*) and George W. Beadle (1967) (*right*). Dodge photo courtesy of the LuEsther T. Mertz Library of the New York Botanical Garden. Beadle photo by Chuck Painter, by courtesy of Stanford University News Service.

perithecia, the fruiting body in which the major events of the sexual phase of the organism take place (fig. 3.2), in cultures of a fungus called *Monilia sitophila*. This fungus had previously been classified as wholly asexual, reproducing only by hyphal growth and conidial formation. The authors described three species within the new group, which they renamed *Neurospora*. These were *N. sitophila*, *N. crassa*, and *N. tetrasperma*, of which the heterothallic *N. crassa* would become the most widely used in biochemical genetics.

Dodge received his Ph.D. in botany and physics in 1912 from Columbia University.[2] He stayed on as an instructor until 1920. During this time, he became a good friend of T. H. Morgan, who was then busy working out the fundamental genetics of *Drosophila* at the same institution. Dodge then went to the U.S. Department of Agriculture in Washington, D.C., where he worked as a plant pathologist under Shear, with whom he published the first *Neurospora* paper.

Dodge followed this first paper with a series of studies of the inheritance of the mating types *A* and *a*. Mating of *N. crassa* or *N. sitophila* strains, which requires fertilization of a protoperithecium by a conidium of opposite mating type, leads to the development of perithecia. Within the growing perithecium, haploid nuclei fuse (*A* + *a*) to form diploid nuclei (*Aa*) in clublike cells, the immature asci. The asci elongate as the diploid nucleus undergoes meiosis. The four nuclear products of meiosis, the tetrad, undergo a final postmeiotic division before spore walls form around each nucleus (fig. 3.2). The eight ascospores form a linear array in the ascus, and the spores can be separated from

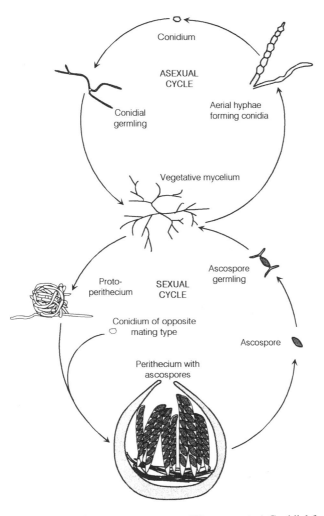

Fig. 3.2.   The life cycle of *Neurospora crassa*. (*Upper portion*) Conidial formation from aerial hyphae growing from the vegetative mycelium; dispersal and germination. (*Lower portion*) The sexual cycle, beginning with formation of the protoperithecium, followed by fertilization, ascus development, dispersal of ascospores, and (after heat or chemical activation) germination and growth of a colony.

one another under a dissection microscope, isolated, and cultivated separately. The ascospores of *Neurospora* species require a heat shock (commonly, for laboratory stocks, 60°C for 30 minutes) to initiate germination.

In his extensive studies with *Neurospora*, Dodge discerned patterns of segregation of the mating types *A* and *a* in individual asci. In each ascus, four *A* and four *a* spores were found.[3] These experiments were not the first to show that Mendelian segregation occurred in an individual meiotic event arising from a single diploid nucleus. But the clarity of this deterministic pattern was a great satisfaction for geneticists at the time, and the phenomenon has been pictured in textbooks ever since. The meiotic products of higher organisms, in contrast, were gathered from many cells undergoing the process and were hopelessly intermingled; segregation and other genetic phenomena were inferred from statistical ratios of progeny. Dodge, in fact, described variants of the segregation patterns of mating-type genes in the strictly heterothallic species of *Neurospora*.[4] Some segregations yielded, in the linear asci, the pattern *AAAAaaaa*, some *aaAAAAaa*, and still others *AAaaaaAA* or *aaAAaaAA*. With *Neurospora* as a major interest, Dodge returned to New York in 1928 as a plant pathologist at the New York Botanical Garden, just as Morgan was leaving for California. Dodge remained at the New York Botanical Garden for the next 20 years.

In 1929 or 1930 Dodge gave a seminar at Cornell University in which he described the different spore-order patterns for mating type in *Neurospora* asci. Although Dodge could not understand these patterns in terms of chromosome behavior, several people attending the talk could. One of these people was George W. Beadle (fig. 3.1).[5] He and his fellow listeners had become familiar with the details of chromosome behavior in corn, in which crossing over was by then known to occur at the "four-strand" stage; that is, after homologous chromosomes had each duplicated into two chromatids and had paired with one another. If a crossing-over event between two nonsister chromatids took place between the mating-type gene and the centromere of the chromosome, the spore order would not be *AAAAaaaa*. Instead, one of several other orders would be seen as indicated above, in which *A* and *a* spores would both be present in the upper and lower halves of the ascus (fig. 3.3). In short, segregation of the two mating types would be deferred to the second division of meiosis, a phenomenon generally known in genetics as "second-division segregation."

This, of course, was not a moment of choice for Beadle or for *Neurospora*, since even the Beadle-Ephrussi experiments were five years in the future. But the Dodge lecture acquainted Beadle with *Neurospora*, and this would play a part in the choice of an organism that began the revolution in genetics 10 years later.

The Development of *Neurospora* Genetics

Dodge, although not a geneticist by training, appreciated at first hand the benefits of *Neurospora* for genetic investigations and pursued them intermittently for some time. His analyses were direct and thorough, and they contributed to the later use of the organism in biochemical genetics. Because Dodge believed that *Neurospora* would be an ideal experimental organism for geneticists, he urged Morgan, as he left for Caltech, to take cultures of *Neurospora* and use them if he could. At Caltech, Morgan assigned an early graduate student, Carl C. Lindegren (1896–1987), the project of developing the genetics of *Neurospora*.[6] Lindegren's doctoral dissertation on the organism and papers published thereafter were major steps in the progress of *Neurospora* to its prime place in genetics at mid-century.

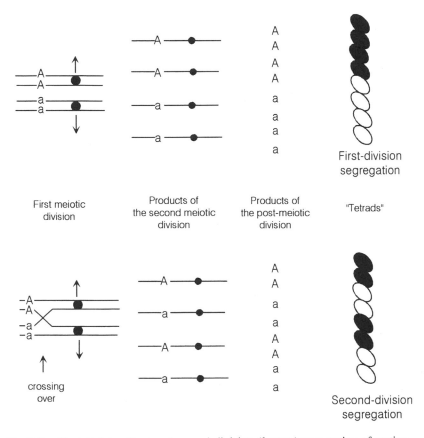

Fig. 3.3. First-division (*top*) and second-division (*bottom*) segregation of mating types *A* and *a*. At left, duplicated, paired homologous chromosomes are depicted during prophase of meiosis. The vertical arrows signify the later separation of homologs at the first meiotic division. The meiotic products, shown in the order in which they lie in the ascus, divide once before spore walls form around them. Second-division segregation occurs after an exchange between the homologs takes place between the centromeres (*dark circles*) and the mating type gene *A/a*.

Lindegren worked with enthusiasm to adapt the fungus for genetic analysis because of its technical advantages, much as Dodge had hoped. Under the direction of Morgan and with the advice of Sturtevant, Bridges, Dodge, and others, Lindegren first standardized the methods of making crosses, isolating ascospore progeny from tetrads, and heat-activating the ascospores. He favored the species *N. crassa* for later work because it had the most uniform temperature requirement for the germination of ascospores. Lindegren remarked on the lucky fact that heat activation of the ascospores killed the parental asexual spores (conidia) that would otherwise contaminate and confuse analysis of the ascospore progeny. From his wild-type stocks of *N. crassa*, he isolated a number of morphological and color markers, some of them induced by X-rays and others by ultraviolet light. He developed inbred strains that had shed some of the worst genetic characteristics of natural isolates, such as spore abortion, lack of vegetative and sexual vigor, and variable growth rate. He confirmed Dodge's findings regarding second-division segregation. He developed inbred stocks that differed from his standard wild type by single mutations, allowing him to map them. He reported many observations now forgotten, and which are often encountered in starting with an organism isolated from nature. These included mutations modifying expression of some of his genetic markers, frequent mutation or back mutation, genetic backgrounds with imperfect or variable sexual development, and the occurrence of recessive mutations in stocks, but hidden in the heterokaryotic state.

Lindegren's work was in retrospect severely limited by the nature of his mutants. All of them had altered morphology or color, and the morphological simplicity of the organism often made it difficult to distinguish their individual phenotypes in multiple mutant strains. Mutations interacted in ways that required inferences and further matings to establish the genotypes of certain progeny of complex crosses. Beadle and Tatum would escape this limitation with their more specific nutritional mutants, which permitted vastly more efficient methods of genetic analysis. Nevertheless, Lindegren performed the extensive spadework that made the later investigations proceed so quickly.[7] We may give credit to Dodge for identifying *Neurospora* and to Lindegren and his advisors for developing it as a model system. But it was to them a model for genetic analysis, not for the investigations for which *Neurospora* became most prominent.

Lindegren, whom we will meet again in connection with his contributions to yeast genetics, started his *Neurospora* work at Caltech in 1928 and continued it at the University of Southern California until 1939. During that time, Beadle and Ephrussi started their transplantation work in *Drosophila* at Caltech. Beadle was therefore well acquainted with Lindegren's *Neurospora* studies, and Beadle would quickly recall them as he embarked on studies of the genetic control of intermediary metabolism.

Beadle's Choice of *Neurospora*

Beadle and Ephrussi, in their investigation of eye-color genes in *Drosophila*, hoped that their findings would set them on a path to understanding gene action. Beadle returned from Paris to Harvard in 1936, but in 1937 he took a position at Stanford University as a professor. He took on Edward L. Tatum, a microbiologist and comparative biochemist, as a research associate. Beadle and Tatum spent the next four years trying to isolate and identify the v^+ substance, a supremely unrewarding effort, as discussed in chapter 2.

Tatum had previously worked on the nutrition and comparative biochemistry of bacteria, concentrating on vitamins. He had adopted the emerging view that bacteria shared with other forms of life the basic biochemical building blocks of proteins, nucleic acids, carbohydrates, lipids, and vitamins. Moreover, the metabolic reactions leading to these precursors also seemed to be widely shared among these and higher organisms.[8] Tatum had developed an interest in the diversity of nutritional requirements among related bacteria and the biochemical consequences of bacterial associations in nature.

Inspiration, chance, and careful choice led Beadle and Tatum to *Neurospora*. In 1940 at Stanford, Beadle listened to a lecture by Tatum on the nutritional requirements of different bacterial species. It occurred to Beadle that if such nutritional differences could arise during the divergence of evolving bacterial species, they might be found as mutational variants within a species.[9] This idea might not have taken hold so easily had Beadle not become familiar, in his earlier Paris sojourn, with the work of André Lwoff (1902–94), a protozoologist at the Pasteur Institute studying the comparative nutritional requirements of ciliates.[10] If metabolic mutants occurred in an organism that could be cultured in a defined medium—commonly used in bacteriological work—they might be identified by their need for specific, additional nutrients. Beadle thereby altered the fundamental question he had been asking. Instead of trying to identify the obscure, terminal steps in eye-pigment synthesis in *Drosophila*, he would turn to a genetic analysis of known biochemical reactions. This strategy had a double virtue. First, the biochemistry of many metabolic sequences had already been determined.[11] Second, if he could show that genes underlay these reactions, there could be little doubt that Mendelian genes were not confined in their role to "trivial" embellishments of development.

Beadle required an organism that could easily display biochemical deficiencies arising by mutation. Such mutants would be "reparable" by addition of nutrients to the growth medium. We see here the moment of choice of a model organism. Ten years separated Dodge's *Neurospora* seminar at Cornell and Tatum's lecture at Stanford, during which Beadle learned much about Lindegren's genetic work on the organism. In a moment in 1940, Beadle could now see that genetics might be used to analyze metabolism. *Neurospora* was thus consciously chosen as a model system to answer a very specific question.

One might ask why Beadle, in considering organisms to use, did not choose a bacterial species, perhaps one with which Tatum was familiar. This question is pertinent in view of the prominence that bacteria would later acquire in the genetic analysis of metabolism. This question has a complex answer. In 1940, the bacteria were quite a mystery to geneticists and cell biologists alike.[12] Many biologists thought that heredity in bacteria was wholly unlike that in familiar, eukaryotic organisms, those with a true nucleus. No chromosomes had been dependably described in bacteria. No obvious sexual or recombinational mechanisms had been discovered. This led some respectable biologists to the view that bacteria lacked genes. This curious position was not as absurd as it now sounds. The gene itself had not been described physically, and it had been argued not so long before that much of the hereditary endowment of cells, including bacteria, lay in the cytoplasm as a persistent or metastable state of metabolism.

If Beadle were to relate genes and physiological reactions, he would have to choose an organism displaying all of the operational features of genes and the chromosome theory. The chromosome theory by that time had become well established. Genes, whatever they might be, were carried by or as part of the chromosomes, whatever their structure or composition might be. Therefore, only a eukaryotic organism would do. The most familiar model systems for genetics—mice, corn, and fruit flies—were the least tractable for fundamental biochemical work, as Beadle, Ephrussi, and Tatum, among others, had found. Other investigations of genes and chemical reactions, sharing the rationales of Beadle and Ephrussi, would seem to have more promise. These were the investigations, mentioned previously, of flower-color mutants at the John Innes Horticultural Institution and the mutational analysis of mating substances in the alga *Chlamydomonas*, carried out by Moewus. But for different reasons, neither of these groups became major, continuing contributors to biochemical genetics.

By the time Beadle sought a suitable experimental organism in 1940, a simple synthetic medium had been devised for other fungi, on which *Neurospora* also grew well. Beadle therefore obtained *Neurospora* stocks from Lindegren and wrote to Dodge for information about other stocks that might be of use. In a letter of February 27, 1941, Beadle wrote:

> Dr. Tatum and I are interested in doing some work on the nutrition of Neurospora with the eventual aim of determining whether these requirements might be dependent on genetic constitution. I have written Dr. Lindegren to see if he can supply us with some stocks. If preliminary experiments prove to be encouraging we will be interested in trying out the available species and also various collections of "wild types" if we can get them.[13]

The First Days of Biochemical Genetics

In December 1941, Beadle and Tatum published a seminal article in the *Proceedings of the National Academy of Sciences* entitled "Genetic control of biochemical reactions in Neurospora."[14] The article described three vitamin requirements that had arisen by X-ray–induced mutation. The mutations were obtained by irradiating conidia, the asexual spores, and using them in matings with untreated female parents of the opposite mating type. The ascospore progeny of the matings were pure in genotype. Thus, after germination, the rare mutant ascospores among the progeny expressed their mutant phenotypes despite the recessive character of the mutations. Beadle and Tatum heat-activated individual spores in tubes containing a "complete" medium supplemented with amino acids, vitamins, and other small organic molecules that the mutants might require. Strains that could not grow without a supplement were identified by testing all ascospore cultures individually on a simple medium containing only the nutrients needed by the wild type ("minimal medium," consisting of salts, sugar, and the vitamin biotin). Finally, tests of the new mutants on individual components of the complete medium identified the nutrient they required. The specificity of the requirements of the first mutants and their orthodox, Mendelian behavior in crosses convinced many that the method promised discoveries about the genetic control of life's most basic functions.

The program would soon be known as a test of the one-gene, one-enzyme hypothesis, but more broadly it sought to explore the role of genes in metabolic reactions. Because the program connected the two fundamental fields of genetics and biochemistry, the 1941 paper of Beadle and Tatum received widespread attention even as the war in Europe preoccupied most of the Western world.

At this time, *Drosophila* began to wane in its role as a model organism. While once it claimed the attention of biologists in general, in 1941 that audience had become more restricted, and for the next 25 years the research would not greatly transcend the field of genetics. In fact, despite the strong advances made in many fields such as neurophysiology, much of biology had settled into routine, specialized work. This was certainly the case with developmental biology, microbiology, and biochemistry. Conservatism prevailed in many institutions and scientific societies. Therefore, even while Beadle and Tatum enjoyed the acclaim of many, they invited the skepticism of traditional biologists. The latter, including some biochemists, resisted the notion that the primary role of genes might easily be discerned, much less that genes had unitary roles. In their view, living cells simply had too many working parts, too many simultaneous reactions, for the mutants of Beadle and Tatum to offer great insight into cell function. The interaction of cell constituents was a web of cause and effect that could not possibly be disentangled. This view was not wholly off the mark at the time; Beadle and Tatum's first three mutants could hardly lead biologists and biochemists to abandon long-standing attitudes.

Beadle and Tatum's work progressed rapidly during the war years, with the goal of generalizing their first findings. Their ability to continue the work during wartime relied in part on government financial support for the development of bioassays for various nutrients, using *Neurospora* mutants requiring particular nutrients for growth. The support not only facilitated the work, but rendered many of the assembled research team exempt from the draft.[15] The dedicated team included Norman Horowitz (b. 1915; trained as an embryologist and geneticist), Herschel K. Mitchell (1913–2000; a chemist), David Bonner (1916–64), Adrian Srb (b. 1917), and Francis Ryan (1916–63; visiting from Columbia). All were authors of papers emerging from the Stanford laboratory during the war. However, a 1945 summary of results acknowledges the technical help of at least 20 additional people. The work continued as Beadle moved in 1946 with Horowitz, Mitchell, and several others to Caltech to become Morgan's successor as chairman of the Biology Division. Tatum moved to Yale in 1945, where he also continued work in biochemical genetics.

The 1945 summary of the group's results lists more than 380 mutant strains, obtained by X-ray or ultraviolet (UV) irradiation, among 68,198 ascospores tested.[16] Requirements for most of the B vitamins, essential amino acids, purines, and pyrimidines are represented. Three areas began to preoccupy *Neurospora* workers at this point, themes that would continue, with others, to the present day. These were (i) metabolic pathways, (ii) enzyme specificity, and (iii) genetic phenomena such as tetrad analysis, genetic mapping, and heterokaryosis.

Metabolism

A project in metabolic pathways concerned a metabolic sequence leading to the amino acid arginine, in which different mutations affected different steps of the pathway.[17] True to the original intent of Beadle and Tatum, this work dealt with a well-known metabolic pathway in mammals, the genetic control of which might be determined. Unlike the first mutations of 1941, the various *arg* mutations all concerned a single pathway; all could grow if arginine was added to the medium. But the mutants could be distinguished by two intermediates in the pathway, namely ornithine and citrulline, that supported the growth of certain mutants. It appeared that some mutants could grow only on arginine, some on either arginine or citrulline, and still others on either arginine, citrulline, or ornithine. Significantly, mutants able to grow on ornithine but not on citrulline were never found. This demonstrated that synthesis of arginine must proceed from ornithine, then citrulline, and not the reverse. Just as important, the results showed that only one enyzme appeared to be missing in each mutant strain, and the others performed their normal roles if an intermediate following the genetic block was provided. Like the limited conclusion of the order of the v^+ and cn^+ substances of *Drosophila*, the arginine pathway gave strong support for the idea that each gene might affect a single enzymatic reaction. It also gave a rationale for studies up to the present for determining the sequence of gene action in any complex process.

An extremely important advance was the use of heterokaryons to distinguish mutations, such as the *arg* mutations, having the same basic phenotype.[18] The significance of the "heterokaryon test," as it was called, was in its ability to determine whether two mutations affected the same or different genes, even if they had the same basic nutritional phenotype (fig. 3.4). Many such sets of mutants arose in early studies, and no simple nutritional test such as growth on intermediates could distinguish them, as they had in the arginine pathway. Indeed, the laboratory was busy using the mutants to identify unknown intermediates. Such experiments could be done by isolating the intermediates that accumulated in mutant, enzyme-deficient cells that failed to metabolize them further. Even in the arginine pathway, three different genes appeared to block the synthesis of ornithine, three genes were needed to convert ornithine to citrulline, and two were needed to convert citrulline to arginine. Even in this case, therefore, one could not know without further study whether each gene controlled a separate enzymatic reaction or cooperated to effect a single reaction.

In the heterokaryon test (fig. 3.4), fusion of two mutant mycelia, each carrying a nutritional deficiency, leads to the formation of a heterokaryon; that is, the two nuclear types lie in a common cytoplasm. If the nuclei carry mutations in different genes, the recessive deficiency of one nucleus is overcome by the normal allele of the gene in the other nucleus and vice versa. Thus growth of the heterokaryon on an unsupplemented (minimal) medium indicates that the mutations lie in two different genes. In contrast, if both mycelia carry mutations in the same gene, the heterokaryon usually does not grow in minimal medium because neither nucleus can perform the function affected. This method

a. non-allelic mutations (*gene-1*⁻ and *gene-2*⁻)

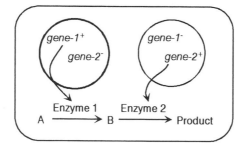

b. allelic mutations (*gene-2ᵃ* and *gene-2ᵇ*)

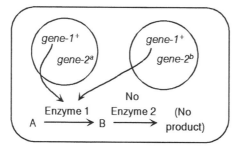

Fig. 3.4.   The heterokaryon test. (*a*) Non-allelic genes (by current definition) affect different functions, in this case enzymes. If nuclei with nonallelic mutations coexist in a heterokaryon, all functions can be performed because what is lacking in one nucleus is present in the other. (*b*) Allelic mutations fail to complement because neither nucleus can perform the vital function of gene 2.

is an example of the so-called complementation test, allowing rapid determination of the functional identity or nonidentity of two mutations. It thus represents a test of allelism. The test is simple and effective, permitting rapid sorting of collections of mutants affecting a single biochemical pathway into sets representing individual genes. The heterokaryon test is now used routinely to sort mutations affecting all kinds of different features of organisms.

Several other technical advances had general significance. One was the discovery that the nonmetabolizable sugar sorbose, if used in the proper ratio with sucrose or glucose, restricts colony growth in agar medium (fig. 3.5).[19] Normally, *Neurospora* grows radially very rapidly, covering a Petri dish with a thin web of mycelium in a single day, only later branching more profusely and deeply into the agar. Sorbose encourages dense branching, so that compact colonies form upon platings of conidia or ascospores. Media containing sorbose allowed investigators to handle *Neurospora* like bacteria and encouraged major programs in mutational analysis, recombination analysis from random spore platings, conidial frequency determinations to measure the nuclear ratios of heterokaryons, and direct selection of mutants after plating.

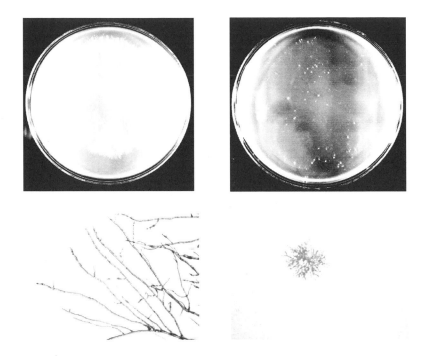

Fig. 3.5.   The effect of sorbose on the growth of *N. crassa*. The upper photographs show (*left*) a mycelium grown for 24 hours at 32°C without sorbose, and (*right*) colonies originating from conidia plated on a sorbose-containing medium as they appear after 72 hours growth at the same temperature. The lower micrographs show details of the hyphae of the two types of colony. Note the small diameter and intense branching, imparted by limiting amounts of glucose and the presence of sorbose. The lower micrographs were taken at similar magnifications. From Davis (2000).

A second important advance was the development of negative selection methods. Beadle and Tatum's work up to and after 1945 depended on "brute-force" methods, in which large numbers of strains derived from mutagenized conidia had to be tested individually for nutritional requirements. In 1952, Srb's group developed a method of enriching mutagenized populations of cells for mutants by a simple mechanical means.[20] Simply stated, ultraviolet-irradiated conidia are allowed to grow in a shaken, liquid minimal medium. The population is filtered through cheesecloth periodically and often over several days. This removes the growing, wild-type germlings and allows mutant, nutritionally deficient conidia, which do not grow in this medium, to pass through and remain in suspension in the liquid. When no further growth occurs in the filtered medium, the population is plated on a medium supplemented with one or more specific nutrients. Among the colonies that grow are mutants that specifically require the added supplement, at hundreds or thousands of times their original frequency in the original, mutagenized population. A later negative selection method, the "inositol-less death" method, is similar: inositol-requiring cells (*inl*, a mutant type isolated in the early days) die if allowed to germinate in the absence of inositol. This does not happen if new mutations impose unsatisfied, additional growth requirements on cells. This provides a method of enriching mutant types in irradiated populations of *inl* conidia, the survivors of a period of inositol starvation.

Enzyme Specificity

The second theme of the studies after 1945 was the notion that each gene specified a given enzyme. In addition to mutants having absolute requirements for nutritional supplements, the laboratory isolated many with partial requirements (whimsically called "leaky" to this day): they could grow, but not at an optimal rate, on unsupplemented medium. These mutants suggested that an enzyme was not lost, but only impaired in function. The idea was reinforced by a number of "conditional" mutations that had been isolated earlier, ones that would express a growth requirement only in part of the pH range (pH sensitive) or temperature range (temperature sensitive) of growth. The leaky or conditional mutations corresponded to Muller's hypomorphic mutations (those with minor damage) in *Drosophila*. Certain mutations of the temperature-sensitive type—those that would not grow well in any medium at the restrictive temperature—would much later become a central tool of the study of "indispensable" genes. These are genes that could not be identified otherwise, since complete deficiencies of such genes are by definition lethal.[21]

In 1945 Beadle described the role of genes thus: "Each of these thousands of gene types has, in general, a unique specificity. This means that a given enzyme will usually have its final specificity set by one and only one gene."[22] The term "specificity" could be used only in a formal sense, and Beadle guarded against a commitment to a mechanism. (In fact, others speculated that genes actually *were* enzymes or their nuclear prototypes.) By 1945, his group had proved the use of a novel mutational method to dissect biochemical sequences as well as extensive, if not conclusive, evidence for the one-gene, one-enzyme hypothesis. The work was the beginning of what is rightly called the molecular revolution.[23]

The *Neurospora* community's major efforts in the area of enzyme specificity turned in the latter half of the 1940s to concerns raised by scientists outside the core groups. Beadle and Tatum, in the 1945 summary paper, predicted the occurrence of mutants unable to grow on complete medium simply because the appropriate supplement could not be provided in the medium; that a supplement, even if provided, could not enter the cell; or that use of one supplement competitively blocked the use of another. Indeed, many "unknowns" were found, and this is why the heterokaryon test was so useful in distinguishing them from one another. But in 1946, the physicist Max Delbrück (1906–81;

discussed in chapter 8) asked whether the methodology of Beadle and Tatum might simply restrict them perforce to mutants of a one-function character. For instance, any gene whose mutants simultaneously caused both an irreparable (lethal) and a reparable deficiency would never be isolated.[24] Conversely, two or more genes might be required to specify a single enzyme, but Beadle and Tatum might find a mutation for only one of them. One could not rule out such possibilities with certainty.

As more detailed evidence for the role of genes in the determination of enzymes accumulated, cases of dual requirements did indeed arise. But some of these mutations were simply shown to be blocked in the formation of a single precursor common to two biochemical pathways. In another case, a mutation imposing a dual requirement for the amino acids isoleucine and valine was shown to block a single enzyme that catalyzed similar reactions in the two metabolic sequences. Indeed, these examples illustrate the power of biochemical mutants to illuminate metabolic systems.

But Delbrück's objection, that if single mutations had multiple effects, one or more of which was lethal, they would never be seen, still stood. Such a possibility rendered the one-gene, one-enzyme hypothesis unfalsifiable and therefore not formally rigorous. Horowitz and Urs Leupold (b. 1922) then showed that a general class of mutants mentioned above could be isolated as an unbiased class: those in which the mutant phenotype appeared only at higher temperatures.[25] They isolated temperature-sensitive mutants, simply as strains unable to grow on minimal medium at high temperature. When they were tested for growth on single amino acids or vitamins at the higher temperature, Horowitz and Leupold found that many of them could do so; that is, they were monofunctional. This argued against the view that such mutants would have multiple, unknown, lethal derangements. Statistical considerations comparing the spectra of mutations of this type and those isolated by normal procedures justified the conclusion that multifunctional defects were at best uncommon.[26]

When we look at the status of the work in 1960, the question was no longer important to most workers, because other findings transcended the initial one-gene, one-enzyme slogan. In 1949, the laboratory of Linus Pauling (1901–94) at Caltech showed a physical difference between normal hemoglobin and the mutant form of this human protein in patients with sickle-cell anemia.[27] During the next eight years, methods of defining the structure of proteins progressed rapidly. It became clear that proteins consisted of one or more unbranched strings of amino acids, which we now call *polypeptides*. Not only that, but some proteins might consist of more than one type of polypeptide. Indeed, in 1957, Vernon Ingram showed that the difference between normal and sickle-cell hemoglobin arose through the mutational substitution of glutamic acid by valine in one position of one of the two types of polypeptide (α- and β-globin) of the protein.[28] This showed for the first time the extraordinarily specific effect of a particular mutation, a discovery made, ironically, in a species wholly unsuited to experimental genetics.

The sickle-cell discovery supported a growing confidence that genes encoded the amino acid sequence proteins, a confidence based less on particular discoveries and more on the pace of molecular insight into the possible role of DNA. DNA structure was determined in 1953; the word "encode" emerged strongly only after that. The word is part of the informational attitude to genes, as opposed to the structural, or template notions entertained previously. In view of the findings with hemoglobin, the one-gene, one-enzyme phrase was modified to "one-gene, one-polypeptide." At about the same time, certain point mutations of *Neurospora* were shown to affect two or more very different catalytic activities, a clear violation of the one-gene, one-enzyme hypothesis.[29] Ultimately, such cases revealed that certain polypeptides might carry two or more catalytic domains and thus catalyze several enzyme reactions. By the time these peculiarities were clarified, few investigators remained adamantly opposed to Beadle and Tatum's general theory. Faith prevailed that all would become clear in a short time. The changeable term for the putative product of a gene (enzyme, protein, polypeptide) reveals advances in specifying what a gene specified, and a relaxed attitude toward minor anomalies and ambiguities. This was normal science for the time: a rush of discovery proceeding without pedantry.

Only in 1948 did clear proof of an enzyme deficiency in *Neurospora* arising through mutation come to light, a year before Pauling's hemoglobin article.[30] But even in 1956, Horowitz could cite only four cases, in three different organisms (*Escherichia coli*, *Neurospora*, and humans) in which an enzyme had been shown to be structurally altered as a result of a mutation.[31]

The narrow focus of Beadle and Tatum on metabolic mutants and the one-gene, one-enzyme question closely resembled the focus and rationales of the old fly room at Columbia University, where Morgan had presided over the adolescence of genetics 20 years earlier. The *Neurospora* workers, concentrated at Stanford University until 1946, shared ideas and resources freely in a small place, led by a charismatic and supportive organizer. They isolated and mapped mutants and developed techniques of culture, mutant selection, and matings as a necessary accompaniment to the main experimental purpose. Where Morgan pursued the chromosomal basis of inheritance, Beadle and Tatum amassed evidence that individual genes controlled single biochemical reactions. The two groups both worked in the face of much conventional wisdom in their respective primes. Morgan wanted to promote a nucleocentric view of inheritance; Beadle and Tatum wanted to demonstrate that Mendelian genes had fundamental, straightforward roles in cell function. These counter-conventional stances required narrowly focused attention to the respective goals, driving home easily grasped points forcefully, without much initial concern for secondary and general implications.

Genetics

The third preoccupation of many *Neurospora* investigators after 1945 was the basic genetics of *Neurospora*. Much had been learned since the days of Dodge and Lindegren, but with all the new mutants and the need for classifying them came intense efforts to define the linkage relationships of the mutants and the number of linkage groups. The chromosomes had been visualized by Barbara McClintock (1902–92) in 1945, and as mutants accumulated, they ultimately distributed themselves genetically into seven linkage groups.[32] The use of chromosomal rearrangements in cytological and genetic experiments identified these linkage groups with the seven chromosomes now known.

In 1954, *Neurospora* research proceeded strongly only at Caltech, Yale, Stanford, Oak Ridge National Laboratory, and a few other places. As the diaspora began thereafter, a Fungal Genetics Stock Center was formed. The first major summary of standard genetic work in 1954 described the available mutants, presented detailed linkage maps, and described rationales of genetic analysis with dissected tetrads and random-spore progeny.[33] This, together with a detailed methods article published previously by Ryan, allowed workers outside of the main *Neurospora* research centers to initiate work independently of one another.[34]

The genetic work soon led to knowledge of how crossovers influenced one another, the distribution of crossovers on single chromosomes, and the behavior of chromosomal aberrations such as translocations of chromosome parts.[35] In addition, tetrad analysis allowed detection of the phenomenon of gene conversion. The term *gene conversion* refers to the appearance of individual tetrads (asci) with aberrant ratios of alleles of a given gene (5:3 or 6:2 instead of the usual 4:4), as though one of the alleles had become converted into the other. The discovery was actually a confirmation of a less dependable earlier claim by Lindegren, working then on yeast (see chapter 5).[36] For a short time, this subject was pursued mainly in *Neurospora* and several related fungi, in which the properties of the phenomenon were largely revealed.

Neurospora Becomes a Tool

At this point *Neurospora* was on its way from being a model system to becoming a model organism. It acquired genetic depth, efficient techniques, and a phenotypically diverse array of mutations. The odd mutants that appeared, often without specific selection, engaged the curiosity of people in the laboratory who studied them further for their own sake. Knowledge of pigment formation, sexual disfunctions, colony morphology, conidial formation, and a host of other attributes could be pursued by a number of workers unfamiliar with or uninterested in biochemistry. The organism engaged even undergraduates learning genetics. The increasing knowledge about the organism provided a rich biological context in which new mutations and observations could be interpreted. Investigators turned to *Neurospora* to investigate phenomena discovered elsewhere, those arising in studies of *Neurospora* itself, and those for which *Neurospora* appeared to be a good place to start.

Several examples illustrate this phenomenon. Formal genetics took an entirely new direction with efficient tetrad analysis. In addition, the large number of spores obtainable from any cross, and selective methods for rare recombinants (the wild-type in crosses of strains carrying closely linked mutations), made recombination studies at high resolution possible. The study of the biological aspects of heterokaryosis, including the control of nuclear ratios and several incompatibility genes that restrict the formation of heterokaryons, has become a model for many other fungi. Tetrad analysis offered a proof of the non-Mendelian nature of some mutations affecting respiration. This observation stimulated study of mitochondrial physiology and biogenesis, ultimately revealing mitochondrial DNA and its role in organelle inheritance (chapter 14). A tool for measuring linear growth rates, the "race tube," was developed in 1943. A horizontal tube half-filled with agar was inoculated at one end and marked each day. Used initially for bioassays of vitamins and amino acids using nutritionally dependent strains, the race tube is now used for leading studies of biological rhythms.

The more detailed analysis of biochemical pathways and the structure of enzymes continued strongly after 1950. Among the most prominent of these efforts were the detailed studies of the tryptophan pathway by Bonner, Charles Yanofsky (b. 1925), and their followers.[37] Contributions in this area were well coordinated. They included the determination of the order of biochemical intermediates in the pathway, one of the early fine-structure maps of a single gene (*trp-3*), the development of a technique to detect catalytically inactive protein (cross-reacting material, or CRM) by immunological methods, and the detailed study of a complex enzymatic reaction, tryptophan synthetase, using mutational techniques.

Yanofsky's group was among the first to abandon *Neurospora* for the bacterium *Escherichia coli*, known best as *E. coli*. After 1954 the group became interested in the fundamental question of how genes and enzymes were related at the level of sequences of nucleotides in DNA and amino acids in proteins. The frustrating fact was that the enzyme of interest, tryptophan synthetase, could not be purified adequately from extracts of *Neurospora* mycelia. The small initial amount of the enyzme and the protein-digesting enzymes liberated upon extraction of *Neurospora* cells rendered the yield of undamaged enzyme too low for further study. As this became clear, *E. coli* had entered the world of genetics and offered a vastly improved biological system for metabolic analysis and enzyme preparation. Yanofsky switched immediately to this organism and remained with bacteria for many years thereafter.[38] He was followed by a number of others working on *Neurospora* at the time.

More important, by 1955 biochemical, genetic, and molecular work on bacteria and bacterial viruses had flourished for more than a decade and had demonstrated that many of life's major secrets could much more easily be uncovered with these organisms. The frontier had moved, with new investigators turning to *E. coli* as the new hub of molecular genetics. By 1960, *Neurospora* had lost its preeminence as *Drosophila* had 20 years before to *Neurospora*. *Neurospora* would begin to serve a subordinate role in biochemical genetics, as a eukaryotic basis of comparison for findings with *E. coli*. *E. coli* would initiate paradigms and *Neurospora* would provide tests of their universality. The change of focus from *Neurospora* to *E. coli*, which I detail in chapter 7, is a prime example of how one model organism displaces another as techniques and the interests of a field develop.

Neurospora was the first microbe to establish a homestead in the analysis of gene action. Within five years a diverse garden of other microorganisms, including *E. coli*, would flourish around it. *Neurospora* soon became just one of many model organisms in the new biology. We will see how these other models were chosen, developed, and discarded by an intellectually aggressive horde of biologists seeking not homes, but frontiers.

4

Aspergillus

I do not like biochemistry.
—G. Pontecorvo

Biochemists of a reductionist persuasion took pride during the late 1940s and early 1950s in the increasing resolution of metabolic sequences achieved by identifying the intermediates and purifying the enzymes that catalyzed their conversions. At the same time in the field of genetics, investigators such as Muller sought to increase the resolution of chromosome structure and even of genes themselves with recombinational mapping and isolation of multiple alleles of individual genes. In Edinburgh, Muller promoted further studies of X-ray mutagenesis. He encouraged Charlotte Auerbach (1899–1994), a resident investigator, to test the possibility of chemical mutagenesis, a test that succeeded. By that time, the mutagenic effect of ultraviolet light had become known. These developments gave hope of progress, by a sort of mutagenic dissection, on questions of gene structure.

Guido Pontecorvo

In 1937, a Jewish Italian expatriate, fleeing Italy as the anti-Semitic climate worsened there, came to the University of Edinburgh, where he found a temporary position at the Institute of Animal Genetics. His name was Guido Pontecorvo (fig. 4.1). He had an improbable research background in breeding silkworms and cattle, but this equipped him for professional work in genetics. Pontecorvo's serendipitous association with Muller began, and Pontecorvo worked under him between 1938 and 1940 on a doctoral thesis on *Drosophila*.

Fig. 4.1. Guido Pontecorvo. Photo courtesy of The Imperial Cancer Research Fund.

In the process, he became imbued with Muller's enthusiasm about the nature of the gene. Upon Muller's return to the United States in 1940, Pontecorvo took a research position for much of the war years in Glasgow, and after a year in Edinburgh in 1944–45 he was offered a tenured position as Lecturer in Glasgow, where he remained until 1968.[1]

In his early years in Glasgow, Pontecorvo worked for a short time on improving the yield of penicillin from the fungus *Penicillium notatum*, an effort sponsored by the British government because of its importance to the war effort. Before the group had achieved any success, the work was taken out of the hands of the British and transferred for reasons of security, safety, and funding to the United States. The work with *P. notatum* was Pontecorvo's introduction to fungi, a step that would mature into a major research program with the related filamentous Ascomycete, *Aspergillus nidulans*. *A. nidulans* would soon achieve the status of a model organism, chosen not for its use in solving an individual problem, but for general reasons. Pontecorvo, a faithful disciple of Muller, would embark on broad-ranging work with *A. nidulans*, laced with speculation about new genetic systems, the nature of genes, their arrangement on chromosomes on the one hand, and gene function on the other. We will see that his style in formulating problems resembled Muller's as strongly as Beadle's single-minded approach resembled Morgan's.

Pontecorvo's Choice

Pontecorvo's choice of *A. nidulans* for genetic work was quite deliberate and was not made definitively until about 1949. His writings before that time cover many of the topics that converged in that era to unite genetics and microbiology. This makes it impossible to assign a single, burning problem for which this model organism was chosen. Rather, the time required to settle on *A. nidulans* reflects the competing influences of many intellectual and experimental innovations in the field. Because of the diffuse determination of the choice of *A. nidulans*, the organism never achieved a status comparable to that of *Neurospora*. However, it joined *Neurospora* as a companion fungus at a time when diverse microbial systems were producing enormous insights about genetic, biochemical, and, finally, molecular phenomena.

Pontecorvo's first study with fungi focused on *P. notatum* and included a genetic proof of heterokaryosis in 1944.[2] Using X-ray–induced mutations, he noted that phenotypically similar mutants of the fungus could fuse and form a heterokaryon with a normal phenotype. The component nuclei thus carried different, recessive mutations that complemented one another in a common cytoplasm (fig. 3.5). However, unlike those of *N. crassa*, *P. notatum* heterokaryons (and, as he later found, those of *A. nidulans*) naturally sectored thereafter into homokaryotic patches as the colony grew radially unless selection for growth on minimal medium were applied to maintain the heterokaryotic state. Here, in the asexual phase of this fungus, Pontecorvo believed that he had found a genetic system somewhat equivalent to that of sexual cycle. Cell fusion brought together different parental nuclei; dominant–recessive relationships of alleles could be established; and segregation took place, albeit of intact, haploid nuclei rather than of genes.

In connection with the last point, I emphasize the difference between *heterozygotes* and *heterokaryons*. Heterozygotes, common in higher organisms such as fruit flies and humans, have diploid nuclei that carry different alleles of one or more genes on homologous chromosomes. Heterokaryons, common in filamentous fungi, are multinucleate hyphal cells that contain haploid nuclei that may carry different alleles of one or more genes. This was not a new subject. Heterokaryosis had been discovered at the turn of the century, and both Dodge's and Beadle's group had worked out the principles of heterokaryosis in *Neurospora*. In fact, Beadle and Coonradt's comprehensive article on their work came out the same year as Pontecorvo's.[3] Nevertheless, in Pontecorvo's words, heterokaryosis deserved "careful investigation in view of its important implications for the theories of gene action and of the evolution of genetic systems."[4] He would make good on his promise soon enough, but in the meantime other interests preoccupied him.

Two years later, in 1946, Pontecorvo reported on an important, early meeting of microbial geneticists, "Gene Action in Microorganisms."[5] He cited several studies of current interest in the field of genetics. These included the work of Beadle and Tatum on gene–enzyme relationships; of Lindegren and

Spiegelman on adaptive enzymes in yeast (attributed to self-duplicating, cytoplasmic gene products; see chapter 5); and Sonneborn's analysis of the self-replicating *kappa* particle in *Paramecium aurelia* (chapter 13). In the last two of these studies, Pontecorvo saw the possibility that the action of constant genomes during embryonic differentiation might be modulated by variable cytoplasmic states or particles established early in development and maintained by feedback to the nucleus. Of special interest to Pontecorvo was the possibility that the nuclear membrane might mediate nucleo-cytoplasmic relationships in specific ways during development. While Pontecorvo does not say so in his review, we may speculate that he felt this issue might be approachable with heterokaryons, where two nuclear membranes separate two genomes. This interest would emerge explicitly in his development of the *cis–trans* test in *A. nidulans*, in which he could compare heterokaryons and heterozygotes. The *cis–trans* test is a formal version of the complementation test. Thus an interest in the action of the genetic material and the role of the cytoplasm, both topics of general interest at the time, merged with experience in his own laboratory.

In the same year (1946), Pontecorvo delivered an elaborate, speculative paper on heterokaryosis at the Cold Spring Harbor Symposium.[6] Here, he advertised the properties of heterokaryons in filamentous fungi in general. His greatest interest centered on the physiological equivalence of heterokaryosis and heterozygosity. But he was intrigued by the possibility that natural selection during growth of heterokaryons would, as Beadle and Coonradt had inferred, establish regions of optimal nuclear ratio. The species Pontecorvo worked with then (*Aspergillus oryzae*) had uninucleate conidia. By using selective media, he could determine the ratio of conidia of two types produced by a heterokaryon. This in turn would allow him to estimate the nuclear ratio of the heterokaryon itself, which he could then relate to the physiological state of the heterokaryon. Such studies would have significance not only in cell biology, but could deepen the study of heterokaryosis in nature. The attributes of heterokaryons were independent of sexual reproduction, and therefore the phenomenon might be not only common, but particularly important in the evolution of asexual fungi. He added that multinucleate organisms like fungi might be thought of as populations of nuclei in which the properties of variation, drift, migration (through fusion with other mycelia), mutation, and selection all prevailed. Although Pontecorvo speaks in retrospect of his "pottering" with heterokaryosis in his work in the mid-1940s, his enthusiasm for it at the time is unmistakable.[7]

Pontecorvo was at least as interested in the structure of the genetic material as he was in heterokaryosis. At the 1946 Cold Spring Harbor symposium, Beadle and Tatum again reported on their work on biochemical mutants of *Neurospora*. In addition, the beginnings of other new approaches with microorganisms appeared. These included Lederberg and Tatum's first results on bacterial recombination and the work of Delbrück, Luria, and Hershey on mutation and recombination in bacterial viruses. These and other presentations demonstrated the ease of using microorganisms for genetic studies of these fundamental processes. The opportunity to detect rare events such as the appearance of wild-type recombinants in matings and of back-mutation of nutritional deficiencies in large populations of fungal cells, bacteria, and bacterial viruses was a huge advantage in the study of the nature of the gene. Indeed, Muller's desire to increase the resolution of genetic analysis had clearly infected Pontecorvo, who would choose *A. nidulans* over *Penicillium, Neurospora,* or the red bacterium *Serratia marcescens* to pursue his interest. The actual basis of the choice is not entirely clear, and in a retrospective essay he implied that *Neurospora* was already claimed by the Beadle group and regretted that *S. marcescens* might have been a good choice, given the fact that bacterial recombination was in its infancy.[8] But Pontecorvo's familiarity with the lower Ascomycetes certainly determined in part his use of *A. nidulans,* even though the most unusual discoveries about the organism had not even been hinted at by findings up to that time. In making his choice, Pontecorvo hoped to solve problems of general interest, not merely to work up the genetics of another organism.

Like *N. crassa, A. nidulans* is an Ascomycete, but of a different and somewhat simpler sort. It has the usual microbial virtues of haploidy, simple nutrition, and large crops of unicellular asexual spores, the conidia. In addition, *A. nidulans* is metabolically versatile, able to grow on many carbon and nitrogen sources, and it grows into compact, green colonies of mutable phenotype (fig. 4.2). Its asexual spores are borne on aerial hyphae. These hyphae form bulbous ends, from which sterigmata protrude. The sterigmata then form characteristic, regular chains of conidia, each containing one nucleus. The many chains of conidia at the tip of the conidiophore give its ends a characteristic brushlike appearance. The sexual cycle (fig. 4.3) culminates in the formation

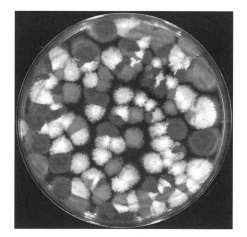

Fig. 4.2.   Petri dish showing *A. nidulans* colonies growing in agar. The colonies are ascospore progeny from a cross of a normal, dark-green parent and a white parent. The segregation of white and the dark colors is clear; among the dark colors, another difference appears due to the segregation of the wild-type green color and a yellow mutant carried by the white parent, but masked by the white mutation. Photo courtesy of Michael J. Hynes, University of Melbourne.

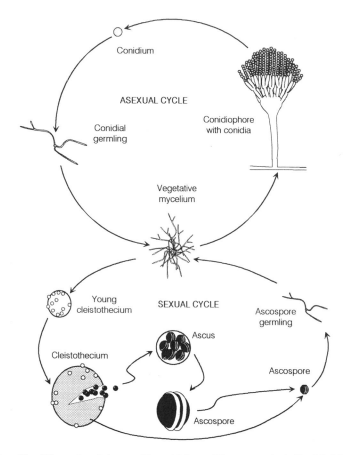

Fig. 4.3.   The life cycle of *Aspergillus nidulans*. (*Upper portion*) Conidial formation on conidiophores growing from the vegetative mycelium; dispersal and
germination. (*Lower portion*) The sexual cycle, beginning with formation of the
cleistothecium (self-fertilized), ascus development, dispersal of ascospores,
germination, and growth into a colony.

of fruiting bodies called cleistothecia, which are structurally simpler than the
perithecia of *Neurospora*. Many asci form within the cleistothecia and, through
the meiotic process, each ascus forms haploid ascospores. *A. nidulans* is homothallic, unlike the heterothallic N. crassa, and therefore nuclei of a homokaryotic
mycelium may fuse in cleistothecia to form the diploid phase and complete
the sexual cycle. As we will see, however, cross-fertilization may be effected
by forming heterokaryons containing different nuclear types.

    Few geneticists would agree that a homothallic organism could easily be
domesticated for genetic analysis. In such organisms, self-crossing is the rule,
while the geneticist requires crosses between known parents of different origin. Nevertheless, from the earliest time, Pontecorvo felt that the advantages
of homothallism outweighed its disadvantages. Homothallism permits crosses

between any two parents, since there is no mating type barrier. Pontecorvo reasoned that if "heterozygous crosses" could be recognized or even favored in some way, the inconvenience of self-crossing could be bypassed. Indeed, yellow, white, and other color mutants derived from the original green wild type were found (fig. 4.2). Hybrid cleistothecia could be recognized in crosses between strains of different color by their mosaic color pattern. In developing this technique, Pontecorvo's group showed that individual cleistothecia arose, as a rule, from only two nuclei. Therefore, selection of hybrid cleistothecia would yield many asci and many thousands of spores (sufficient for genetic analysis at high resolution) from two known parents. And this, after all, was one of the goals of the program: high-resolution genetics.

The first work on *A. nidulans* genetics in 1948 and 1949 showed the practicality of the system, and for the next few years mutant isolation, culture and crossing techniques, and linkage maps were developed.[9] These efforts were, of course, necessary and yielded little that was novel, but the goals were always in mind: what was the significance of heterokaryosis as a genetic system, and what did *A. nidulans* have to offer in studies of the structure and arrangement of genes? In both pursuits, the efforts were richly rewarded. A major summary of the major lines of work of the group on the organism that appeared in 1953 became the introduction and technical guide for several generations of *A. nidulans* researchers thereafter.[10]

The Parasexual Cycle

Alan Roper, a key co-worker of Pontecorvo's in these domestication efforts, sought diploid strains of *A. nidulans*. Using a genetic approach and the ready availability of heterokaryons, he sought not only diploids, but diploids heterozygous for two or more complementing, recessive nutritional markers.[11] In designing a regime to detect diploids, Roper followed a route paved by others. To detect events that might be rare, selection of the product of these events by nutritional or easily observed genetic markers was necessary. In *A. nidulans*, the conidia form profusely on the mycelium as it matures. If autonomous conidial color markers (e.g., white and yellow, which would reveal themselves in the phenotype of individual conidia) are incorporated into a heterokaryon, the conidia impart a mosaic appearance to the colony. However, sometimes patches of wild-type green color appeared. These sectors contained heterozygous, diploid nuclei and produced uninucleate conidia having the wild-type color. The normal phenotype of a single cell with a diploid nucleus arises through complementation of the different mutations used in constructing it. Diploids arose rarely, but their frequency increased greatly after treatment of colonies with camphor vapors. The discovery of diploids was an important step in Pontecorvo's laboratory because it represented still another attribute shared by the asexual and sexual systems of fungi.

Soon thereafter, Roper and Pontecorvo discovered that heterozygous diploid strains yielded recombinant strains in which the original recessive markers reappeared. They could easily follow such events by observing the emergence of mutant color sectors in colonies that were diploid and green at the outset. Two processes, in the end, underlay this phenomenon.[12] The first was *mitotic crossing over*, a genetic phenomenon that occurs very rarely in vegetative, diploid cells and fully characterized by Muller and others in *Drosophila*. In this process, rare crossing-over events between homologous chromosomes during mitosis renders all markers from the point of exchange to the end of the chromosome homozygous, although the daughter nuclei remain diploid (fig. 4.4). Therefore, recessive markers masked in the heterozygote gain expression in sectors of the mycelium that become homozygous for those markers. In the second process, *haploidization*, a diploid nucleus loses one or more chromosomes during mitosis and becomes aneuploid (having more than one, but less than two complete sets of chromosomes). Perhaps in several successive divisions thereafter, further chromosome loss renders some nuclei haploid. Thus chromosome loss—a messy process—is an additional mechanism by which recessive traits originally masked in the diploid reveal themselves.

Pontecorvo named these processes collectively the "parasexual cycle." Although the characterization of parasexuality would take several more years, Pontecorvo and Roper recognized that this was a final, and perhaps a most significant, element shared by the sexual and asexual systems of this fungus. The interest in the phenomenon lay not only in the biological versatility it imparted to *A. nidulans*, a sexual species. More exciting was the possibility that parasexual recombination might be a widespread, if cryptic, process in wholly asexual fungi. In seeking to generalize the findings, Roper and others found parasexual recombination in the first five asexual species in which it was sought. For industries dependent on improvement of fungal strains that produced pharmaceuticals and organic acids, and for plant breeders and agriculturalists interested in the variation of fungal pathogens, the significance of parasexual genetics could not have been clearer. Pontecorvo is best remembered for this contribution. In parasexuality, he saw the possibilities of performing genetic analysis in human tissue-culture cells, thereby bypassing the need for sexual crosses. His later career was devoted to that effort.

Genetic Analysis

Starting about 1950, the field of genetics began to see microorganisms as representative, in fundamental ways, of living forms in general. Although the genetics of many lower eukaryotic organisms had been studied in some detail, Beadle and Tatum's work initiated a much more widespread use of microbes. There were several reasons for this. First, the demonstration that fundamental cell biochemistry was a province of normal Mendelian genes suggested the universality of Beadle and Tatum's findings. Second, their work offered a method of genetic analysis that depended on isolation of nutritional mutants. Such mutants could in turn be used for selective methods of detecting rare recombinants, mutations, and back-mutations and for analyzing gene interaction

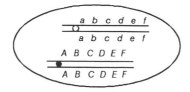

Diploid cell heterozygous for linked genes *A/a, B/b, C/c, D/d, E/e,* and *F/f*
(after chromosome replicaiton)

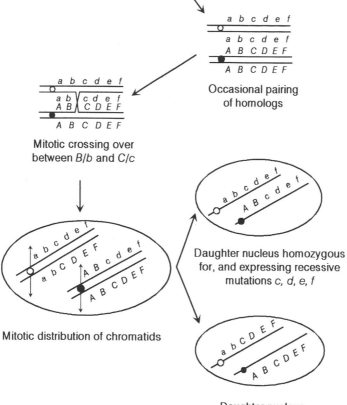

Occasional pairing
of homologs

Mitotic crossing over
between *B/b* and *C/c*

Mitotic distribution of chromatids

Daughter nucleus homozygous
for, and expressing recessive
mutations *c, d, e, f*

Daughter nucleus
homozygous for *C, D, E, F*

Fig. 4.4. Mitotic crossing over in *A. nidulans*. Rare pairing and recombination of duplicated homologs in a vegetative diploid nucleus are followed by their dissociation. At this point, a normal mitosis separates sister chromatids from one another. The daughter nuclei, still diploid, will each be homozygous for the genes previously exchanged.

and a host of other phenomena biochemically. They did not press the issue of the nature of the gene with their material, nor were they particularly interested in genetic systems (with the possible exception of heterokaryosis) per se. Only their followers such as David Perkins (b. 1919) would fully explore the potential of *Neurospora* as a genetic system.

Thus, after 1946, many biologists became aware of new microorganisms being developed as genetic systems. In all cases, investigators standardized techniques, obtained mutants, characterized their phenotypes, made chromosome maps based on recombination, and then diverged to other studies of particular kinds. Following Beadle and Tatum's work, Delbrück's phage group took shape in the early 1940s, with an ambition to define the gene and its replication in entirely new ways (chapter 8). Tatum, at Yale University in 1945, took on Joshua Lederberg, and together they demonstrated recombination between different mutant strains of *E. coli* (chapter 7).[13] Pontecorvo could see in the lower Ascomycetes the possibility of using genetic methods to study the structure of genes and their organization in eukaryotic chromosomes, a desire imparted earlier by Muller.

Pontecorvo's group took up two specific questions. The first concerned the divisibility of genes and their distribution on chromosomes, a problem embodied in the term "pseudoallelism." Muller, as early as 1938, had performed extremely detailed studies of the complex *scute* region of *D. melanogaster*. By sophisticated studies of X-ray–induced chromosome inversions with cytologically visible breakpoints in this region, he found that breakpoints seemed to be confined to only four positions. This suggested that chromosomes were discontinuous: genetically inert regions might separate genes or groups of genes. Indeed, despite their similar phenotypes, different *scute* mutations within these demarcated regions recombined, showing (by the definition of the day) that several genes might lay within each region. Curiously, however, even mutations lying in different locations separated by these inert "spacers" did not fully complement one another. In other words, two different *scute* mutations located on different homologs of a heterozygote did not wholly resemble the wild type, even though these mutations could recombine with one another. Perhaps, Muller thought, a special, positional relationship existed between the mutations such that the normal gene copies must all be together on one of the two homologs in order to function properly.[14]

This conundrum is more easily appreciated using genetic symbols. In a diploid *aB/Ab*, where the lowercase letters represent the two *scute* mutations, each on a different homolog, meiosis in germ cells would occasionally yield a normal gamete (haploid), *AB*. Such rare wild-type recombinants, which Muller detected in crosses, defined *A* and *B* as different genes by the usual standard. Yet in the somatic cells of the same diploid (where no recombination occurs), the *scute* phenotype prevailed. This could only mean, to orthodox geneticists, that the *a* and *b* mutations affected the same gene. The term *pseudoallelism* was later coined to describe this contradiction. The function of the *scute* region appeared to require an intact copy of both the *A* and the *B* alleles on one of the homologous chromosomes of the heterozygote. The question, which had occurred to many geneticists at the time, was whether the functional "gene" consisted of smaller units—alterable by mutation and perhaps resolvable by crossing over.

Before proceeding, I note that it is almost as hard now to remember why this matter was so difficult as it was to understand it at the time. The problem is in part a cognitive one, residing in a long-respected operational criterion of genes—separable by recombination—coming into conflict with observations that such separable units could not always complement one another. The field had grown used to the unspoken assumption that only alleles of the same gene failed to complement. There was simply no place in the mindset of many geneticists for mutations that could recombine but could not complement. A later terminology, initiated by Edward Lewis (b. 1918) and exploited by Seymour Benzer (b. 1928; see chapter 8) describes the arrangement of mutations in organisms carrying two copies of genetic information. The arrangement of mutations when they are on different copies of the genetic region in question is termed the *trans* arrangement (*Ab/aB*, where the numerator and denominator represent the two homologous chromosomes). The other arrangement, *AB/ab*, is called the *cis* arrangement. In the example above, the *trans* arrangement yields a mutant phenotype, and the *cis* arrangement is normal due to the intactness of the *AB* region on one homolog. The phenotypic discrepancy prevails despite the fact that the same genetic information is present in both cases. Equally important, the phenotype endowed by the *cis* arrangement demonstrates that the two mutations are recessive and that the mutant phenotype of the *trans* arrangement cannot reflect the dominance of one or both mutations over their wild-type counterparts.

Pontecorvo thought that he could analyze both the possibility of inert spacers between genes as well as the substructure of genes with much greater precision in *A. nidulans* than Muller had been able to do in *Drosophila*. Pontecorvo had his own basis of comparison: a student in his group did thesis work demonstrating recombination of *white* mutations of *Drosophila*, mutations that behaved much like Muller's *scute* mutations.[15]

Pontecorvo's second ambition was to study the chromosomal locations of genes of related function, a question emerging naturally from the problem of pseudoallelism. Lewis, in seminal work in *Drosophila*, had described a set of mutations in the chromosomal region known as *bithorax* that transformed segments of the adult body.[16] Where the normal fly has wings on the second segment of the thorax, and balancers (halteres) on the third, the extreme *bithorax* mutants have wings on both segments and lack halteres. These so-called homoeotic mutants as a group showed complex functional defects that appeared to control a gradient of developmental information through the body segments affected. Interestingly, the mutations appeared, in recombinational studies, to be distinct and organized in a sequence on the chromosome that was the same as the order of the affected body segments. Despite their recessive nature, heterozygotes carrying mutations in *trans* failed to complement fully. The data suggested again that this chromosomal region must be intact on one homolog in order to function properly, consistent with Muller's data on the *scute* locus.

Lewis proposed that a sequence of "millimicromolar" reactions normally took place on the chromosome at the *bithorax* locus (fig. 4.5). The tiny amounts of the products of these reactions could not diffuse to a homologous chromosome of the diploid nucleus. If the sequence was to be completed properly, all the wild-type alleles had to lie on the same homolog (the *cis* configu-

a. Muller-Pontecorvo model (gene divisible by recombination)

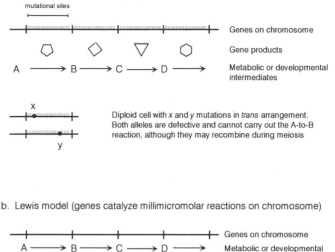

b. Lewis model (genes catalyze millimicromolar reactions on chromosome)

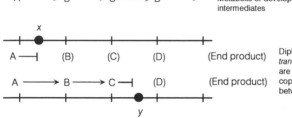

Fig. 4.5.   Two views of pseudoalleles. (*a*) The Muller-Pontecorvo model, in which contiguous genes each produce a product (e.g., an enzyme) that catalyzes an individual biochemical reaction. In a diploid, allelic mutations *x* and *y* cannot complement in *trans*, although recombination can occur between them. (Altered mutational sites are shown as filled circles within a gene.) (*b*) The early Lewis model of the bithorax region of *D. melanogaster*, in which genes direct reactions taking place on the chromosomes themselves. According to the model, all genes on a single homolog must be free of mutations to complete the reaction series. Interrupting the sequence by mutation of different genes on the two homologs would make the cell unable to complete the sequence. (Intermediates that cannot be formed in the *trans* arrangement are in parentheses in the lower portion of the second model.)

ration). This is why the *trans* heterozygotes appeared abnormal. The question could again be asked: is the *bithorax* region one gene or not? Remember that no one at that time had any idea of what the gene might be, nor what its primary product might be, nor where its primary biological function might be carried out. Lewis's work and his interpretation extended the pseudoallelism debate until these matters became clearer with the discovery of DNA structure and Benzer's work with a bacteriophage (chapter 8).

Thus another question occurred to Pontecorvo: if gene products formed on the chromosome itself, it would be unsurprising to see some groups of genes of related biochemical function adjacent to one another.[17] The category of such groups that would show a *cis–trans* position effect would be those in which only small traces of gene product (or metabolite) are made or needed. Pontecorvo and Roper therefore determined the linkage relationships among *A. nidulans bi* mutations (not to be confused with Lewis's *bithorax*, above) which imposed a requirement for the vitamin biotin, needed in exceedingly small amounts.[18] Because he could detect low frequencies of recombination, Roper could demonstrate recombination of certain independently isolated *bi* mutations, even though they were very closely linked. He then made heterokaryons and diploids carrying different *bi* mutations in the *trans* arrangement (fig. 4.5). The mutations failed to complement in both situations. This observation sharpened the pseudoallelism debate into two opposing hypotheses: (i) Genes such as *bi*, of which an increasing number began to appear in the literature, might be single genes that must be intact in order to function. In such cases, the *trans* arrangement, placing different defective copies on the two homologs of a heterokaryon or diploid, would produce a deficient phenotype (fig. 4.5a). (ii) Mutations might occur in different genes, whose function must be coordinated by their position on one homolog of a diploid (fig. 4.5b).[19] There was, in fact, no reason to believe that examples of both cases could not be found. In the end, the unitary gene model was realized in many cases of pseudoallelism such as the *bi* locus, but the nature of the *bithorax* and other such genetic regions has transcended the model offered by Lewis, and much developmental biology now concentrates on their structure and function. Pontecorvo thus contributed to the ultimate definition of the gene by using the *cis–trans* test to highlight the question of allelism or nonallelism of mutations and thus of the functional identity of genes.

Because the two models of gene action (unitary gene vs. coordinated, localized function of several genes) were still valid, Pontecorvo hoped in 1952, vainly as it turned out, that a millimicromolar scale of successive reactions localized in the nucleus might better be revealed by the phenotypic difference between a heterokaryon (where physiological transactions are internuclear) and the corresponding diploid (where the transactions are intranuclear).[20] He built upon this idea in the hope that he could use genetic analysis to determine whether the higher levels of organization of the eukaryotic cell (macromolecular aggregates and compartments) were encoded in some way in the chromosomes. It appears fanciful now for him to have entitled one of his reviews, "Genetical analysis of cell organization," in which he thought that with his approaches "we shall obtain some clues as to certain models of morphogenesis."[21] Such was the promise at that time that "pure" genetics could yield such straightforward answers to such complicated questions. And Pontecorvo could without apology maintain his dislike of biochemistry.

Role of *A. nidulans* as a Model Organism

How influential was *A. nidulans* as a model organism? To the extent that it contributed to the question of gene organization, it had a decided impact. However, the question of pseudoallelism—the basic question of the definition of the gene—prevailed in the minds of almost all geneticists from the time of Lewis and Muller, who formalized the issue, to the acceptance of Benzer's luminous work in bacteriophage T4 in about 1957 (chapter 8). Pontecorvo, a trenchant and imaginative scientist, was only one of a large, competitive, and diverse group that focused on the problem. They argued it out until the segment of a chromosome devoted to encoding a polypeptide and governing its expression (the *cistron*, which became synonymous with *gene*) was distinguished from its constituent DNA nucleotides, which we see as mutable sites. The claim of *A. nidulans* as a model illuminating universal phenomena is therefore weak. It came into play 10 years after *Neurospora*, and 4 years after the introduction of bacterial sexuality. Both of the fungi slipped slowly into more parochial roles, and *A. nidulans* took its place as a model for recombination studies and for parasexuality. *A. nidulans* would go on to make unique contributions to the study of fungal conidial differentiation, mitosis, and many areas of general metabolism. We will leave it, as we left *Neurospora,* at this "pre-DNA" stage until after we have explored how bacteria and their viruses set an entirely new agenda for genetics.

5

Yeast

The truth is better served by error than by confusion.
—Francis Bacon

The baker's and brewer's yeast known now as *Saccharomyces cerevisiae* has become the most prominent model organism of all eukaryotes.[1] We might call it a supermodel, a counterpart to *E. coli,* which is discussed in chapter 7. *S. cerevisiae* is unicellular; laboratory strains are self-sterile (heterothallic), permitting controlled crosses; and the organism can multiply indefinitely in the diploid or the haploid state. A simple minimal medium supports its growth, and populations of single cells can be plated like bacteria on agar medium, forming small, tight colonies. With a disk of paper or velveteen, the colonies can be replica-plated onto new plates, a technique developed in the early days of bacterial genetics (see chapter 7). Cell populations can be synchronized in their growth in liquid culture, allowing biochemical analysis of the cell cycle and meiosis. Although tetrad dissection is required for genetic analysis, yeast is alone among eukaryotic microbes in having high enough rates of recombination to study the comparatively rare events of gene conversion by this method.

Another attribute of yeast has rendered it a model for the study of the physiology, genetics, and biogenesis of mitochondria, the seat of respiratory activity in eukaryotic cells. This attribute is the ability of the organism to survive even the complete loss of respiratory function. This reflects the ability of yeast to extract energy from simple carbohydrates through anaerobic fermentation, the very feature Pasteur remarked upon when he demonstrated, with yeast, *la vie sans air.* These characteristics, among many others, have made yeast by far the most useful eukaryotic microorganism in basic molecular biology since 1975. Thousands of investigators now work on the organism or use it as a tool in connection with other studies. With this knowledge, we may look with amazement at the fact that the first international yeast meeting took place only in the summer of 1961, in Carbondale, Illinois. Eleven people attended, five of them resident investigators at Southern Illinois University.[2]

Knowing its extraordinary advantages, we must ask why *S. cerevisiae* became popular only in the early 1970s. An overview of this matter is instructive as a guide to what follows. Serious investigations of yeast began in the 1800s, notably by Pasteur. He did not seek an organism to do what we now call basic research. Instead, he was commissioned to work toward improving beer production in France. Yeast strains (variants of *S. cerevisiae*; see hereafter) had been selected empirically for baking and brewing for many centuries, each strain appropriate to the particular industry. Most of the production strains were *polyploid*, a condition in which the nuclei contain not one (haploid) or two (diploid) sets of chromosomes, but three or more. At the advent of Mendelism in the early 1900s, the life cycles of yeasts were unknown, and up to the 1950s, many people were uncertain about the difference between yeasts and bacteria. Because most strains were diploid or polyploid, yeasts would not have yielded simple, recessive, mutational variants for use in genetic studies in any case. In the 1930s and early 1940s, the life cycle was finally worked out. Both homothallic (self-fertile) and heterothallic strains became known, and certain strains displayed both attributes in a most confusing way. Finally, yeasts seemed to be quite plastic in their behavior, as the successful selection ("training") of strains for several industries had demonstrated. In short, students of basic genetics had no reason whatever to become interested in these peculiar organisms at a time when *Neurospora* and *E. coli* had the spotlight.

Winge

By the early 1900s, the brewing industry had collected a variety of yeasts, many used as the basis of their working stocks. The first person to do straightforward genetic work on yeast was Øjvind Winge (1886–1964; fig. 5.1), a well-known Danish geneticist and microbiologist. Winge had been appointed director of the Physiology Department of the Carlsberg Laboratories, associated with the Carlsberg brewing company in Copenhagen.[3] He worked on the genetics of yeast, hops, and barley, among other organisms, hoping in the case of yeast to improve knowledge of the life cycle. Working on strains left at the Carlsberg Laboratories by previous investigators, he defined the alternation of the haploid and diploid phases, punctuated by cell fusion and meiosis. As a geneticist, he naturally introduced genetic rationales into the analysis of the life cycle.

Winge's strains were homothallic, in the sense that one or two divisions after emerging from meiosis of diploid cells, the haploid meiotic products could mate with one another in any combination. This was not because the cells lacked mating-type identity, as in the case of *A. nidulans*, but because they switched mating types frequently during cell division. Therefore, even after physical separation of the four haploid products of meiosis, the daughter cells of each

Fig. 5.1. Øjvind Winge (*left*), Boris Ephrussi (1951) (*middle*), and Herschel Roman (1970) (*right*). Photos of Winge and Ephrussi courtesy of the Cold Spring Harbor Laboratory Archives. Photo of Roman taken at National Science Foundation Genetic Biology Panel, 1970.

one also had to be separated from one another to prevent fusions to form zygotes, the initiation of the diploid phase. The immediate formation of a diploid from a single spore at times was a further complication. Winge attributed this to the fusion of nuclei in binucleate spores. These complications enforced great care in making matings, which were done laboriously with a micromanipulator, bringing together haploid spores from two different strains to form a diploid zygote of known genetic constitution. The analysis of the ascospores emerging from the diploid itself after meiosis also had to be done with a micromanipulator. By this tedious means, Winge defined the life cycle of his strains of yeast (fig. 5.2). With genetic variants such as giant colony and the ability or inability to ferment various sugars, he established the Mendelian character of several traits. Winge's work may be contrasted with that of B. O. Dodge, who, in his first publications, defined the life cycle of three *Neurospora* species. He simultaneously demonstrated the Mendelian behavior of the mating types with startling clarity, using handheld needles to remove ascospores from asci and separate them to individual tubes. Dodge also related his findings to parallel cytological observations and thereby easily captured the interests of geneticists when they turned to microorganisms.

Winge worked with strains known then as different species. *S. ellipsoideus*, *S. chevalieri*, and *S. oviformis* were later considered variants of *S. cerevisiae*, and, with some success, they interbreed. However, their homothallism enforced Winge's means of genetic analysis. Although Winge discovered heterothallism in another yeast, *Saccharomycodes ludwigii*, he did not exploit the advantage it afforded—namely, that haploid cells of a single strain would not automatically fuse to become diploid. It fell to Carl C. and Gertrude Lindegren to find and exploit heterothallic strains of *S. cerevisiae*, beginning in 1943. The Lindegrens had developed the genetics of *Neurospora* in the 1930s, first under T. H. Morgan, then independently after Carl Lindegren left Caltech.

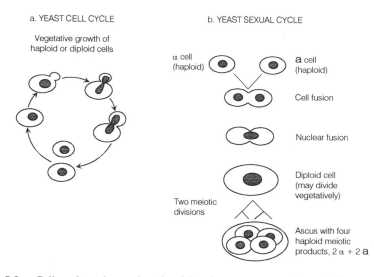

Fig. 5.2.   Cell cycle and sexual cycle of *Saccharomyces cerevisiae*. (*a*) Propagation of yeast cells by budding of daughter from mother cell. The nucleus, haploid or diploid, positions itself at the neck of the bud, and chromosomes are distributed accurately to mother and daughter cell. As in other fungi, the nuclear membrane does not break down during mitosis. (*b*) The sexual cycle requires cells of different mating type (α and **a**). Cells fuse in pairs to form diploid cells, which may propagate indefinitely by mitosis. When starved for carbon and nitrogen, the diploid cell will undergo the two meiotic divisions, yielding four haploid cells.

Lindegren

Carl Lindegren left Caltech for the University of Southern California in 1933 and continued working on *Neurospora* until, after a year at the University of Missouri, he moved to Washington University in St. Louis in 1941. At that time, he switched to studying yeast. The change to yeast reflects several possible influences. The first may have been the source of funding for the work, the St. Louis-based Anheuser-Busch brewing company.[4] We do not know whether the choice of yeast or the availability of funds came first. A second consideration, the attraction of the organism itself, may well have been compelling because it was even simpler than *Neurospora*. With Winge's prior work and Lindegren's appreciation of the potential of yeast for fundamental studies, Lindegren may have felt he could do more easily for yeast what he had already done for *Neurospora*. A possible third influence was a desire to separate himself from the Morgan school, a process that began before Lindegren left Caltech. (Lindegren later described himself as "thoroughly indoctrinated in Gene Theory in Morgan's laboratory.")[5] A last consideration may have been the knowledge that Beadle and Tatum would use *Neurospora* in their first studies of biochemical genetics. (Lindegren, in fact, appears to have given his *Neurospora* strains over to them willingly.) Whatever his reaction to the last development,

Lindegren may have wanted to stake out a scientific pursuit in which he, a born maverick, would be alone.

The Lindegren laboratory started work in 1943 with a variety of yeasts, including a diploid (strain EM93) that produced stable heterothallic strains of two mating types when the diploid underwent meiosis (or, as yeast workers say, "sporulated"). Lindegren named the mating types **a** and α. Using other strains varying in fermentation capabilities, the laboratory derived a number of stocks carrying single-gene differences from their standard wild type. This repeated the strategy of Lindegren's early *Neurospora* work. The discovery of heterothallic strains allowed the Lindegrens to pursue genetic studies much more easily. They could perform matings simply by mixing **a** and α haploid populations (mass matings), rather than by the spore-to-spore technique of Winge. This technique, unfortunately, permitted some illegitimate matings of cells derived from the same parent, and this would eventually be the source of some confusion. The Lindegren laboratory developed procedures for mutant isolation, methods of nutritional and genetic analysis, and early mapping data. Lindegren sent the strains derived from strain EM93 to others, who in turn isolated mutants from them. These became the foundation of the stocks used thereafter by the American yeast community. The recipients of the initial strains included investigators at Yale, Seymour Pomper and Sheldon Reaume (the latter in Tatum's laboratory); Herschel Roman (1914–89) at the University of Washington; and Robert Mortimer (b. 1927) at the University of California, Berkeley.

These investigators developed more rigorous methods of mating. Instead of using the products of mass matings as such, they mated cells carrying complementing, auxotrophic markers, and selected a prototrophic colony derived from a single diploid cell for further analysis and sporulation.[6] (The terms *prototroph*, denoting a strain able to grow on the simple minimal medium, and *auxotroph*, a strain that requires an additional nutritional supplement, were introduced during the development of bacterial genetics [chapter 7].) In this case, the prototrophic diploid arises through complementation of different nutritional deficiencies, much as in a *Neurospora* heterokaryon. The groups led by Roman and by Mortimer would become the first American laboratories to perform rigorous genetic work with yeast and render it suitable as a basis of for modern studies. However, Mortimer credits Lindegren for bringing yeast into the genetic traditions of America and for advertising its advantages.[7] While Lindegren and his team amassed many findings on yeast genetics and cytology that would initiate American yeast genetics, he also initiated a variety of serious controversies. At the outset, the genetics community looked with great interest at Lindegren's findings, but as time passed, confusion and difficulties arose with his work that finally made him something of an outcast.

Lindegren, described by Beadle as enthusiastic, energetic, and, for better or worse, imaginative, became restive working under Morgan. Their scientific temperaments differed radically, with Morgan a skeptical empiricist and Lindegren a constant boil of theories and novel interpretations of experiments. He seems to have adopted, almost instinctively, alternative views of many widely accepted genetic principles. To be sure, this was not always unjustified because many of these new principles had been accepted by many biologists by default. Genetics was too inferential to provide clear biochemical and molecular information about the very issues (the gene and its importance to the cell phenotype) that preoccupied the field. Unlike the equally imaginative Muller, however, the ambitious Lindegren was as uncritical of his own data as he was critical of others'. His publications and public discussions entangle his data with his models and seek to prove his predictions, rather than to test them. His premature interpretations often survived in his own mind after rigorous experimental contradictions from other laboratories, and sometimes even from his own.

Lindegren moved to Southern Illinois University in 1948 and remained there until his death in 1987. In the late 1940s, his and other laboratories diverged on a number of controversial matters. Two publications, one a 1946 Cold Spring Harbor Symposium presentation and the other a comprehensive book published in 1949, summarize his early work.[8] Then and later Lindegren challenged, among other things, the stability of genes and the randomness of chromosome assortment in meiosis. Well after 1953, he opposed the notion that DNA actually carries genetic information. His cytological interpretations were startlingly off the mark. Still imaginative and controversial, he published an anti-Morgan–Mendelist tract, *The Cold War in Biology*, in 1966.[9] This book was written in the midst of the superpower era in which East and West faced each other with nuclear threats and competed in many areas of science and politics. This was a time when Trofim Lysenko had for some years dominated Soviet agricultural policy with his theories of crop improvement based on the inheritance of induced, acquired characteristics. In his *Cold War* book, Lindegren explicitly favors this view, disputes the dominant theories of evolutionary change, speculates about the nature and evolutionary affiliations of different cell types, and offers a theory of the origin of life. It is therefore not surprising that Lindegren can be credited both with the beginnings of American yeast genetics and with its slow development for some years thereafter.

Four major controversies deserve brief descriptions.[10] These have to do with gene conversion, the cytogene theory, yeast cytology, and the integrated cell. In the first, Lindegren documented a number of cases of aberrant segregation in tetrads arising from diploids heterozygous for fermentation genes. Instead of the regular 2:2 segregation expected of alleles, he observed 3:1 and 4:0 segregations, in both of which either the dominant or the recessive allele might be in the majority. Lindegren eventually integrated these observations into his theory of how genes were disposed on chromosomes. He believed that in rare instances, one allele replaced the other, simply by having a copy move to the homologous chromosome of the diploid in meiosis, due to what he felt were weak attachments of the genes to the chromosomes. The peculiar background picture he uses to justify his expectations is less important than the dispute about the observations themselves. In fact, the Lindegrens' data sufficiently interested both Winge and Roman that they published critical experiments attempting to repeat and explain the observations. Roman demonstrated that some products of the Lindegrens' mass mating technique were polyploid.[11] Aberrant segregation would be expected among the meiotic products of such cells. Winge, on another point, demonstrated that after meiosis, nuclei in rare tetrads could divide once, followed by random degeneration of some of the daughter products.[12] The remaining nuclei would be included in four ascospores, but in non-Mendelian ratios. Still another problem concerned the multigenic control of certain phenotypes. Where strict Mendelian segregation (2:2 ratio) is expected of a trait controlled by one gene with two alleles, other ratios are expected if two or more genes contribute to a particular phenotype. Hawthorne, in fact, demonstrated the latter condition with one of Lindegren's own problematic diploids.[13]

These challenges moderated Lindegren's insistence on his hypothetical picture of gene conversion, but he actually presented believable data in one case in 1955 for the reality of gene conversion.[14] It was frustrating for him to see Mary Mitchell (1911–86) and Roman's group demonstrate clear cases of the phenomenon in *Neurospora* and yeast, respectively, in 1955 and 1956.[15] The study of gene conversion is the starting point for our modern vision of the process of genetic recombination. Understandably, Lindegren continued to insist on his priority (and that of his predecessor, Winkler) in the early discovery of gene conversion for the rest of his life, despite the inadequacy of the data on which it was based.[16] As Mortimer put it, "Although Lindegren was the first to draw attention to this phenomenon, he did so mainly on the basis of faulty interpretation of his data. When he did present a valid example, he was largely ignored."[17]

The second controversy had to do with Lindegren's cytogene theory. Well into the 1960s, Lindegren's picture of the gene was that of a protein attached to a scaffold (which he later identified with DNA), the latter enabling proper mitotic and meiotic distribution of the informational material. At the time he proposed the cytogene theory (around 1945), observations on yeast and bacteria had shown that many enzymes appeared in cultures only when the substrates of the enzymes were added. These are called "adaptive" enzymes, and no one had any idea at the time of the mechanisms by which they were synthesized or controlled. Lindegren's theory held that the proteinaceous gene could duplicate and send a copy (cytogene) to the cytoplasm, and the latter could duplicate further. This interpretation was initially supported by the claim that yeast cells continued to synthesize melibiase, an adaptive enzyme that broke down the sugar melibiose, without the presence of the chromosomal copy of the gene if melibiose was present to stabilize it.[18] This situation was arranged by sporulating a diploid population of cells heterozygous for a melibiase-deficiency mutation (mel^+/mel^-), grown in the presence of substrate. The haploid meiotic products of one ascus, half of which would lack the wild-type gene, were maintained in the presence of melibiose and allowed to multiply. Solomon Spiegelman (1914–83) and Lindegren claimed that all of these cultures continued to produce the enzyme indefinitely. However, if the populations were transferred from melibiose medium to one containing glucose, the capacity to make melibiase was lost in half the cultures and was not restored by returning them to melibiose medium. In the words of the authors, "Therefore, as far as melibiozymase is concerned the function of the mel+ gene would be limited to the initiation of its synthesis."[19] Thus, in this picture, the cytogene is not wholly autonomous: it is derived from the gene, and it is stabilized by substrate. A failure to repeat the results, except to show that the enzyme-positive phenotype can be carried over at least transiently to genotypically *mel⁻* progeny, did not deter Lindegren from propagating a cytogene theory for several years thereafter.[20]

The third controversy concerns Lindegren's views on the cytology of yeast. These are spectacularly at variance with our current knowledge, and to a great extent with that of his contemporaries. Lindegren's review of this work identifies the vacuole as the nucleus, the nucleus as the centrosome, the nucleolus as the "centrochromatin," and various vacuolar inclusions (many of which are probably polyphosphate) as chromosomes.[21] He can surely be forgiven for some of these claims, given the primitive techniques then available. He cheerfully acknowledged that others had identified as the vacuole what Lindegren called the nucleus. The confidence with which he named the parts is characteristic of his assertions in other areas, given above. It would be some years before the modern picture of yeast cytology was well established.[22] A scanning electron micrograph of a freeze-fractured yeast cell is shown in fig. 5.3.

Finally, Lindegren's tone carried over to a later review of the "integrated cell," in which he began by chastising those who believed that the genes are

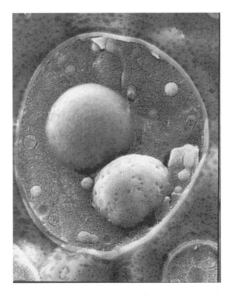

Fig. 5.3.   Scanning electron micrograph of a freeze-fractured yeast cell, giving a three-dimensional impression of certain internal structures. The cell wall at the periphery surrounds a large vacuole (with a smooth membrane), the nucleus (with a membrane pocked with nuclear pores), and several smaller bodies, some of which may be mitochondrial elements. Fractured mitochondria on the left may be identified at higher magnification by their invaginated inner membranes. Micrograph courtesy of Elizabeth W. Jones, Carnegie Mellon University.

the cell's most fundamental elements.[23] He noted that biological duplication is a property of the cell, not of any particular molecule. This reasonable point, however, was a springboard to the view that a mutation, while it may have a quite specific effect, does not reflect the specific and unitary action of the wild-type allele from which it arose. Lindegren insisted, as others had, that the chromosome behaves in an integrated fashion, having no discretely demarcated units of function. (This view was also held by another maverick, Richard Goldschmidt [1878–1958]. It is wholly at odds with Lindegren's own picture of the cytogene, a discrete entity capable of autonomous function and replication.) Lindegren went on to write that the chromosome is integrated functionally with other components of the cell to assure cell duplication. His speculations on the nature of the gene, the origin of life, the development of the eukaryotic cell, and the process of differentiation contained some solid conjectures but extended them to extremes unjustified by the information then available. The article concludes with revealing, anti-Morganist (and ungrammatical) passion: "On this basis the gene is a secondary item in evolution rather than a primary one, and although it now appears to control a wide range of cellular activity it seems clear that it has achieved this position by a creeping bit by bit onslaught until the cell, originally independent of genes, has now become hopelessly dependent on them."[24]

Roman and the American Yeast Community

Lindegren had brought the field of yeast genetics to a troubled state and had sequestered himself in his institute at Southern Illinois University with a loyal group of students. He made forays into the limelight with combative zeal for some time thereafter. He had done little to convince general biologists that yeast was a suitable organism for genetic research. During the late 1940s, however, his work had intrigued Roman (fig. 5.1), then at the University of Washington in Seattle. Roman and several other workers would soon rescue yeast from its tarnished reputation.[25]

Having arrived at the University of Washington botany department in 1942 after receiving his training in corn genetics, Roman realized after a few years that he could not easily pursue this field in the climate of Seattle. At that time, biologists showed considerable interest in gene conversion and the cytogene theory as Lindegren promoted his views. Roman invited Lindegren to give two weeks of lectures at the University of Washington in 1947 to see just what sort of data Lindegren had. Lindegren's 1949 book is based on these lectures.[26] During the visit and thereafter, Roman brought up explanations for Lindegren's findings that Lindegren had not thought of or felt he had eliminated. The state of the work convinced Roman that he might clarify some of the ambiguities by conducting investigations of his own.[27]

Fortunately for Roman, the microbiology department of the University of Washington had a capable microbiologist, Howard C. Douglas (b. 1910), who was already working on yeast. Douglas and Roman shared a student, Donald

Hawthorne (b. 1926), soon to become a key contributor to modern yeast genetics in its formative period. In 1949, Hawthorne was an undergraduate, charged by Roman and Douglas with the duty of learning and perhaps improving upon Lindegren's techniques of mating and ascus dissection, work he continued when he became Roman's Ph.D. student. Soon the laboratories of Roman and Douglas became proficient in mating techniques, mixing parents with biochemical mutations (auxotrophs) to form prototrophic diploids and then isolating diploid colonies derived from single cells for use in meiotic analysis. They isolated more mutants and began mapping the yeast genome, extending Lindegren's work. They also demonstrated that many of Lindegren's reports on gene conversion could be explained by more orthodox mechanisms involving polyploidy, multigenic determination of certain traits, and so forth. Douglas developed knowledge of galactose fermentation and enzymology in conjunction with these studies and clarified, with Hawthorne, the genetic determination of this system in several strains.

Roman's laboratory would eventually demonstrate that gene conversion did indeed occur in yeast. He did so with adenine-requiring mutations and rigorous controls that were much more convincing than Lindegren's experiments. These studies, summarized in 1956, contributed not only to questions about allelism, pseudoallelism, and the nature of the gene, but also certified yeast as a favorable organism for work in general genetics.[28] Roman would go on to establish a Department of Genetics at the University of Washington, with a large enough group of yeast geneticists to give visibility to the field and to his leadership of the American school in the 1960s and 1970s. Part of his success in the enterprise lies in the recruitment of Leland Hartwell (b. 1939) to the department in 1968 and the conversion of later recruits such as Benjamin Hall (b. 1932) and Walton Fangman (b. 1939) to the use of yeast in their studies of gene expression and DNA transactions.[29]

In 1949, Mortimer, trained in physics, began studies of radiation inactivation of yeast, working as a student under Cornelius Tobias (b. 1918–2000) at the radiation laboratory of the University of California, Berkeley. He was asked to test the proposition that the greater radiation resistance of diploid yeast cells, compared to haploids, could be explained by the ability of diploids to sustain recessive lethal mutations. Although his studies yielded results at variance with expectation and thus failed to confirm the simple hypothesis, Mortimer was committed by then to developing yeast genetics in parallel with the Seattle group. He joined the Berkeley faculty in 1953 and began an ambitious program of mutant isolation, genetic mapping, and study of the mutational process. Many of the strains of yeast used thereafter by the American yeast community originated in his laboratory during the 1950s, all derived from a single haploid strain, S288C (a close relative of strain EM93), with favorable characteristics for genetic analysis. (Rare mating-type switching allowed this strain to generate both **a** and α mating type strains; see chapter 15.) Mortimer isolated a large array of mutants that he, in collaboration with Hawthorne, used to map the genome. These mutants were the germ of the yeast stock center at Berkeley, and the beginning of yeast as a model organism. In addition, a major advance was made with the use of a snail-gut enzyme preparation to digest the tough ascus wall of yeast for tetrad analysis. This greatly accelerated genetic work on yeast, making it attractive to a growing number of researchers. The various groups of biochemical mutants isolated during this time were given freely to various American and European laboratories. Significant research programs began, and many continue to this day.

Ephrussi and the French School

In the late 1940s, another major program in yeast genetics arose in Europe in the laboratory of Boris Ephrussi (fig. 5.1). As discussed in chapter 2, Ephrussi had been trained as an embryologist in France and had gone to Caltech in 1934 to learn more about genetics, a subject that was poorly developed in France at that time. Ephrussi retained an intense interest in development, the role of the cytoplasm in differentiation, and the possibilities of self-replicating determinants in the cytoplasm. Nevertheless, he acquired a strong grasp of and respect for Mendelian genetics as he performed his groundbreaking work with Beadle on eye-color mutants of *Drosophila*. This distinguished him from the majority of biologists in France, many of whom resisted the Darwinian view of evolution and Morgan's representation of the Mendelian system, which they considered unphysiological, metaphysical, and, ironically, speculative.[30]

Ephrussi spent much of the Second World War in the United States at Johns Hopkins University but returned to France as the war in Europe concluded. He became a prominent modernizing force in French genetics thereafter. He received the first chair of genetics in France in 1946 and was charged with organizing research in this area under the auspices of the Centre National de Recherche Scientifique (CNRS). His own and other laboratories, which he organized with financial help from the Rockefeller Foundation, were devoted to teamwork approaches on certain well-defined problems.[31] Ephrussi, following Beadle's lead, chose yeast as an experimentally accessible, eukaryotic microorganism. Ephrussi, after all, stood on the very ground where Pasteur had explored this organism in the interests of rendering French beer superior to the German. (Ephrussi remarked that "in spite of Pasteur's efforts, German beer remained much better than French beer.")[32] By the time Ephrussi chose yeast, its life cycle had been defined by Winge, Lindegren, and others. Moreover, he was aware of a possible case of cytoplasmic inheritance in yeast published by Winge in 1940.[33]

In his early studies, Ephrussi noted that plating clones of yeast derived from single cells on agar medium led to the appearance not only of normal colonies, but always of a low percentage (ca. 0.2%) of small ones (*petites colonies*).[34] While populations of the large-colony yeast, whether diploid or haploid, continued to throw off petites upon plating, the petites themselves were stable in further growth and plating. Matings of a haploid petite with a normal haploid of the opposite mating type yielded a normal (large-colony) diploid population. These diploids, when sporulated, yielded asci in which the petite character failed to reappear; asci had a 4:0 phenotypic ratio of large to petite. The result suggested that the petite variant arose through the irreversible loss of a replicating, cytoplasmic determinant that could be restored only by mating the petite to a normal cell. The findings resonated strongly with Ephrussi's preoccupation with the possible role of the cytoplasm in cell determination during embyrogenesis.

The petite phenotype was correlated with slow growth in aerobic culture and a deficiency of cytochromes and other respiratory enzymes. Although Ephrussi was cautious about the conclusion, further work demonstrated that the cytoplasmic particles were in fact mitochondria, a conclusion reached well before mitochondrial DNA was discovered. Ephrussi also discovered that that the acridine dye euflavine directly caused the loss of the cytoplasmic particles (as opposed merely to selecting for cells that had already lost them).[35] Thus Ephrussi's well-focused program was from the start distinct from the efforts of the American school. Much of Ephrussi's laboratory moved from Paris to Gif-sur-Yvette, a small town lying a short distance south of Paris, in 1956. Both Hawthorne and Roman visited Ephrussi's laboratory for substantial periods during the mid-1950s and thereby established an important, early connection between the American and European schools of yeast genetics. Nevertheless, these schools, both in their organization and in their research programs, remained quite distinct. In both cases, their initial work (on gene conversion in the United States and on cytoplasmic inheritance in Europe) was novel and, with Lindegren's continued presence, confusing even to geneticists working on other organisms. The glory days of yeast were therefore delayed until both areas came to be perceived as foundations of major advances in biology.

Again, I close this account prior to the discovery of the structure of DNA so that I can summarize in the next chapter the state of microbial genetics, dominated in this early period by the three fungi I have discussed. In particular, I compare *Neurospora*, *Aspergillus*, and *S. cerevisiae* with respect to their different origins as model organisms and their different technical attributes. This comparison facilitates an understanding of how the bacteria and their viruses would displace these model organisms, at least temporarily, in investigations directed to the structure and action of the gene—investigations the three fungi had done so much to initiate.

6

Leaving the Fungi

Scientists don't discover. They make *discoveries—and then maintain them.*

Background

Humans have always been aware of the fungi, owing to their visibility and importance as foods, animal and plant pathogens, and agents of spoilage and fermentation. One of the prime features of fungi leading to their initial use in biochemical genetics was their status as microorganisms. This status is justified by the fact that fungi grow as single cells (yeasts) or as hyphae (filamentous fungi), which are simply extended cellular units. Unlike bacteria, which remained mysteries to most biologists well up to the 1950s, the fungi were visible and familiar. Some background is necessary to appreciate the manner in which *Neurospora* and *A. nidulans* became model organisms.

Researches on fungi began with taxonomic work (and its amateur counterpart, mushroom hunting), a phase that concentrated on the morphology of the fruiting bodies of the Basidiomycetes and Ascomycetes particularly, but which included the Phycomycetes (a primitive group now considered derivatives of algae) as well. Experimental work on fungi began in the mid-1800s by Heinrich de Bary (1831–88), who defined the life cycles of many groups, starting with the plant-pathogenic rusts. By the 1880s, these studies were extended by cytological descriptions of sexual processes. The sexual cycle was of particular interest, and rapidly the sexual stages of many theretofore asexual ("imperfect") fungi, known only through their mycelia and asexual spores, were discovered. In 1904, Alfred F. Blakeslee (1874–1954) discovered heterothallism in Phycomycetes (*Rhizopus stolonifera*, the common black bread mold, is a member of this group), determined by two alleles of a mating type locus, a phenomenon that could be interpreted with the then-new concepts of Mendelian genetics.[1]

73

Blakeslee also showed that homothallism, the capacity for self-fertilization, prevailed in other species.

Strains of hermaphroditic fungi, having well-developed male and female organs and gametes and with identical morphology, failed to self-cross due to the requirement for two different mating types to consummate fertilization. Some of these species even displayed relative sexuality, in which the magnitude of differences in affinities between strains determined whether mating would proceed. Therefore, mycologists began to appreciate the difference between morphological attributes of male and female on the one hand and genetic differences between mating types on the other. The higher Basidiomycetes (mushrooms and puffballs) have no male and female sexual structures at all; mating is accomplished by fusion of vegetative hyphae. The ability of pairs to fuse is governed by elaborate mating-type systems with several genetic loci, each with many alleles. It is therefore understandable why academic interests in fungi as a whole should have become so concentrated on their sexual and mating-type systems. Dodge, Winge, Pontecorvo, Roper, and Lindegren made early and important contributions to this field as they discovered the sexual systems and developed the genetics of *Neurospora*, *Aspergillus*, and yeast.

The study of fungi was driven as much by practical considerations as by academic curiosity. Long-standing interests of industry included fermentation products, edible mushrooms, food spoilage, and, later, antibiotic production. More recently, fungi have been used for preparing catabolic enzymes for industrial use. Plant pathology had a long-standing presence in colleges of agriculture, seeking understanding and control of such diseases as wheat rust, corn smut, chestnut blight, and Dutch elm disease. Students of forestry were interested early in the requirement of many tree species for symbiotic *mycorrhizae*, the fungi that inhabit root systems and deliver many nutrients from the soil to their hosts. In medicine, medical mycology is a long-standing field, covering allergens, commensal forms, and pathogens such as *Candida*, the agent of thrush, certain infectious *Aspergilli*, and an important opportunistic pathogen of AIDS patients, *Pneumocystis carinii*, recently revealed as a member of the fungi.

The Choices of Fungal Geneticists

Let us look at the choice of fungi that became model organisms in microbial genetics in midcentury. Did those who chose them have no alternatives? The insiders' accounts of the choices of model organisms are as suspect as the founder myths of molecular biology. We must be detached in looking at the process because clear purpose or fortuitous circumstances might drive the choices. Hindsight is somewhat unreliable in reconstructing the basis for these choices.

By the early 1930s, the Morgan school had certified the Mendelian patterns of inheritance in studies with *Drosophila*. Studies of the mouse, corn, and a number of other organisms established the near universality of Mendelian genetics. *Drosophila*, mice, and corn, due to their extensive use and resources, permanently trapped many younger investigators trained in the use of these organisms. At the beginning, many hoped they could use them to solve the important genetic questions emerging in the 1930s: the structure of genes and chromosomes and how genes determined both discrete and continuous (quantitative) traits. When the need to escape the limitations of these organisms became obvious, many investigators found them to be prisons: to begin with another organism required a huge investment of time and effort in the process of standardizing conditions, mutant hunting, and gene mapping. Indeed, even Beadle moved only between well-developed models as he abandoned corn for *Drosophila* and *Drosophila* for *Neurospora*. Nevertheless, he had to acquire considerable expertise at each move, and therefore the moves were made with daring, effort, and imagination; each move put his career at stake. Beadle and Tatum's success after 1941 emboldened many other geneticists to make similar moves later in the decade.

To geneticists at that time, the only way of investigating gene structure required crosses—that is, breeding analysis. As Morgan had insisted, the gene was a theoretical entity inferred only from the behavior of phenotypes in crosses. His approach, based on mutation, segregation, and recombination, would continue to yield vital information for decades. Other approaches to the physical basis of heredity had not yielded experimentally testable hypotheses. Speculation centered on protein as the substance of the gene, and biochemists and biophysicists had just begun to describe the most fundamental properties of proteins. Therefore, geneticists could only say that mutation and recombinational studies might "resolve" gene structure, much as the contemporary physicists were hoping to resolve atomic structure with the cyclotron. In both cases, the raw-sense data were somewhat removed from the objects of interest, and in both cases, a good deal of faith underlay progress.

The challenge of understanding gene structure and gene function had to be met by organisms that would breed easily and inexpensively, produce large progenies, grow rapidly on a simple medium, and submit to biochemical analysis. Beadle's choice began with a conscious effort to find an organism that met all these criteria, and he found himself at Stanford University, 400 miles north of Caltech, where some years before he had witnessed Lindegren's early work with *Neurospora*. Other fungi were demonstrably favorable for genetic analysis, though not nearly as well domesticated. In addition, most of these other fungi, unlike *Neurospora*, were known only in the genetically unsophisticated provinces of mycology or microbiology.

Pontecorvo's choice of *A. nidulans* demonstrates that Beadle and Tatum's choice was not the only one that could have been made, and the review that began this chapter shows that other fungi were indeed available for use. However, Pontecorvo's choice of *Aspergillus* seems, according to his own retrospective comments, almost whimsical. He spoke of wanting to explore alternative genetic systems, or an organism that would improve the resolution of genetic analysis. When he made the choice, there is no evidence that he was influenced by anything more than his familiarity with the *Aspergillus-Penicillium* group of Ascomycetes and his confidence that a new genetic system, once explored in detail, would yield surprises. Thus this choice had a fortuitous element, rather than being natural and obvious. While Beadle and Tatum had specific hopes for what finally made them famous, Pontecorvo could not anticipate what eventually made *A. nidulans* an important organism in mid-century genetics. This illustrates the point that a field, or even an entire science, starts as often with a surprise as a plan: the unexpected phenomena of parasexuality followed Pontecorvo's choice of *A. nidulans*.

Winge's and Lindegren's separate choices of yeast reflected their involvement with the brewing industry, Winge as director of the Carlsberg research enterprise and Lindegren as recipient of support from Anheuser-Busch. Lindegren's research was driven by his quest for scientific autonomy, his abandonment of *Neurospora*, and his desire to apply his already extensive experience with *Neurospora* to a new microrganism. He saw as clearly as Beadle, Tatum, and Pontecorvo the merits of biological simplicity. A fortuitous element did enter the picture thereafter: Roman's surprise in meeting Lindegren and finding such extensive grounds for skepticism about his work. Thus a personal challenge to Roman and his need for a new organism to study genetics in Seattle, among other things, drove the second phase of yeast studies. Ephrussi's choice of yeast was complementary to Winge's, Lindegren's, and Roman's, reinforced by the surprising behavior of petite yeast, if not based upon it.

The cluster of new fungal geneticists had restricted intellectual origins. Morgan had trained Muller, who in turn trained Pontecorvo. Morgan and his associate Sturtevant trained Lindegren, Ephrussi, and Beadle. Rollins Emerson (1873–1947) trained Beadle in his early days as a corn geneticist, and he also trained L. J. Stadler (1896–1954), Roman's mentor. This tight pedigree underlies the similar genetic aims of the early fungal geneticists and the process of choosing organisms with which to pursue them.

The Early Uses of the Fungi

We are left to ponder the fact that the choices summarized above committed at least three generations of investigators to continue work on these fungi. As noted previously, the structure and function of the gene preoccupied those who chose fungi as experimental organisms. Beadle and Tatum asked whether genes had a role in a fundamental class of biological activities without asking initially about the details. Their 1941–45 work, the revolutionary discovery of the connection between genes and enzymes, was celebrated as much for its promise as for its findings. Little was revealed at the time about the molecular products of genes, and nothing could be inferred about the nature of genes themselves. Pontecorvo celebrated heterokaryosis as a new genetic system in almost mystical terms. This promotional activity continued until the truly novel parasexual cycle emerged, to be admired without so much manufactured enthusiasm. And Pontecorvo's salesmanship finally focused on his cultural inheritance from Muller: would high-resolution genetics tell us about the structure of the gene? It did, but only in the familiar, formal terms of a geneticist. Finally, gene conversion—the surprise of yeast genetics, first suspect but later confirmed—would remain for some time only a promising starting point toward our present understanding of recombination in eukaryotic organisms.

These remarks make a point embodied in the dismissive judgment often made in the 1960s: "Genetics is only a tool." It is difficult to place one's self in the time between the publication of the early observations made with the three model fungi and the time when the observations began to be understood in molecular terms. In retrospect, one is amazed at the excitement that the formal genetic observations in microbes generated. It was the promise, not the facts, that lured the large number of investigators to enter the field of microbial genetics in the late 1940s. Formal genetic analysis was the only workable tool for understanding the gene up to the late 1940s. But even before DNA and protein structure became clearer, the time had come for biochemical approaches to answer questions posed by geneticists. Beadle made one of his greatest contributions in promoting a fusion of these fields through reviews, lectures, and eventually, his collaboration with Pauling in renewing the Biology Division at Caltech. This effort, started in 1946, had as great an impact on molecular biology as the organization of the new genetics at Cold Spring Harbor, which is discussed in later chapters.

The first job of the early workers in *Neurospora*, *Aspergillus*, and yeast was the domestication of the organisms. Only several years into each program could the benefits and limitations of the organism in question be appreciated. Let us look at some of these before we go on to the choice of other organisms used later in molecular genetics.

All three organisms grow on a simple minimal medium and all have small genomes of 16–45 megabases (Mb, or millions of nucleotide pairs of DNA), with yeast having the smallest genome. Each has its technical advantages and disadvantages, as shown in table 6.1. With respect to advantages, each organism could be exploited for certain purposes. For instance, the uninucleate conidia of *Aspergillus* made identifying mutations straightforward, and allowed nuclear ratios in heterokaryons to be easily estimated. The size of the yeast genome is among the smallest of any free-living eukaryote. This almost assured that yeast would be the first eukaryote whose genome would be sequenced. *Neurospora*, with comparatively large asci (fig. 6.1), was the first organism to be fully exploited in tetrad analysis, and the first in which gene conversion was unequivocally observed. Disadvantages were overcome in each case. The rangy growth of *Neurospora*, which can cover a Petri dish in a day, could not be plated for colony formation until the sugar, sorbose, was found to restrict growth to tight colonies (fig. 3.5). The homothallism of *A. nidulans* was a disadvantage essentially nullified by the use of strains that favored hybrid matings, and that afforded easy detection of hybrid cleistothecia. The small size of yeast tetrads (fig. 6.1) and the necessity of dissecting them was dealt with by using snail gut enzymes and inventing simple, effective micromanipulators to dissect tetrads. Ironically, tetrad analysis in yeast can now be done more quickly and with vastly greater efficiency than in any other organism.

Table 6.1. Technical advantages and disadvantages of *N. crassa, A. nidulans,* and *S. cerevisiae*[a]

Properties	N. crassa	A. nidulans	S. cerevisiae
Advantages	Heterothallic	Uninucleate conidia	Unicellular
	Rapid growth	Stable diploids	Replica plating simple
	Pure crop of progeny easily obtained free of parental conidia	Colonial growth allows easy plating	Homo- or heterothallic, depending on genotype
	Freehand ascus analysis	Colors mutable as genetic markers	Stable diploid and haploid vegetative phases
	Heterokaryons relatively stable	Versatile nutrition governed by easily analyzed regulatory systems	Diploids sporulate rapidly
	Colonial growth inducible	Parasexual assignment of genes to linkage groups	Red *ade* strain usable as visible marker in many genetic experiments
	Young mycelial growth suitable for accurate physiological and biochemical analysis	Sexual cycle prolific	Strictly homologous integration of trans-forming DNA
	Repeat-induced point mutation as a mutagenic technique	Morphogenesis can be synchronized, and easily analyzed by molecular techniques	Autonomous nuclear plasmids of several types
		Frequent homologous integration of trans-forming DNA	Survives loss of mitochondrial DNA
		Autonomously replicating nuclear plasmid	Very small genome
Disadvantages	Crosses take 2.5–4 weeks	Homothallic	Tetrad analysis must be used for most genetic analysis
	Older mycelia contain proteases that compromise enzymological work	Progeny must be freed of parental contamination or selectively plated	Tough cell wall complicates biochemical work on organelles
	Filamentous growth prevents indefinite, homogeneous growth	Tetrad analysis very difficult	Little morphological complexity, limiting developmental studies to cell type, cell polarity, and cell division
	Most conidia are multinucleate	Many degradative enzymes present	
	Transforming DNAs rarely integrate in a homologous location, most integrate randomly into chromosomes	Tight, buttonlike growth in liquid culture compromises homogeneous growth	
	No autonomous nuclear plasmids known	Growth is slow compared to *Neurospora* and yeast	

[a]Some items refer to subjects discussed in later chapters.

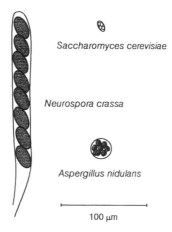

Fig. 6.1.   Asci of the three model fungi, drawn approximately to scale. (100 μm =
0.1 mm.)

Some of the disadvantages of filamentous fungi could not be ignored by
investigators anxious to pursue certain goals. Yanofsky found it impossible to
purify tryptophan synthetase from *N. crassa* dependably and transferred his
affections to *E. coli.* The filamentous growth habits of *N. crassa* and *A. nidulans*
made study of homogeneous cell populations in physiological steady state or
during culture transitions cumbersome or impossible in comparison to yeast
or bacteria. And as work progressed into the 1950s and 1960s, formal genetic
approaches to gene action pursued with the filamentous fungi and yeast diminished in importance as the biochemical and molecular questions using bacteria
and their viruses began to preoccupy microbial geneticists. The days of fungi
as universal model organisms darkened as *E. coli* and bacteriophage emerged
into the light. The eclipse of filamentous fungi as model organisms lasted a
long time. But it is often forgotten by biologists and historians alike that the
fungi initiated truly revolutionary developments in genetics in the 1940s.[2] We
will see in chapter 15 how yeast became the prime focus of eukaryotic molecular biology after 1975 and in chapter 16 how *Neurospora* and *Aspergillus*
reemerged in the 1980s as model filamentous fungi rather than models of life.

7

Escherichia coli

But science is enclosed in its own explanatory system, and cannot escape from it. Today, the world is messages, codes and information.

—F. Jacob

History and Historians

The rapid growth of molecular biology has evoked intense interest among historians and sociologists of science, journalists, and the scientists themselves. Many accounts of the development of the field, focused on different aspects of the science or its context, have been published. The early newspaper coverage of events as they were happening promised enormous breakthroughs in our knowledge of living systems and the benefits to human health that would ensue. More sophisticated treatments, usually by the scientists, appeared in popular magazines devoted to advances in science, among which *Scientific American* was the most influential. In fact, the artwork in this magazine has influenced the representations of molecular biological phenomena in the popular imagination and in textbooks written since that time.

The discovery of the structure of DNA in 1953 and recognition of its probable genetic role imparted almost transcendent insights into the phenomena being uncovered in microbial genetics. The new insights evoked specific hypotheses regarding the divisibility of the gene, the nature of the genetic code, protein structure and synthesis, enzyme regulation, and molecular (inherited) diseases of humans. Historians soon attempted to capture the currents and crosscurrents of this period, in which the world witnessed an unprecedented pace of discovery and unification of biological knowledge. Numerous descriptions of research accomplishments by the first generation of molecular geneticists appeared as the story unfolded. Among these are the early collection of memoirs, *Phage and the Origins of Molecular Biology* (1966) and the most widely known personal account, James Watson's *The Double Helix* (1968).[1] More comprehensive histories by historians and journalists in the 1970s include Olby's *The Path to the Double Helix* (1974) and the popular and accessible *The Eighth Day of Creation* by H. F. Judson (1979).[2] Most of these accounts focus on the intellectual background and subjective experiences of the founders of the field, as the founders were the source of much of the anecdotal information. The authors of the first-generation books are therefore subject to strong bias by the original investigators, and a heroic picture understandably emerges from these histories. Later treatments analyze the events in a larger context, calling attention to the source of research funds, the influence of institutional structure, the role of instrumentation, the neologisms and changes of language, and the change of our attitudes toward living things.[3] These histories stress conceptual changes from structural to informational aspects of genes and the attendant changes of scientific approaches from holistic to reductionist to cybernetic.[4] The more revisionist accounts focus on some of the "losers" in the "struggle for authority" in biological thinking or on the now-shadowy roles of wives, forgotten members of laboratory teams, and other "also-rans" left in the wake of the founders as they went on to other things.[5] These latter writings are part of the growing field of the history and philosophy of biology, increasingly self-contained, and unfortunately ignored by working scientists. (Richard Feynman is reputed to have said that a working scientist needs the philosophy of science like a bird needs ornithology.)[6]

Thomas Brock's *The Emergence of Bacterial Genetics* describes in depth the origins and history of the genetics of bacteria and their viruses (called *bacteriophages*, or *phages*), largely in scientific terms.[7] I have used Brock's book as an invaluable resource and defer to the work as a systematic treatment. Here I focus, as in earlier chapters, on the choice of a model organism, in this case the bacterium *Escherichia coli*, in the development of molecular genetics. As we will see, this choice was made inconspicuously just three years after Beadle and Tatum published their 1941 paper on the first biochemical mutants of *Neurospora*. The initial discovery of bacterial recombination was announced in 1946.[8] In the case of phages, the physicist-turned-biologist Max Delbrück insisted in 1944 on the choice of the T-series as models of virulent phages (see chapter 8). While the choice of model temperate phages would be delayed until their lifestyle was understood, a group at the Pasteur Institute in Paris had begun seminal work on this category of virus even before 1940 (chapter 9). From the mid-1940s, bacteria and their phages would dominate progress in both biochemical and molecular genetics.

As *E. coli* gained prominence in microbial genetics, two displacements are notable. One, referred to in the last chapter, is the eclipse of the fungi as models of life processes. The second is the related, decreasing attention to enzymes and metabolism—the world of biochemistry—and the increasing preoccupation with nucleic acids promoted by the new molecular biologists and geneticists after 1953. Despite the acceptance by most biochemists of biochemical genetics, the biochemical and molecular communities remained for a time separate and institutionally competitive. As a result, historians of molecular biology have tended to ignore the role of biochemistry in its origins. I argue in chapter 17 that biochemical techniques and thinking played a constant and significant role in the evolution of the new field.

Bacteriology: A World of Its Own

Let us look at the state of bacteriology in 1945, as described in the influential treatise of René Dubos (1901–82), *The Bacterial Cell*.[9] The publication of this book followed Oswald Avery's (1877–1955) characterization of the transforming agent of *Pneumococcus* as DNA in 1944 and preceded Lederberg and Tatum's announcement of bacterial recombination in 1946, which I will describe below.[10] The field of bacteriology had developed by then largely in the areas of academic medicine and agriculture, and in industry, all at least somewhat separate institutionally from traditional biology and genetics. In the early 1800s, the prevailing opinion regarding bacterial variation (*pleomorphism*) held that only a few bacterial types existed, and that these could, in various environments, generate all the forms found in nature. The idea that bacterial species (as we would call them now) are interchangeable succumbed in the 1870s and 1880s, largely through the efforts of Robert Koch (1843–1910) and Ferdinand Cohn (1828–98). These efforts were considerably abetted by the earlier studies of Pasteur in his researches on beer, wine, vinegar, and diseases of silkworms, cattle, and humans.

Koch's work established the use of semisolid media (using gelatin) and pure-culture techniques, which demonstrated a substantial stability of bacterial strains during serial propagation through single-colony isolation. With this demonstration, Koch could propose that each bacterial type was specific for a given disease or constellation of biochemical activities. "Koch's postulates," a set of rules to identify the agent of bacterial diseases, embodied this view. The identification required infection by and reisolation of the agent from a host that shows, in the interim, characteristics of the particular disease. Unfortunately, the principle emerging from this work, termed "monomorphism," was so useful in guiding progress in bacteriology that the idea began to retard progress at a later date. *Monomorphism* held that bacterial species were invariant and that they divided solely by fission. This led bacteriologists to think that variation in bacterial cultures was probably due to contamination or a vague category of pathological change that could be named ("involution," "dissociation," *Anpassung*) but not understood. The presumed place of bacteria among living organisms in 1910 was at the border between the living and the nonliving. Taxonomically, they were classified as lower fungi, the Schizomycetes.

Despite the principle of monomorphism, variation in bacterial cultures in the laboratory was common and had to be faced in practice. As Brock points out, bacteriologists thought of bacteria as cultures, rather than as cells or clones.[11] This quasi-unitary picture of what were in fact populations varying in response to the environment allowed several interpretations. Bacteriologists, most with nongenetic—indeed, non-Darwinian—attitudes, believed that bacteria might simply be unorganized flecks of protoplasm. It seemed reasonable that bacteria reacted holistically, changing from one metastable state to another as they were "trained" to adapt to one or another environment. Changes from motile to nonmotile, the acquisition and loss of enzymatic capabilities, alteration of antigenic character, and the appearance of variant colony morphology were readily assimilated to this view. Thus, a limited pleomorphic attitude and Lamarckian interpretations of the phenomena that underlay it returned and persisted well into the 1950s.

Two of these variations were common during propagation of bacterial populations. In the first, bacteria acquired a particular enzyme upon the presentation of the substrate of the enzyme. For instance, the "adaptive" enzyme we now call β-galactosidase appeared in populations of bacteria only when the sugar lactose was present as the sole source of carbon. The enzyme enabled the bacteria to use the sugar for growth, but the change was reversible. The second sort of variation is exemplified by avirulence developing in laboratory populations of a normally pathogenic species, often with accompanying morphological changes. Many such changes were later recognized as irreversible, but not in the hands of investigators who failed to purify the variants from single-cell colonies after plating a population on solidified medium. By 1945, however, some sophistication had penetrated the field. Dubos, in a down-to-earth way, characterized variations as either physiological changes in the state of all cells of a population (adaptive enzymes), or the selection of rare, discontinuous mutational variants arising during the growth of the culture.[12] Certain variations of both sorts (adaptive or discontinuous) could even be predicted in certain species. Dubos therefore held that if one must use the term monomorphism, it must refer to a spectrum of potentials rather than to a fixed type. Dubos had to make a detailed point of these arguments, given the state of the field at the time. The time was atypical, with major discoveries coming in quick succession. The 1943 paper of Salvador Luria (1912–99) and Delbrück on the randomness of mutations to bacteriophage resistance (a case of discontinuous variation, discussed hereafter) was too new to be mentioned in Dubos's text.[13] The 1944 work of Avery's group on DNA-mediated transformation of *Pneumococcus* is noted briefly as "potentially very significant."[14] It might not have been cited at all if Dubos and Avery had not both worked at the Rockefeller Institute.

Dubos's discussion of bacterial cytology shows that microscopic observation had not told bacteriologists much about the organization of the bacterial cell. Even the electron microscope revealed little more at the time than the light microscope because whole-mounts were too thick to show structures within the cell. In fact, microscopic observations had evoked many proposals regarding nuclear structure and a life cycle, sexual or asexual, from many wishful workers. The diversity of these observations and their interpretations guaranteed that there was a great deal more to learn. Only at the time that Dubos's text was written did staining methods improve, largely through the efforts of Carl F. Robinow (b. 1909), to the point where the DNA-rich nuclear region (now casually called a nucleus) could be easily visualized.[15] But even here, the DNA appeared as a clump, either an amorphous mass or in the form of one or more rodlike "chromosomes," as some pronounced them.

Debates regarding the subcellular organization of bacteria continued into the early 1950s, well after bacterial recombination had been discovered. Only in 1953 would observations from the laboratories of George Chapman (b. 1925) and Keith Porter (1912–97) demonstrate the lack of a nuclear membrane in bacteria with thin sections observed using the electron microscope.[16] And, curiously, later genetic techniques were the first to suggest that the bacterial genome was a circular molecule of DNA. As late as 1962, Roger Stanier (1916–82) and Cornelis B. van Niel (1897–1985) felt compelled to publish a white paper that defined bacteria as a coherent and distinct biological group.[17] Their definition stressed the absence of a nuclear membrane (fig. 7.1) and of chloroplasts and mitochondria; the ability to carry out large numbers of metabolic reactions; and the distinctive character of the cell wall and various motile apparatuses. In using these criteria, they excluded viruses, fungi, and other microorganisms and included the familiar bacteria, the blue-green algae, and the Actinomycetes (of which *Streptomyces* is a member) in a group we now call prokaryotes. It was about time. The definition still serves us well, even if more recently the bacteria have been divided into two "domains," the true bacteria (Eubacteria) and a newly recognized taxon, the Archaea (see chapter 18).

In the late 1930s and early 1940s, a number of investigators claimed that bacteria fused, or produced "gonidia," or displayed other behavior suggestive of a sexual cycle. These rudimentary observations could hardly prevail against the equally poor and unsystematic morphological evidence that bacteria reproduced strictly by fission. Dubos came down firmly in the middle: not proven on either side. In his words, "any attempt to classify bacteria in the present state of ignorance [about sexual structures] is likely to duplicate the confusion which

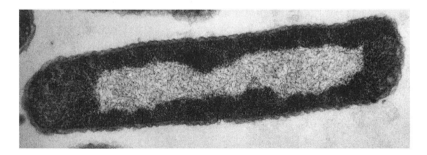

Fig. 7.1.   Transmission electron micrograph of thin section of *E. coli* K-12. The light central nuclear region ("nucleus") contains a fibrillar precipitate of DNA. The nuclear region is not bounded by a membrane. The denser region surrounding the nuclear region is packed with ribosomes and the proteins that carry out most of the metabolic reactions of the cell. The thin cell membrane and outer cell wall are barely distinguishable in this photograph, but may be seen more clearly in figure 8.3. The cell measures approximately 3 × 0.7 microns. Photo courtesy of Conrad Woldringh, University of Amsterdam.

prevailed over the group of *fungi imperfecti* before their sexual forms were identified."[18] He then reminded readers of the discovery and analysis of sexuality in yeast by Winge and by the Lindegrens, a quite recent development at the time (see chapter 5). Dubos suggested that new experimental methods would be needed to test critically for the existence of sex in bacteria.

Mutation and Transformation in Bacteria

Dubos's book was written just as two major advances, alluded to above, had taken place in the study of bacteria. The first, Luria and Delbrück's characterization of the process of spontaneous mutation in 1943, is often called the beginning of bacterial genetics. This was followed a year later by Avery's discovery that DNA alone could carry genetic information.

The first study concerned the origin of mutations in bacteria. Luria had received medical training in Italy but soon became interested in physical approaches to the gene. He observed that bacterial populations, exposed to a bacteriophage, were killed by the infection, but a few resistant bacteria, immune to the bacteriophage, remained. The resistance to bacteriophages persisted upon subculture of these bacteria and therefore appeared to be a heritable characteristic. Luria asked whether bacteriophage caused these mutations in a small fraction of the population or whether such variants preexisted the challenge. At the time, this question represented a general one: were bacteria "trained" by changed environmental circumstances to adapt to them in some Lamarckian way? In another context, did the immune system of a mammalian host induce virulence in a few bacteria that otherwise might be cleared from the circulation? Luria, according to his retrospective account, imagined that if mutations occurred before the bacteriophage challenge during the growth of the culture, resistant cells would propagate as a *clone* (a population originating from a single cell) within the larger population.[19] If so, tests of many small cultures, each started with a few bacterial cells, should reveal occasional "jackpots" of bacteriophage-resistant cells derived from mutational events occurring at an early stage of growth. Most cultures, however, would have none or only a few mutants— those arising very late in the growth of the culture when many more cells were available for rare mutational events.

The alternative scenario, in which a bacteriophage caused the mutations, would have the resistant bacteria arising in the cultures only after the cultures were fully grown, when they were exposed to the bacteriophage. The cultures would be expected to show similar numbers of mutant cells, distributed statistically among the cultures according to the Poisson distribution, which describes such cases of rare, random events. Luria and Delbrück tested these hypotheses, using the bacterium *E. coli*, and determined that the first case prevailed: wide fluctuations in the number of mutants per culture were found. They published a rigorous study that proved that the mutations to resistance were spontaneous, and that, in this respect, bacteria obeyed the rules of higher organisms.[20] Mutant cells occur randomly, without environmental induction, and are merely favored, or selected for, by their subsequent indifference to the selective agent, in this case bacteriophage. The paper influenced many geneticists to think that bacteria had conventional genes that might be studied at least through mutational analysis.

The second study was that of Avery, who worked in the 1930s and early 1940s to analyze the process of bacterial transformation in the murine pneumonia bacterium, *Pneumococcus*. The bacterium, a model system for studies of this disease, was well domesticated. Culture of the organism from various sources had yielded antigenically different types (I, II, III, etc.) distinguished by slightly different polysaccharides that form the slimy capsule around each cell. These capsules are secreted by the cells and account for the virulence of the bacterium: the capsules defend the cells against the immunological defenses of the host. This became clear with the occurrence of *Pneumococcus* strains that had lost the sheath by mutation and were consequently nonvirulent. The virulent cells were called S, for the smooth, glistening character of colonies that appeared after plating cells on agar. The nonvirulent cells were called R, for rough. The phenomenon of transformation was first described in 1927 by Fred Griffith (1877–1941) in studies of mice infected with the bacterium. Transformation consisted of the ability of heat-killed, virulent cells (S) to transform living, nonvirulent cells (R) to the virulent form when the two were injected together into mice. The acquisition of virulence reflected the appearance of the polysaccharide capsule on some of the R cells. The transformation was quite specific. R cells derived from a strain having polysaccharide Type I, when transformed by a heat-killed Type III strain, yielded transformants with the Type III polysaccharide.

By 1940, Avery's group could transform R cells in a test tube with a soluble extract of S cells. The phenomenon not only displayed specificity, but the transformation was clearly permanent—heritable—during further culture of transformants. Avery sought to identify the transforming agent, the effective chemical entity present in S cell extracts. After much labor, Avery's group, which included Colin MacLeod (1909–72) and Maclyn McCarty (b. 1911), showed that the transforming agent was not the polysaccharide itself, nor did it appear to be protein.[21] It was, as far as the chemical analyses could be trusted, pure DNA. Thus a hereditary character, the ability to make a specific polysaccharide, could be imparted by another type of molecule to bacteria lacking the character. In effect, what was lost by mutation in the origin of R mutant cells could be restored by S-cell DNA. One was left to wonder how the character might be embodied in the DNA molecule. The phenomenon was so unusual at the time that relatively few people could assimilate its implications, and it flew in the face of the near-universal belief that genes were made of or at least contained protein. Many geneticists, for instance, thought this might be a case of a "directed mutation," although the specificity of transformation seemed at odds with the prevailing understanding that mutation was random. (The habit of using language to subdue mystery is not uncommon.) But the more imaginative geneticists at the time, as well as the taciturn Avery, realized that DNA might be the hereditary substance, directing the formation of a phenotypic character.

While Avery would not publicly promote this idea himself, his data, in the view of some, clearly supported this theory.

Lederberg

Columbia University in New York had a long tradition of interest in cytology, genetics, and evolution imparted by Edmund Beecher Wilson (1856–1939), T. H. Morgan, and others early in the century. The neo-Darwinian conception of evolution had become accepted by most geneticists with the theoretical work of Ronald Fisher (1890–1962), J. B. S. Haldane (1892–1964), and Sewall Wright (1889–1988). This work was bolstered greatly by experimentalists such as Theodosius Dobzhansky (1900–75), at Columbia, and the naturalist-systematist, Ernst Mayr (b. 1904), at the nearby Museum of Natural History in New York. Working exclusively on animal systems (*Drosophila* and birds, respectively) Dobzhansky and Mayr propounded the view that evolutionary change consisted of the occurrence of Mendelian mutations which, in sexual systems, yielded an enormous variety of genotypes by recombination and segregation. The resulting variation would be filtered by natural selection to yield genotypes in each generation equally or better suited to the habitat. The occurrence of variability in populations through mutation and sexual recombination became an accepted principle. In more advanced taxonomic thinking, species came to be defined in biological terms. Mayr formulated the principle that a species comprises actually or potentially interbreeding populations separated reproductively from other such groups. The principle contradicted the practices of traditional taxonomy, which defined species by comparison to type specimens (i.e., typologically) on morphological criteria. (The latter habits and the associated mindset lingered on because it was impossible to apply Mayr's definition in a practical way.)

In their two syntheses of genetics and evolution, Dobzhansky and Mayr both called attention to the question of asexual species: what was their genetic structure, and how did they evolve?[22] Was mutation the sole source of variation? This question was especially pertinent to the ubiquitous bacteria, which up to that time had not revealed any means of gene exchange. Because the sexual life cycles of other microorganisms, particularly yeasts, filamentous fungi, and flagellates, had been discovered and described in the 1930s and 1940s, a biologist could wonder whether the so-called Schizomycetes could also reproduce by sexual means. A young medical student at Columbia University named Joshua Lederberg (b. 1925) did more than wonder about this question at the time: he went on to answer it.

Lederberg (fig. 7.2) entered Columbia College as a premedical student in 1941 with a passionate interest in biology and a specific interest in the behavior of chromosomes.[23] He enlisted in the Navy when he was 17, still an undergraduate, and entered an accelerated program toward the M.D. degree in 1942. During the next four years, he continued biological work in medical programs to which he was assigned. The research programs of Avery's group and of Luria

Fig. 7.2.   Joshua Lederberg (1958). Photo courtesy of Joshua Lederberg and the
University of Wisconsin Archives.

and Delbrück were part of the climate in which Lederberg began his work. To
these we may add the excitement surrounding the Beadle and Tatum work. Their
rationales were brought from Stanford University to Columbia by a young
instructor, Francis Ryan, introduced in chapter 3. Although Ryan was trained
as an embryologist, he had sought a postdoctoral fellowship at Stanford in 1941
to work on a simpler system (an alga), but he begged to join Beadle's *Neurospora* laboratory when he got there. He was Beadle's first postdoctoral fellow
and played an important methodological role in the earliest work in *Neurospora*.
A year later, he set up a program of *Neurospora* research as a new instructor in
the Columbia zoology department. As part of the war effort, the Beadle-Tatum
lab had devised bioassay methods for nutritional substances, using *Neurospora*
mutants that required them. Ryan continued bioassay efforts in his own laboratory as a sideline, specifically for the amino acid leucine. The funding of the
work by the government allowed Ryan to hire Lederberg as a part-time assistant in 1943.

Still another research program, initiated by Tatum, would have considerable influence on the course of Lederberg's work. My description of the state
of bacteriology makes clearer why Beadle and Tatum had no intention of using
bacteria for their initial studies on the role of genes. They could hardly turn to

organisms that seemed to lack discrete genes and were sexually innocent. But Tatum, with Beadle's encouragement, sought to extend the search for biochemical mutants to bacteria shortly after his and Beadle's first paper on *Neurospora*. Why? The reason is simple enough if one understands that another aim had emerged. The genetics of bacteria may have been obscure, but bacteria, like fungi, had simple nutritional requirements, and Tatum had studied bacterial nutrition extensively. The success of the *Neurospora* work therefore prompted him and others to seek biochemical mutations in bacteria. Their motive, arguably, was to use mutations to certify the discreteness and specificity of genetic information in these organisms. Auxotrophic mutations had just been certified as mutations of orthodox genes in *Neurospora*. Tatum's laboratory therefore isolated mutants of *E. coli* that required certain vitamins and amino acids, just like those of *Neurospora*. The success of the effort led Tatum to interpret the results "as indicating an analogy with the induction by irradiation of true gene mutations in *Neurospora* with similar biochemical changes."[24] The use of the word "true" here indicates the novelty of the idea of bacterial genes.

In 1945, Lederberg read the paper of Avery et al. that, although it did not advertise the possibility that genes in general were made of DNA, allowed many geneticists to think so. Lederberg was excited by the possibility that the genetic material could be characterized chemically. He believed that if Avery's findings could be reproduced in a more familiar eukaryotic organism, the gene would indeed be open to chemical analysis. Others entertained the same hope, stimulated by Avery's findings and the developments in microbial genetics over the next few years. In Ryan's laboratory, Lederberg was quick to consider transforming the *leu⁻* (leucineless) *Neurospora* mutant to the Leu⁺ (leucine-independent) phenotype.

The possibility of transforming *Neurospora* represents a key intellectual step: the use of an auxotroph to select, on minimal medium, for what might be a rare event leading to the appearance of a prototroph. The experiment could not be carried out, however, because the *leu⁻* mutant Lederberg used reverted at a low rate (perhaps 1 in 100,000 cells) to prototrophy. (The term *prototroph* was used first here, after the word *auxotroph* had been in use for several years.) The inability to distinguish Leu⁺ transformants, should they arise, from Leu⁺ back-mutants doomed Lederberg's initial hope.

Avery's work would continue to linger in Lederberg's mind. Lederberg thought that bacteria could not tell him much about the genetic significance of the transforming principle as long as these organisms could not be subjected to genetic analysis. Lederberg believed that the question of whether the Schizomycetes had a sexual stage was not closed, a point reinforced by Dubos's text. Lederberg decided he would find a new way of looking for sexuality in bacteria. Thus, instead of bringing transformation to *Neurospora*, he would bring genetics to bacteria. He reasoned that with nutritional mutants of bacteria, he could mix populations with different nutritional requirements and then select for prototrophs by plating the mixture on minimal medium. Lederberg did not attempt to repeat Avery's transformation experiments in *E. coli* with purified

DNA because at the time the bacterium had only marginal advantages over *Pneumococcus*.[25] But his alternative procedure of mixing mutant strains could theoretically reveal the production of tiny numbers of prototrophs, whether they arose by transformation, by forming a heterokaryon or diploid, or by exchanging and reshuffling their genetic material.

Still in Ryan's laboratory, Lederberg chose *E. coli* for his work. He was encouraged to do so by his familiarity, as a medical student, with the wide variation in the antigenic serotypes of the related pathogenic bacterium *Salmonella typhimurium*. Lederberg thought that serotype variation might reflect sexual recombination and selection in natural populations, a possibility raised at the time by Dobzhansky and Mayr. The closely related *E. coli*, because it was nonpathogenic, was more suitable for the work than *S. typhimurium*, though the latter bacterium would figure prominently in Lederberg's later studies. Lederberg succeeded, like Tatum, in obtaining mutants through mutagenesis of wild-type *E. coli*. He was encouraged in this effort by his *Neurospora* work and by Luria and Delbrück's demonstration of the mutational origin of bacteriophage-resistant variants of *E. coli*. He obtained mutants that required single metabolites, but tests for recombination in mixtures of different strains foundered on the appearance of revertants (back-mutants) to prototrophy, the same problem that had foiled his transformation experiments in *Neurospora*. He realized that a successful test would require double mutants, in which simultaneous reversion of two mutations in a single cell to yield a prototroph would be vanishingly rare. Before he could embark on the arduous task of obtaining such strains (by the brute-force methods that were then quite laborious), he discovered that Tatum was about to move to Yale University, and that Tatum had obtained mutants and double mutants of *E. coli*.

Tatum's mutational work on *E. coli* had made another important point.[26] Its intent, according to the introduction to the 1945 paper, was to demonstrate that single mutations can be used to mark a strain so that the provenance of additional mutations is certain. Thus, if a threonine-requiring strain were to be mutagenized and a methionine-requiring strain were derived from it, the product should require both nutrients. Not only was the strain in which the second mutation occurred clear, but isolation of the double mutants showed that the mutations could occur independently or together, as in higher organisms. This in turn would show with little doubt that the genetic material of bacteria is resolvable into different and quite specific units; its system of inheritance was not holistic. The paper claimed "presumptive evidence for the existence of genes in bacteria, perhaps contained in the nuclear structures which have been observed in a number of bacteria."[27]

Ryan suggested that instead of waiting for Tatum to honor Lederberg's request for appropriate strains, Lederberg should seek to join Tatum's new laboratory. This was arranged in early 1946, with Lederberg on leave from medical school. (There is some hint, according to Lederberg, that Tatum contemplated bacterial recombination tests before Lederberg requested a place in his laboratory.) Within six weeks, a mixture of two strains, one carrying deficiencies for threonine and proline (Thr⁻ Pro⁻), the other carrying deficiencies for biotin and methionine (Bio⁻ Met⁻) yielded very rare (perhaps 1 in 1 million) Thr⁺ Pro⁺ Bio⁺ Met⁺ cells on minimal medium. Neither parent would do so if plated by itself. Additional experiments included a bacteriophage-resistant marker in one of the auxotrophic strains. In such matings, both resistant and sensitive strains were found among the rare prototrophs. This segregation reinforced the probability that the prototrophs arose by recombination and a return to the haploid state. This in turn lessened the doubts about whether "recombinants" were actually diploids or persistent associations of parental cells that cross-fed one another.

The Choice

Much has been made of Lederberg's peculiar luck in choosing strains derived from *E. coli* strain K-12.[28] This strain was isolated in 1922 from a person recovering from diphtheria and had been part of a bacterial strain collection at Stanford University since 1925.[29] Lederberg stated that *E. coli* was probably the most studied bacterial species in the area of biochemistry and metabolism in the early 20th century.[30] Tatum had used the K-12 strain in his laboratory course at Stanford. From this strain, Tatum had selected mutants and double mutants, as we have seen, by X-ray mutagenesis—the same mutagen that he and Beadle used in their *Neurospora* work. *E. coli* K-12, however, represents a special class of *E. coli* strains, soon to be called F⁺ (for fertility). Such strains, unbeknownst to anyone at the time, carry a small, accessory circular DNA molecule called the F plasmid. This plasmid, which replicates independently of the bacterial chromosome, carries genes that enable F⁺ cells to conjugate with cells lacking it (called F⁻). Only about 5% of *E. coli* strains are the F⁺ type. Moreover, during the sequence of mutagenic treatments applied by Tatum in the development of the Thr⁻ Leu⁻ strain, the F factor was lost. Therefore, the matings Lederberg performed took place between F⁺ and F⁻ cells, optimal at the time for detecting bacterial gene transfer.

The real choice made by Lederberg, then, was not of a bacterial stock to work with, but of Tatum himself, who had the material for experiments he wanted to do. Tatum's choice was rather casual, as indicated above. The choice of *E. coli* K-12 was made without reference to its unique suitability, recognized retrospectively, for the recombination experiments. The widespread acceptance of Lederberg and Tatum's initial work, despite Delbrück's skepticism based on the lack of kinetic analysis of the mating process, indicates how ripe the scientific community was for it.[31] Lederberg and others have speculated that the discovery of bacterial recombination was "postmature"; it could have been made with materials available much earlier.[32] Such speculation illuminates certain possibilities and the intellectual climate of earlier times. But it does not change the way bacterial genetics was born: with several fathers, a lot of luck, and a future no one could imagine in 1946.

The Adolescence of Bacterial Genetics

Lederberg had made a striking discovery, soon confirmed in other laboratories with the strains he had used. A rather long period of difficulty in understanding the process of recombination followed the discovery. Lederberg chose to study the mating process by genetic means, seeking evidence for linkage as one would in higher organisms. His initial presumption was that rare pairs of parental cells fused and formed a diploid. The genetic material of the two parents would then recombine and yield recombinant progeny, presumably haploid. Unlike genetic analysis in higher organisms, Lederberg's method of detecting bacterial recombination relied on the specific selection of cells having recombinant phenotypes. He did this by using medium on which only cells with certain nonparental nutritional characteristics could grow. This selective technique of revealing a rare biological event is arguably Lederberg's most important technical contribution. However, with this technique it was impossible to express recombination frequencies as a percentage of the total progeny, given the unknown frequency of mating, the inability to distinguish most progeny from parental cells, and the incomplete recovery of recombinants in the final, plated population.

To make his method more practical, Lederberg used selective markers by which recombinant progeny could be isolated, followed by determination of the frequency and associations of additional, unselected markers carried by the parents. Among the convenient selective markers was the streptomycin-resistance mutation of *E. coli* K-12 isolated early in this work and resistance to the bacteriophage T1.

Lederberg set up his early crosses such that one parent was resistant and the other sensitive to bacteriophage T1, using this difference as an unselected marker. But he used a rationale that could easily be understood by geneticists: he did the cross in two ways. In one cross, one strain with selective markers A$^+$B$^+$C$^-$D$^-$ carried T1 resistance and the other strain, with markers A$^-$B$^-$C$^+$D$^+$, carried T1 sensistivity. He found that recombinants A$^+$B$^+$C$^+$D$^+$ were predominantly T1 resistant. If, however, the parent A$^+$B$^+$C$^-$D$^-$ carried T1 sensitivity and the other were resistant, recombinants A$^+$B$^+$C$^+$D$^+$ were predominantly sensitive. To a geneticist, this means that the form of the T1 marker (resistant or sensitive) is the one more closely associated with A$^+$ and B$^+$ than with markers C$^+$ and D$^+$. The degree to which the parental association persists in the selected recombinants is a measure of linkage. Using this rationale, Lederberg developed an extensive set of linkage data that initially promised a linear map, as though the bacterium had one linkage group comparable to the genes on a single chromosome of a eukaryote. Complications soon appeared that made it difficult to rationalize the data in this way, and a branched chromosome was proposed at a later stage.

The problems of genetic analysis could not have been disentangled without more knowledge of the biology of mating. Only in 1950 was it generally accepted that cell contact, referred to thereafter as *conjugation*, was required for recombination.[33] Recombination did not reflect a transformationlike event, nor was it mediated by bacteriophage. (The latter process, as we shall see, was discovered by Norton Zinder [b. 1928] in Lederberg's laboratory in 1952.)[34] In 1952, William Hayes (1918–94) discovered that bacterial matings were polarized: one of the two parents (the "male") could be killed with streptomycin, but could nevertheless donate genes to the other (the "female"). The latter could then go on to yield recombinant descendants. The reverse arrangement, in which the other parent was streptomycin sensitive, did not lead to recombinants. Hayes's laboratory and Esther M. Zimmer Lederberg (b. 1922) inde-pendently identified F⁻ strains, unable to mate with one another.[35] This was, in fact, the first demonstration that there was a fertility factor of some sort that endowed the carrier to act as a male, donating genes to the female. This factor (later shown to be the F plasmid) had the peculiar property of being infective and was transferred from F⁺ to F⁻ strains during conjugation. The plasmid carries genes which, when expressed in the *E. coli* host, promoted recognition, intimate contact, and the formation of a cytoplasmic bridge between F⁺ and F⁻ cells. The plasmid genes then enabled the F factor to transfer a copy of itself into the recipient cell (fig. 7.3). In a mixture of F⁺ and F⁻ cells, all of the F⁻ cells soon became F⁺ cells. However, the F factor did not carry bacterial genes with it as it entered the recipient. How, then, could conjugation lead to exchange of bacterial genes?[36]

These confusing results were not easily understood until a new development occurred. Still another new type of strain was discovered independently in the laboratories of Luca Cavalli (later, Cavalli-Sforza, b. 1922) and Hayes. These strains acted as males, but donated bacterial genes to the recipient with high efficiency. Surprisingly, they did not transmit a fertility factor; the recombinants recovered were unable to confer fertility to other strains and thus remained F⁻. The super-male strains were called Hfr (high frequency of recombination) and were later shown to have integrated the F plasmid DNA (F DNA) by recombination into the chromosome of the bacterium (fig. 7.4a). The F DNA of these strains, as in the F⁺ strains, could initiate conjugation of cells but transmitted only part of itself to the recipient, before any of the bacterial genes. But because the F DNA was part of the bacterial DNA, it dragged bacterial genes with it (fig. 7.4b). The part of the F DNA entering the recipient did not endow the recipient cell with fertility. These exconjugants remained F⁻, rather than becoming Hfr or F⁺ in their sexual capacity. By 1955, it was clear that bacterial mating had little in common with eukaryotic sex. It did not involve full fusion of cells to form a diploid. Only parts of the bacterial chromosome were transmitted in most matings. Finally, progeny arose by a messy process by which segments of the donor (male) chromosome became integrated into the recipient's chromosome after they had been transferred. This makes it understandable why investigators had such difficulty in interpreting the initial empirical results in bacterial recombination studies. With genetic analysis alone, no clear picture of the mating process could have emerged. The initial insight about the donor–recipient nature of the mating process radically changed the working hypotheses. We must admire the work of the few laboratories participating in research that made such headway with such indirect methods and with unseen sexual partners.

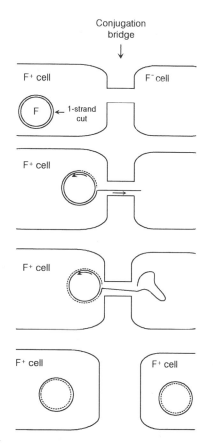

Fig. 7.3.   Conjugation and transfer of the F plasmid from an F+ cell to an F- cell. The transfer begins by cutting one of the two strands of DNA, the passage of one cut end into the F- cell as it is peeled from the intact, circular strand. The process is coupled with replication of the plasmid in the F+ cell, maintaining the double-strandedness of the initiating plasmid (the "rolling circle" mode of replication). After one complete genome is transferred, the donated DNA is circularized and replicated to form an intact F plasmid; the recipient is now an F+ cell.

An elaborate study of many aspects of bacterial matings (indeed, the domestication of the *E. coli* genetic system for its later service in molecular biology) was initiated at the Pasteur Institute in Paris by Élie Wollman (b. 1917) and François Jacob (b. 1920) in 1953. This was the year that the structure of DNA (see appendix 2) was announced, a structure on which our modern picture of the bacterial mating process depends.[37] Wollman and Jacob summarized their studies in the English version of a comprehensive book in 1961.[38] Wollman and Jacob are discussed later, and we will see how they helped make the bacterial mating system a major tool of molecular biology in the 1960s.

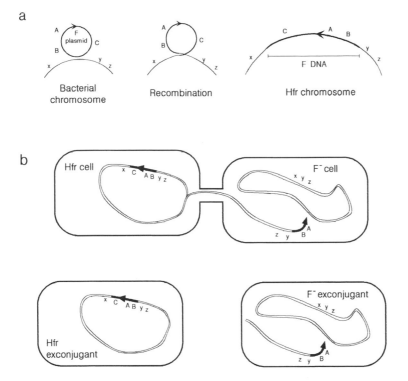

Fig. 7.4. Conjugative chromosomal transfer in *E. coli*. (*a*) The F plasmid may occasionally integrate into the bacterial chromosome by a single recombinational event, resulting in an Hfr bacterium. No free F plasmid remains in the cell. The arrow point will become the "origin of transfer" in conjugation, followed by the plasmid genes *A* and *B* and the bacterial genes *y* and *z*. (*b*) Conjugative transfer of the bacterial chromosome is initiated in the manner of F plasmid transfer (fig. 7.3). Transfer continues until the conjugation bridge breaks. The bacterial genes transferred may be integrated into the chromosome of the F⁻ exconjugant by the recombination machinery of the cell. The incomplete F plasmid DNA is destroyed, and the cell remains F⁻.

The early studies of the Lederberg laboratory included several noteworthy technical advances. The first was the replica plating technique, devised by J. Lederberg and E. M. Lederberg.[39] The technique allows one to sample of all the colonies of a Petri dish at once and identify interesting variants. The process uses a cylinder (called a replica block) slightly smaller than a Petri dish, with velveteen secured over the flat top of the cylinder with a ring. The velveteen has a nap that, when pressed onto the surface of a "master plate," samples the cells. The entire sample is then printed onto a fresh plate, where they may or may not grow according to the composition of the new medium. Comparison of the master and the derivative plate permits one to identify interesting mutants (particularly auxotrophs) and to culture them from the master plate.

The technique yielded a spectacularly clear demonstration that mutants arise independently of a selective agent to which the parent population is sensitive. A large population of E. coli was spread on a Petri dish and allowed to grow into a thick lawn of cells. The cells were replica plated to a medium containing streptomycin. In a few positions on the new plate, colonies appeared. The corresponding positions on the lawn of cells on the master plate were identified and sampled. The samples were grown in liquid culture and then spread on a plate without streptomycin. The resulting second lawn of cells was replicated once more onto streptomycin medium. This time, many more colonies appeared than the first time; the sample appeared to be enriched in streptomycin-resistant mutants. (Similar samples from other positions of the master plate yielded only a few resistant colonies when put through the same procedure.) By going through this process several more times, a pure, streptomycin-resistant strain was isolated. Tracing the pedigree of the final culture showed that a bacterial mutation to streptomycin resistance had arisen and had been purified without the slightest exposure to streptomycin. The streptomycin-containing medium had been used only to determine the positions on the master plate at which clones of mutants would be found in high numbers. This demonstrated that contemporaneous Lamarckian interpretations of the origin of bacterial antibiotic resistance, in which streptomycin causes the mutation, were egregiously off the mark.

The second technical advance, called the penicillin enrichment method, arose in the laboratory of Lederberg and Norton Zinder and independently in the laboratory of Bernard Davis (1916–94).[40] The technique greatly improved the efficiency of isolating auxotrophic mutants. Penicillin kills growing cells because it blocks the formation of the tough outer cell wall of bacteria. If bacteria grow without maintaining their cell walls, the cells soon burst, due to the decreasing strength of the cell wall against the normal and substantial turgor pressure within the cell. Lederberg and Davis reasoned that auxotrophic cells could be greatly enriched in a mutagenized population (after treatment with ultraviolet light, for instance) by first allowing the cells to grow in the presence of penicillin in minimal medium. Under the proper conditions, the prototrophs of the population grow and die, while the rare auxotrophs, unable to grow, survive. The remaining population, washed free of penicillin and plated on a medium containing nutritional supplements, yields many auxotrophs. The replica-plating and penicillin enrichment techniques greatly increased the pace of investigations of bacterial genetics and biochemistry.

E. coli as a Model Organism

We may now appreciate the series of commitments made in developing E. coli as a model organism. Tatum and Lederberg independently chose E. coli as a nonpathogenic bacterium with which to seek mutants. The mutants isolated led to the lucky discovery of recombination in the particular strain, K-12, used. Recombination was rare and difficult to analyze. But at that point, there was no turning back. To study the phenomenon in any detail, there was no question of adopting another organism: the phenomenon was unique, and there was no practical reason to do so. Slowly knowledge accumulated, and the investigators devised new techniques which would be effective with many other bacteria. As work went on, other laboratories took up *E. coli*. The bacterium by then had yielded nutritional mutants to study metabolism, and the technical means of applying the selective-marker rationale of matings. The work quickly stirred hopes of mapping bacterial genes and understanding a wholly novel sexual system. The commitment to the organism was made early and almost insensibly, well before it was clear what was in store.

Before leaving *E. coli*, I must note other, reinforcing influences that solidified the model status of the organism for biological work. The first is the independent research of Delbrück and his collaborators on the virulent bacteriophages of *E. coli*. The next chapter introduces this program, designed to study the replication of genetic information in the simplest possible form: during the multiplication of a bacterial virus. In this program, Delbrück used *E. coli* strain B as a host for bacterial viruses. As studies on viral replication progressed, considerable biochemical information about the infected hosts also accumulated.

E. coli B does not mate as a male because it lacks a fertility plasmid. With difficulty, it can accept the F factor of strain K-12, and then it will mate with an *E. coli* K-12 F⁻ recipient, but this property was not known or exploited at the beginning.[41] However, *E. coli* B became an early model for studies of radiation-induced mutation and initial studies of DNA damage and repair.[42] Systematic work on bacterial growth and macromolecular synthesis through the use of radioactive tracers were initiated by a group of investigators trained in physical methods. Their systematic description of the flow of carbon, nitrogen, sulfur, and phosphorus in logarithmic cultures is a classic study, which laid the foundation for more detailed studies of gene regulation thereafter in *E. coli* K-12 (see chapter 11).[43] These studies contributed more immediately to the knowledge of the composition of *E. coli* cells and of precursor-product relations in general metabolism. Strain B became useful thereafter in studies of replication of DNA and cell division due to its ability to yield well-dispersed cultures of separate cells and its tractability in synchronizing the growth and division of cells of a population. Therefore, although this strain contributed little directly to bacterial genetics per se, many parallel or comparable researches were done on both strains, providing some indications of variation in certain areas such as enzyme regulation.

The introduction of *E. coli* altered the course of molecular genetics. Up to that time, the fungi—in particular *Neurospora*—were the favored microorganisms for biochemical genetics. And biochemistry was a favored focus of genetics at that time. By the time that DNA structure was discovered in 1953 and relatively quickly accepted by a majority of geneticists, the standards of biochemistry had increased greatly. Many questions posed by geneticists in the one-gene, one-enzyme program could be solved only by solid biochemical techniques. Curiously, biochemistry at that time was quite conservative, and the impetus and much of the experimental work in biochemistry after 1950 appeared in the new field of microbial genetics. Few geneticists doubted that biochemistry could be more easily done in bacteria—now with respectable genetics—than in fungi. As soon as this was clear, *E. coli* and its bacteriophages also began to assert their hegemony over the new field of molecular genetics—the study of the properties, behavior, and role of DNA in living systems.

8

The T Bacteriophages

We define as living those things that can be killed.
—Anonymous

Bacteriophages

If the nature of the bacterial cell was obscure until the 1950s, one can hardly be surprised that the nature of bacterial viruses was even more so. Bacteriophages or phages, as bacterial viruses are called, manifest themselves most vividly by making a turbid, liquid bacterial culture clear. The bacteria of the culture, after infection with a virus preparation, simply break down after about 45 minutes; the bacteria lyse, as we now say, with the liberation of vastly more of the infectious agent. The particulate character of bacteriophages was demonstrated in the earliest studies. Samples of a sufficiently diluted infectious suspension would either cause an infection or not, depending on whether the sample happened to have an infectious particle. Moreover, such dilute solutions, when applied to a turbid lawn of bacteria spread on agar in a Petri dish, yielded circular areas of clearing, or *plaques*, in the lawn (fig. 8.1). Plaques represent an area of infection, spreading from a point where an initial bacterium has lysed, and in which large numbers of infectious particles can be found. These attributes appeared to justify naming the infectious agent bacteriophage (bacterium eater). Their key peculiarities were their ability to pass through filters fine enough to stop bacteria and their inability to propagate alone on any culture medium. They grew well in bacterial hosts, and for a time they held great promise as antibacterial agents in medicine.

The discovery of these entities independently by Frederick Twort (1877–1950) in 1915 and Félix d'Herelle (1873–1949) in 1917 initiated a long con-

99

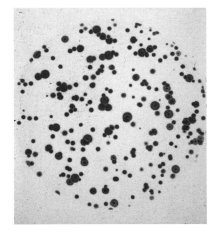

Fig. 8.1. Phage T4 plaques. The variation in plaque character reflects the phenotypic variation of progeny from a multifactor cross involving *m* (minute), *r* (rapid
lysis), *tu* (turbid), and a modifier of the turbid phenotype. The image was made by
placing the Petri dish into a photographic enlarger and making a direct print. The
plaques, lacking or deficient in bacterial cells, transmit light and thus appear dark;
the background lawn of cells appears light. Photo courtesy of Franklin W. Stahl,
University of Oregon.

troversy about their nature.[1] Were they intracellular parasites, a stage of the
bacterial life cycle, or symbionts with the bacterium (which might be likened
to the alga–fungus relationship in lichens)? Or were they perhaps infective
entities below the level of the living, or lethal substances that triggered bacterial cells to make and release more of the same substance? d'Herelle considered them similar in some respects to known submicroscopic viruses of higher
organisms from the outset, entities like Pasteur's rabies virus or the filterable
Rous sarcoma virus, discovered in 1910.[2] These considerations reflect the
uncertainty at the time not only about the nature of bacteriophage, but about
the nature of the bacterial cell itself.

Muller and Delbrück

Herman J. Muller, who was discussed in chapters 2 and 4, rarely ignored an
opportunity to comment on major discoveries in biology. In 1922, he proclaimed
that viruses might be analogous to genes, calling attention to their properties
of variation, replication, and submicroscopic dimension.[3] Although this analogy drew little attention at the time, it demonstrated Muller's interest in physical approaches to the gene. Muller's ambition to study the structure of the gene
led him finally to the discovery of X-ray–induced mutation in 1927.[4] This process yielded not only a greatly increased repertory of mutants to use in studies
of gene action and crossing over, but a method of studying the susceptibility

of the gene to an external, damaging agent. Both biologists and physicists had turned to studies of the effects of radiation on biological material. Muller had previously devised an elaborate, highly effective means of measuring spontaneous mutation frequencies in *Drosophila*, and he used it in his discovery of X-ray mutagenesis. The method, called the *ClB* method, takes advantage of females, one of whose X chromosomes is marked with a recessive lethal mutation; the dominant *Bar eye* mutation; and a chromosomal rearrangement (inversion) that suppresses the formation of viable recombinants upon crossing over between her two X chromosomes. Recessive lethal mutations that are induced in the other X chromosome of the female by X-rays will render crosses with normal males devoid of male progeny. This outcome is inevitable because both her X chromosomes (of which her sons receive only one or the other) now carry a lethal gene. Crosses in which an individual female carries a new lethal gene, carried out in a small vial, are easily recognized simply by looking through the glass wall of the vial and noting the absence of male progeny.

During his year in Berlin in 1932, Muller collaborated with a member of the genetics unit of the Kaiser-Wilhelm Institute, Nicolai Timoféeff-Ressovsky (1900–81). Timoféeff-Ressovsky, like Muller, had broad interests that included gene transmission and evolutionary genetics. He hoped that other disciplines such as physics might be applied to biological problems, and, indeed, Muller's radiation-induced mutation using the *ClB* method lent itself to quantitative analysis. Before he could proceed far with this study, Muller left for the Soviet Union (see chapter 2), but Timoféeff-Ressovsky continued with the project. The study benefited from an informal seminar group involving him, a radiation biologist, Karl Zimmer (1911–88), and the atomic physicist, well versed in quantum theory, Max Delbrück (fig. 8.2). (Muller had met Delbrück in Berlin in 1932, and they would meet again later in Europe.)[5] A joint paper published in 1935 by Timoféeff-Ressovsky, Zimmer, and Delbrück took advantage of the developing discipline of target theory, which related the dosage of ionizing radiation to its effects on biological materials, in this case mutation frequency.[6] This relationship did not resemble the effect of toxic chemicals on viability, for instance. In the latter case, small amounts of the chemical have no effect until a threshold is reached, and the effect rises steeply to a maximum thereafter. In contrast, ionizing radiation appeared to have no such threshold.[7] Variation of the dose or the rate of application of a given dose allowed Delbrück, following earlier formulation of the theory, to conclude that the mutational effect of ionizations caused by X-rays obeyed one-hit kinetics. With some now invalid, but clearly stated assumptions, he estimated that ionizations that cause mutations must take place within a diameter of 10 atoms of the target, or within a volume of 1000 atoms, the empirical target size of the gene.[8] Although this figure is not applicable to the gene as we now know it, the paper introduced a way of thinking about biological phenomena in physical terms, exploring them with physical methods, and analyzing data with mathematical tools.

Delbrück proposed that the gene was a molecule subject to discontinuous changes among several stable states. With a sufficient input of activation energy provided by ionizing radiation, the gene could mutate in a quantized fash-

Fig. 8.2.   Max Delbrück (1963). Photo courtesy of Cold Spring Harbor Laboratory Archives.

ion. The collaborative paper of 1935 (often called the "Three-Man" paper in recognition of its three authors), though published in an obscure journal, was widely distributed by way of reprints sent out by the authors. It marks the point at which the gene was certified by physical methods as a molecular entity that might be analyzed further in molecular terms. The findings and their implications excited Niels Bohr (1885–1962), with whom Delbrück had studied, and Muller, among others. Bohr organized a conference on the application of physics to biology, in which Delbrück's work was featured. From that point on, Delbrück would commit himself to genetics and would be encouraged and sought after by those who could foster his interests.

The Three-Man paper would later become influential to physicists who wanted to enter biology because of its detailed discussion by Erwin Schrödinger (1887–1962) in his 1944 essay, *What Is Life?*[9] Schrödinger, a Nobel Prize-winning physicist, had developed wave mechanics and further rationalized quantum theory. Like Delbrück, he speculated about the gene as a physical entity. Schrödinger's essay explored a physical view of life and the gene, concentrating on the curious fact that the gene, presumably a molecular entity represented by very few copies in cells, was endowed with an extraordinary chemical stability. He conceived of the gene (and other components of the living cell) as an "aperiodic crystal" which, later in the book, he said might be a "code-script." The latter idea is said to have inspired early molecular biologists in their attempt to discover the molecular basis of inheritance.[10] Although the book may have inspired molecular biologists, the scientific impact of Schrödinger's ideas on the progress of studying the gene, according to Edward

Yoxen, was minimal.[11] He favors the view that geneticists may have appreciated Schrödinger, like Garrod, much longer after the fact than at the time of their publications.

Delbrück and Ellis

Max Delbrück, born in 1906 in Berlin and trained in atomic physics by Max Born (1882–1970) and Niels Bohr, was respected as a highly promising professional early in his career.[12] He owed his interest in biology to Bohr, who had maintained an enthusiasm for biology from his boyhood. Bohr and Werner Heisenberg (1901–76) had contributed to the quantum revolution and the idea of uncertainty and acausality at the quantum level. The uncertainty principle and the idea of complementarity are embodied in the impossibility of measuring both the momentum and the position of an electron in a single experimental arrangement. Moreover, the need to use probabilistic analysis in describing quantum states was unsettling to more deterministically minded forbears in physics.

In the early 1930s Bohr believed that life also might have an irreducible mystery, one that could not be penetrated by known methods at its most fundamental level of organization.[13] He suggested that just as the quantum of action is both the mystery and the foundation of atomic physics, life might still have something to hide at the level of molecules. As one probes beneath the level of still-living cells, he thought, one reaches the realm of molecules and atoms, which by themselves have no vital qualities. Bohr did not promote vitalism here, but only a way of thinking: perhaps new physical principles as elegant and equal in importance to those of quantum theory would be discovered if one's mind remained open to mutually exclusive models. Bohr inspired Delbrück by suggesting that biology might reveal paradoxes that could be described only in complementary, nonintuitive terms. Delbrück found such a pursuit more attractive than a career in atomic research which, after the development of quantum theory, seemed as if it might be routine. His mind was after big things, ones that might be embraced by a natural philosopher, universal principles that had so far eluded biologists.

Delbrück decided that after his work with Bohr, he would go, still a physicist, to the laboratory of the atomic physicist Lise Meitner (1878–1968) in Berlin. He chose this location because the laboratory was situated near Kaiser-Wilhelm Institutes in which biology was thriving. His personality and background made him welcome in the genetics division in which Timoféeff-Ressovsky worked and especially in the discussion group that led to the Three-Man paper on genes defined by radiobiological techniques. Delbrück's interest in genetics deepened when Wendell Stanley (1904–71) announced the crystallization of the tobacco mosaic virus (TMV) in 1935, showing that a preparation of replicating particles consisting of protein could be crystallized like a conventional chemical. Here was a demonstration that a chemically defined entity, having a fundamental property of genes, could be characterized by physical methods. It confirmed the impression that genes, too, were molecules. The next year, Frederick C. Bawden (1908–72) and his associates showed that RNA was also a constituent of this virus, but the significance of this remained unclear for some years.[14]

TMV was vastly simpler than complex organisms with Mendelian genetics; its replication simply happened within suitable host cells as an autocatalytic process. The essence of an autocatalytic process was embodied, in most biochemists' minds, in the example of the protein-digesting enzyme pepsin and its precursor, pepsinogen. An active molecule of pepsin can cleave pepsinogen, liberating more pepsin; as more pepsin is formed, the process proceeds more quickly as long as the pepsinogen lasts. With little understanding of protein and nucleic acid structure, biologists felt genes might be enzymes that also might undergo autocatalytic duplication in some fashion. In 1937, Delbrück finished his work with Meitner, and with some knowledge of Stanley's work on TMV, committed himself to biology, starting with a visit to the United States supported by a Rockefeller Foundation fellowship.[15] His background in physics assured him of the interest of the Rockefeller Foundation because it strongly promoted interdisciplinary research and had for some time supported work in genetics. The director of the foundation, Warren Weaver, sought to introduce more physical methods into biology and would coin the term "molecular biology" in 1938.[16]

Delbrück needed a system amenable to highly quantitative analysis. He hoped to continue work on *Drosophila* by going to Morgan's laboratory at Caltech. But even in his travels to Caltech from the East Coast of the United States, Delbrück's visits to various laboratories revealed what he viewed as baroque complexities of working with higher organisms. Mutational analysis and the laboratory routines involving *Drosophila* or plants (or even TMV, which required a plant host) or any other higher organism rendered them unsuitable for efficient, quantitative work. This impression solidified when he arrived at Caltech in 1937 and underwent further orientation in *Drosophila* genetics. But before he succumbed to despair, Delbrück sought out the biochemist Emory Ellis (b. 1906), who worked in Morgan's laboratory on bacteriophage, a subject that by then had exhausted its claim on most scientists' attention.

Ellis worked on cancer, specifically cancers that might be caused by viruses.[17] He held a postdoctoral fellowship endowed by a family interested in research on radiation and cancer therapy. Ellis began by looking at carcinogenesis in mice caused by chemicals, then by viruses, the latter originating much earlier with the discovery of the Rous sarcoma virus. But he needed, in the interests of time, a model for viral infections because animal and TMV experiments were both costly and time consuming. He therefore turned to bacteriophage, hoping to confirm the early findings of d'Herelle regarding the infection process. D'Herelle, in a book published in 1926, had described how to handle bacteriophage in a microbiological laboratory.[18] He summarized the key stages of bacteriophage infections as attachment to the host, multiplication, and release of progeny. These stages probably prevailed for other viral infections, and therefore bacteriophage appeared to provide an inexpensive and workable model for Ellis.[19] Although Ellis favored this picture, a debate still prevailed about the nature of bacteriophage itself: was it an extrinsic parasite or was it wholly derived from the bacterium? D'Herelle had done some quantitative work, inspiring Ellis further in his choice of a model system. Starting from scratch, Ellis simply took an "available bacterial stock," *E. coli* strain B (obtained, curiously, from Carl Lindegren in Morgan's laboratory) and used it to detect and to isolate phages from ultrafiltrates of sewage.

We can appreciate from how Ellis went about selecting a phage that the tractability of a model phage outweighed the particulars of its biology. He simply selected a phage whose plaques were large enough to be seen with the naked eye and small enough to allow many plaques to develop on a single Petri dish without becoming overlapping and confluent. The phages chosen were virulent, without a latent, provirus form. This was lucky in hindsight because the temperate phages (which are discussed in the next chapter) have their own peculiarities that could not have been dissected at the time.

Ellis sought to understand the adsorption step, by which phage particles collide with bacterial surfaces and attach to them before their multiplication within the bacterium. This is a problem in kinetics, complicated by the differing efficiencies with which different phage types attach. Even with a single type of phage, the times at which different cells of the bacterial population become infected vary. The primary difficulty in studying a population of infected cells rigorously with time, which d'Herelle had noted, is that before all phages are adsorbed, others have already begun multiplication within their host cells. This causes an overlap in the periods of attachment and multiplication that confuses the temporal analysis of each process.

Delbrück teamed with Ellis in 1937, as Ellis's work was beginning. They analyzed various parameters of the separate stages of growth mathematically. In doing so, they sought to improve on d'Herelle's report of discontinuous production of bacteriophage during infection of a liquid culture of bacteria. Delbrück determined the rate of adsorption of phage and its variation, and the fraction of viable phage that would form plaques. The latter was analyzed by a Poisson-based method later to be used in his work with Luria. With these parameters in mind, a rapidly adsorbed phage, and reproducible conditions, Ellis and Delbrück defined the "one-step growth" process. Here, bacterial populations are infected quickly at low multiplicity (that is, with many fewer phages than bacterial host cells) followed by large dilution to prevent further adsorption of remaining phage to bacteria. The culture is then sampled and assayed over the next several hours until the number of progeny phage particles, the first of which appear in about 30 minutes, rises to a plateau as all the initially infected bacteria lyse. Further experiments showed that even if the average number of progeny phage per infected bacterium is quite reproducible, individual infected cells vary greatly in their yield of phage particles or "burst size."

With these observations, Delbrück began to explore the process by which single cells, infected with a single phage particle, can produce more than 100 copies of the phage in 45 minutes. This finding of what looked like pure and simple replication was perhaps the most fundamental quality of life and form of inheritance, divorced from sex, recombination, embryogenesis, and most of the complications of genetics as it was known at the time. Delbrück's first publication on phage appeared in 1939—his only joint paper with Ellis—and became a guide for all further work on virulent phages.[20] Ellis was forced by the terms of his fellowship to move back to the study of animal viruses and cancer, leaving to Delbrück the promise of phage biology.[21]

The Choice

Before Delbrück came to Caltech, he had a plan, but, perhaps, like Pontecorvo's plan as he entered fungal genetics, it was more a philosophical dream. Delbrück's early interest in biology had turned on his desire to explore universal properties of life, including replication, its most important attribute, by physical rationales. He had already expressed his ambition in his model of the gene in the Three-Man paper, an ambition further solidified by Stanley's crystallization of TMV. He looked for a living system embodying simplicity and replication, much as Morgan had looked for an organism for efficient research in genetics, and much as Beadle and Tatum would find, in *Neurospora*, a tool for biochemical genetics. Delbrück's choice of bacteriophage as a model system was based not on a surprising observation made with an organism already familiar to him. Instead, he was simply excited by the possibility of quantitative analysis of plaques, visualized on Petri dishes, and was confirmed in his desires by his domestication of the phage growth cycle.[22] Chance had favored a highly prepared mind, one that saw in phage "atoms in biology."[23] Up to this time, no one had ever seen a bacteriophage.

I note in passing that Delbrück's scientific orientation resembled that of Morgan. Both men were empiricists, interested in the operational attributes of the subjects of their interest, neither of which could be seen. Both eschewed speculation that could not be pursued into the "black box" of their biological system, as Delbrück called it. Both thought of the gene as a theoretical entity, whose behavior, by rigorous quantitative analysis, might reveal its nature within the particular frame of discourse of the investigator. Morgan's high regard for Delbrück led him to recommend his staying an additional year at Caltech. Delbrück wanted to remain in the United States even after that to continue his work as the war began in Europe. He could not remain at Caltech after that extra year, however, and he had to take a position in physics at Vanderbilt University in Nashville, Tennessee. There he developed a phage laboratory and, in conjunction with his work at Cold Spring Harbor, made phages the center of a growing scientific area that attracted biologists, biochemists, chemists, and physicists.

Luria

Salvador Luria, discussed in chapter 7 in connection with bacteriophage-resistant mutants of bacteria, started as a medical student in Italy.[24] But his heart was not in the profession, and he sought to learn enough physics and mathematics to enter work in what we now call biophysics. In a course on spectroscopy, his teacher called his attention to the paper of Timoféeff-Ressovsky, Zimmer, and Delbrück. Inspired to apply target theory to biological systems, he met by coincidence a bacteriologist, Geo Rita (b. 1911), studying dysentery bacilli and using susceptibility to a particular bacteriophage as a taxonomic criterion. Immediately, Luria felt he could apply target theory to bacteriophage, an ambition realized after he fled to France in 1938. A paper describing his results appeared as a collaborative work with Eugène Wollman (1890–1943), then at the Pasteur Institute in Paris. It was an extraordinary coincidence, remarked on by Luria, that both he and Delbrück independently (and both, in a sense, by chance) encountered bacteriophage as an optimal system for study of fundamental properties of living things: replication and the nature of the genetic material.

Luria, hardly settled in France, had to flee again in 1939 as the German army entered Paris. He finally came to the United States, supported by a Rockefeller Foundation fellowship in the laboratory of Frank Exner at the Columbia University College of Physicians and Surgeons. He met Delbrück in New York at the end of 1940, and they quickly made plans to collaborate in the summer of 1941 in Cold Spring Harbor. The Cold Spring Harbor Laboratory, a privately endowed biological laboratory on Long Island in New York, had just come under the direction of the *Drosophila* geneticist Milislav Demerec (1895–1966). Demerec had been up to that time the assistant director of the Department of Genetics of the associated Carnegie Institution of Washington, also at Cold Spring Harbor, and he would continue in this role.[25] Demerec supported the collaboration between Luria and Delbrück during summers at Cold Spring Harbor for some years thereafter, and he soon became an influential bacterial geneticist himself.

Luria moved to Indiana University in 1943; Delbrück had already moved from Caltech to Vanderbilt University. They collaborated thereafter by working at Cold Spring Harbor in the summers and by long-distance communication during the academic year. Their studies on spontaneous mutation, using phage resistance as a selective factor, appeared in 1943 (see chapter 7). At the same time, studies of phage flourished as the "phage group" formed around them.

If there were a surprise for Delbrück and Luria, it would be the unexpected complexity of phage biology. One of their first difficulties in studying phage growth came when they tried to perform a mixed infection using distinct phages. Expecting to see each phage replicate independently, yielding a mixed burst of phages, they saw only one or the other phage multiply: the two phages were mutually exclusive in the infection of a given cell. Luria commented:

> Yet our experiments on mixed infection of bacteria with two phages proved to have a seminal role in the rise of molecular biology. They focused attention on the possible mechanisms of phage multiplication—and more generally of virus reproduction—and shifted the interest of virus workers from the problem of cell damage by viruses to the life cycle of viruses themselves. Not without an eye to this possible fallout, Delbruck and I in our first set of published papers referred

to bacteriophages as bacterial viruses, and were gratified by the interest evinced by specialists in virus diseases.[26]

Gradually, with the recruitment of biochemists such as Seymour Cohen (b. 1917), Lloyd Kozloff (b. 1923), and Alfred Hershey (1908–1997) and of the electron microscopist Thomas F. Anderson (1911–1991), the study of phage became less physical and increasingly biological and biochemical.[27] The hope of a paradox or an abstract appreciation of new physical principles began to evaporate, and a wet biological problem developed in its place. Nevertheless, Delbrück continued to impose one of the most rigorously maintained choices of organism in genetics. The dominant personality of the field, Delbrück would promote, proselytize, criticize, teach, and inspire a generation of phage geneticists, recruiting many of them from the field of physics. The phages he chose, the T (for "type") phages, became models of model organisms.

The Phage Group

Delbrück had moved in the rarefied higher circles of both physics and biology by the time he left Caltech and had won the respect of both. By then he had demonstrated a broad scientific background and acumen, the ability to collaborate, an adaptability in changing fields, and leadership crucial to the development of work on his model system. Cold Spring Harbor was an ideal place for collaborative research. The Cold Spring Harbor Laboratory became increasingly oriented toward microbial genetics under Demerec's leadership. Demerec, a *Drosophila* geneticist interested in mutation, turned briefly during the war to penicillin production by the mold *Penicillium*, and soon to bacterial genetics, with interests in antibiotic and phage resistance.[28] In addition, Demerec oriented the summer symposium series increasingly toward genetics, thereby bringing many geneticists to Cold Spring Harbor, promoting crucial and productive interchanges thereafter. Finally, the collaboration of Delbrück and Luria on the study of bacterial resistance to phage T1, culminating in the classic paper on the nature of spontaneous mutation, was a major step in the evolution of both bacterial and phage genetics.[29]

Many important steps in the development of the T phages as model systems took place during the war, much of it at Cold Spring Harbor. One, as noted above, was the collaboration in several aspects of phage multiplication with Luria, which continued during the academic year, with Luria doing laboratory work and Delbrück doing the theoretical and mathematical analysis. Early in this period, Luria and Anderson obtained the first electron micrographs of phage, showing a surprising morphological complexity (fig. 8.3), a harbinger of complications to come.[30]

Delbrück's commitment went well beyond his own scientific accomplishments in the laboratory. He believed strongly that phage work had to involve many people and many disciplines if it were to realize its promise in revealing fundamental aspects of life. Delbrück therefore instituted a summer phage

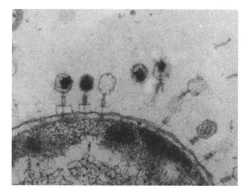

Fig. 8.3.   Phage T4 particles attached to the surface of an *E. coli* cell. The "tail" attaches with tail fibers to the cell surface, and, acting as a syringe, permits the DNA in the head to be injected into the cell. In this micrograph, one phage (third from left) has discharged its DNA and the protein "ghost" remains. Others to the left retain a dense bolus of DNA. Reprinted with permission of Eye of Science/Photo Researchers, Inc.

course at Cold Spring Harbor Laboratory in 1944, selecting individuals with sufficient mathematical skills to perform at a sophisticated level. In the course, he provided the phage, the bacterial host, the Petri dishes, and media. He trained students in the basic steps of growth and assay of phage and bacteria. Those enrolled were expected to replicate major prior experiments and to do others they might devise themselves at the time. The course included lectures, often by visitors as the years went on, supplemented by the regular Cold Spring Harbor symposia.

The phage course also embodied the decision to standardize materials and procedures. The T phages T1–T7 replaced the haphazard collections of phages studied at time. The T phages, all virulent forms that could grow on the chosen "substrate," *E. coli* strain B, were used by Luria and Delbrück in their early work (T1, T2) or were isolated from mixtures obtained from other investigators in the early 1940s by Demerec and Ugo Fano (1912–2001) (T3–T6).[31] This assured that those completing the phage course would take with them standard materials and would be equipped to compare their subsequent work effectively. The *Phage Information Service*, an informal newsletter that maintained contact among phage workers between summers, and professional meetings facilitated communication. By these mechanisms, Delbrück established a scientific following and a burgeoning area of research. He moved back to Caltech in 1945, where he spent the rest of his career. There, he offered versions of the phage course to students for a time, which amplified his influence.[32] Delbrück's recruiting efforts resulted in many parallel studies by ex-students on the T phages, including the biochemical characterization of phage, the biochemical attributes of the infected cell, phage mutation and recombination, the population-oriented analysis of replicating (and recombining) phage, and the means of infection. The work developed continuously, and the major milestones are recounted in every textbook in molecular genetics.

The transition between the more theoretical and the truly biological aspects of the work may have come with the discovery in 1946 of recombination between host-range and rapid-lysis variants of phage T2 by Hershey.[33] The host-range variant arose through Luria's isolation of *E. coli* B cells resistant to the standard T2 phage (h^+). The bacterial mutant, *E. coli* B/2, was then used to find phage mutants (called *h*) able to overcome the resistance and infect it. The rapid-lysis mutants (*r*) could be detected visually because they created abnormally large plaques (fig. 8.1). With a minimum of two pairs of alternate characters, geneticists may test for recombination. Unlike the mutual exclusion of different T phages, mixed infections of mutants of the same phage were successful. In a mixed infection of *E. coli* B by r^+h^+ and r^-h^- phage, both these parental types and a number of r^-h^+ and r^+h^- phage appeared. This, and similar results reported by Delbrück at the same symposium, were attributed in the 1946 study to induced mutation (of one phage by the other, a model resembling transformation) during phage growth.[34] Although both investigators noted that the results could be interpreted as evidence for recombination, Delbrück only later and reluctantly subscribed to this interpretation. A detailed account of recombination in phage T2 by Hershey and Raquel Rotman appeared in 1948, eliminating any doubts about the matter.[35]

A more obvious stage of the transition to biological approaches came in 1951, with improved methods of mounting specimens for electron microscopy. It was not obvious at the time how the infection of bacteria was accomplished. Phages were known by then to consist of nucleoprotein, with about equal amounts of protein and DNA. But electron micrographs showed an astonishing fact: phages did not appear to enter bacteria during infection! In the electron micrographs, one could not distinguish normal phage particles from empty protein coats, now called "ghosts" (fig. 8.3). Therefore, Anderson, who had previously shown that the phage particles would lose their nucleic acid when subjected to osmotic shock, could not know that the phage attached by their tails to the bacterial cells had injected their DNA into the cells.[36] Only when Roger Herriott (1908–1992) showed that the ghosts on the surface of infected bacteria were made entirely of protein could he suggest, as he did to Hershey, that "the virus may act like a little hypodermic needle full of transforming principles; that the virus as such never enters the cell; that only the tail contacts the host and . . . then the nucleic acid of the virus head flows into the cell."[37]

The observations were more fully rationalized by a classic experiment in which Hershey and Martha Chase (b. 1927) used radioactively labeled bacteriophage preparations.[38] They labeled one preparation with radioactive sulfur to label the protein coat and the other with radioactive phosphorus to label the DNA. Bacteria, infected with one or the other preparation of phage, were subjected to the shear forces of a Waring blender after the initial adsorption step. The treatment was not strong enough to kill the bacteria, but it did remove the phage ghosts on their surfaces. By centrifuging the bacteria and analyzing the supernatant fluid and the sedimented bacteria thereafter, Hershey and Chase showed that labeled sulfur had been largely removed from the bacteria, but the labeled phosphate largely remained with the bacteria until lysis. The demonstration that DNA was the molecular species with continuity through the infection suggested that DNA, not protein, was the substance of phage genetic material. The Hershey–Chase experiment could be criticized on the basis of incomplete association of the phosphorus with bacteria and incomplete removal of sulfur from the bacteria. Despite this, the experiment supported the implications of Avery's work on *Pneumococcus* transformation and became a strong spur to the work of James Watson (b. 1928) and Francis Crick (b. 1916), already underway in England, that revealed the structure of DNA and its suitability as the substance of genes.

Delbrück had hoped for years to reveal a productive scientific paradox in the analysis of the replication of phage, whether they were living things or not. He hoped that his group's studies would lead him to this paradox and perhaps to a new physical principle underlying life. But Kay remarks:

> It is ironic, therefore, that despite his professional acclaim in biology, in his own epistemological quest Delbrück did not achieve his goal; for the gene was ultimately explained in 1953 by the conventional application of physics and chemistry, without invoking new knowledge of physics. With this in mind, the evaluation of Delbrück's success as a physicist in biology must be tempered, and his important contribution to genetics must not be confused with his own intellectual aims.[39]

9

Temperate Phages and Transduction

Lysogeny: a gracious truce that tends to break down in hard times.

Temperate Bacteriophages

Not all varieties of phage invariably destroy their hosts after entering cells, as virulent T-series phage do. Even early in the study of phage, some isolates of bacteria appeared be *lysogenic*; that is, culture filtrates of one bacterium carried an agent capable of lysing other bacteria, and the lysogenic cultures would themselves lyse under certain environmental conditions, producing the infective agent in much larger quantities. A controversy developed between d'Herelle, who had co-discovered phage, and the eminent immunologist Jules Bordet (1870–1961), who took up the question of the nature of these unusual bacteria, for which he introduced the term "lysogenic."[1] The question was whether, as d'Herelle thought, lysogenic cultures were simply contaminated with phage, or, as Bordet believed, the agent produced upon lysis was produced by the bacterium as an intrinsic, hereditary character. The phagelike agent isolated from lysogenic cultures that caused lysis of Bordet's cultures (of *E. coli*) was, he thought, simply a trigger that induced the affected bacteria to make the more of the same material until they autolyzed under its influence. Survivors of the lytic agent among the test bacteria, Bordet observed, had become lysogenic themselves. Bordet could find no means of curing his lysogenic cultures of their properties despite strong serial-transfer measures designed to rid them of any contaminating phage. Nor could he cure the bacteria of their lysogenic properties when he applied antibodies made against the agent. Bordet was not troubled by the fact that an agent introduced from the outside could induce a hereditary property in the affected bacteria: he was a Lamarckian in France at a time when Lamarckian mechanisms were accepted.[2]

While Bordet's experiments addressed the suggestion of d'Herelle regarding contaminating phage, Bordet, like d'Herelle, could not have suspected an unusual property of certain phages. The DNA of these temperate phages, as we now call them, can integrate themselves into the bacterial chromosome in a noninfective form called a *prophage*. Moreover, the prophage may occasionally and spontaneously leave the chromosome, multiply, and kill the host. This accounted for the low level of phage particles in cultures of lysogenic bacteria, phage that d'Herelle took to be contaminants. It also accounted for the inability of Bordet to cure his cultures of these viruses. One might ask why, once phage were liberated by spontaneous induction, they did not go on to infect other cells of the culture. Much later, it became clear that the lysogenic bacteria produce a repressor protein, encoded in the prophage DNA. The repressor both prevents the prophage from leaving the chromosome and multiplying and blocks the multiplication of phage DNA injected by particles in the culture medium. Cases in which an entire lysogenic culture clears (i.e., lyses) reflects an environmental shock that leads to the loss of the repressor in most cells of the population.

While the issue between d'Herelle and Bordet became polarized between an infective and a hereditary mechanism, it would become clear that the agent was indeed a phage and that all cells of a lysogenic culture had the potential to produce the phage in question. Later studies on a lysogenic strain of *Salmonella typhimurium* in the laboratory of F. MacFarlane Burnet (1899–1985) showed that no further phage of the type produced by this strain appeared in the medium after cultures had lysed after infection with a second, virulent phage.[3] Burnet proposed instead that all bacteria of a lysogenic culture contained a noninfective element that under suitable conditions caused the production of mature and infectious phage. Here we see the convergence of concepts of infectious and hereditary elements. This convergence was realized much later in the notion of a prophage carried as a segment of the bacterial chromosome. Burnet went on to confirm Bordet's observation, though with surprisingly modern interpretation—the appearance of the phage-resistant, lysogenic cells that survive the infection of a sensitive bacterial strain. At the time, however, Burnet's work could only add weight to the identity of the infectious and hereditary element as a form of phage.[4] Nothing of the molecular mechanisms surrounding the observations was known until the 1950s and 1960s.

Experiments with a lysogenic strain of the large bacterium *Bacillus megaterium* finally put to rest the question about the nature of the lysogenic state. Until the final experiments with this bacterium were done in 1950, the idea of lysogenic bacteria encountered serious opposition on the part of many microbiologists and others. Eugène (1890–1943) and Elizabeth Wollman (1890–1943; both were to perish in a concentration camp during the war), working at the Pasteur Institute in the 1930s, had both the lysogenic normal strain of *B. megaterium* and a nonlysogenic, sensitive derivative of the bacterium that could be used for phage assays. A critical experiment was much like that done earlier by Burnet. The walls of lysogenic cells were digested with the enzyme lysozyme. The resulting lysis of the cell liberated no phage detectable in the assay system.[5] The question, then, was how lysogenic strains managed to produce the phage. Some investigators thought phages were secreted at a low level by all cells; others thought that phages were produced in large numbers by a tiny minority of the population. Delbrück remained skeptical of the whole idea of lysogenic bacteria, taking d'Herelle's view that contamination could explain most of the observations. Delbrück's authority in this matter in the mid 1940s was such that the principal work in this entire area attracted little interest in the American community and instead continued mainly in Europe.

In France, the most compelling experiments were done by the protozoologist André Lwoff (1902–1994). Lwoff was an associate of the Wollmans at the Pasteur Institute, and he continued the work with *B. megaterium* after World War II. The large bacterium lent itself to work with a micromanipulator. Lwoff, starting with a single cell of the lysogenic type, isolated 1 daughter cell of 19 successive divisions to microdrops of culture medium.[6] All of these daughter cells yielded lysogenic cultures when propagated, and no phage particles could be detected in the fluid of the microdrops. However, even after 19 divisions, an occasional cell would lyse. The microdrop in which this occurred contained more than 100 plaque-forming particles. Lwoff concluded that lysogenic bacteria carried this property as a hereditary character, and occasionally single cells would yield a burst of phage. Even Delbrück was convinced by this experiment: there could not possibly have been enough contaminating phage on the original bacterium to persist in all microdrops for 19 divisions. At this point, lysogeny had been or would soon be recognized in its modern form in a number of bacterial species, including *E. coli*. The use of *B. megaterium*, the size of which was its major advantage in the temperate phage studies, was then abandoned.

The history of work on lysogeny was long and filled with controversy and is ably summarized both by Brock and by Stent.[7] Lysogeny forms the milieu for the discovery of *transduction*, a process of phage-mediated transfer of bacterial genes. Before going on, some terms and properties of the phages I have discussed should be given systematically and emphasized. The term *lysogenic* applies to the bacterium carrying a *prophage*, the DNA of the phage within the bacterium. If a bacterium is lysogenic, it is immune to further infection by phage of the same type. When such phages infect host cells lacking a prophage of the same type, however, the injected DNA may have one of two fates, depending in part on the physiology of the cell. It may either become a prophage or it may undergo multiplication and lyse the cell. These processes are called the *lysogenic* and *lytic* responses, respectively. Phages that can adopt a prophage state are called *temperate* phages, in contrast to the *virulent* forms exemplified by the T series of the Delbrück's phage group. When sensitive bacteria (lacking prophage) are mixed with a small number of temperate phage and plated together, the bacteria grow into a lawn of cells in which plaques appear, as they would with virulent phage. However, the plaques are turbid, owing to the growth of bacteria that have become lysogenized rather than lysing. In lysogenic bacteria, the repressor protein (discovered much later) produced by the prophage DNA maintains the prophage in its dormant state.

The prophage of lysogenic bacteria can be induced to undergo multiplication within lysogenic bacteria by treatment of the bacteria with ultraviolet light (UV) and other DNA-damaging agents, a discovery of Lwoff and his associates in the early 1950s.[8] The phenomenon of induction reflects the destruction of the repressor protein, an indirect consequence of DNA damage.

Transduction

Lederberg's meteoric ascent to the first rank of microbial geneticists can be more fully appreciated if one remembers that he joined Ryan's lab as an undergraduate and Tatum's lab at the age of 21 in 1946. He was awarded the Ph.D. degree in 1947 "after the fact" on the basis of his short period of work with Tatum. He became an assistant professor in the School of Agriculture at the University of Wisconsin the same year, at the age of 22. With Beadle and Tatum, he would be awarded the Nobel Prize in Physiology and Medicine in 1958 at the age of 33.

The tradition of precocity continued as Lederberg's first graduate student, Norton Zinder, joined the laboratory in 1948, at the age of 19. He had been recommended by his undergraduate mentor, none other than Francis Ryan. Lederberg's interest in medical microbiology had survived his diversion from medical school. Indeed, he was now committed to scientific investigations in medical microbiology. He had long before felt that the large amount of antigenic variation among strains of the pathogen *Salmonella typhimurium* signified some sort of genetic recombination in their natural setting (see chapter 7). Lederberg hoped to explore this possibility, as it promised considerable medical benefit. Therefore he urged Zinder, for his doctoral project, to test for genetic recombination in *S. typhimurium* by the methods Lederberg had devised in Tatum's laboratory. Here we see another choice of a model organism, *S. typhimrium*, made on the general basis of its medical interest, but without foreknowledge of its potential for bacterial genetics. *S. typhimurium* soon took its place as a companion of *E. coli*, to which it is closely related. Its current place in basic microbial genetics emerged from the discovery by Zinder and Lederberg of a new and surprising form of genetic recombination that Lederberg called transduction.[9] (Lederberg applied this word to a wider range of phenomena at the time, but the term now has the more restricted meaning of phage-mediated transfer of bacterial genes.)

At the time of this discovery, the laboratory was deeply involved in the genetics of *E. coli* K-12. The mapping studies, all done with $F^+ \times F^-$ crosses, had yielded increasingly complex data that were still being interpreted according to the cell-fusion and recombination model.[10] The polarity of mating had not yet been discovered. Lederberg's information on bacterial recombination was summarized in a 1951 Cold Spring Harbor symposium paper renowned for its length and opacity.[11] However, the authors, one of whom was Zinder, reported observations on recombination in *S. typhimurium* at the end of the paper. From a collection of strains of this bacterium, Zinder had prepared multiple mutants that allowed him to test rigorously for genetic recombination between certain pairs. Upon detecting recombination in one case, he used a U-tube apparatus to determine whether recombination required cell contact, as it did in *E. coli*.[12] In this experiment two bacterial populations are placed, one in each arm of a U-tube, the two separated by a bacteriological filter at the turn at the bottom. Applying air pressure alternately to each of the openings at the top can mix the media of the two sides (fig. 9.1). Zinder showed that, in contrast to the experience with *E. coli*, recombination in *S. typhimurium* did not require cell contact. Moreover, adding a DNA-digesting enzyme (DNAase) to the medium did not reduce the frequency of recombination. This made DNA-mediated transformation as described in *Pneumococcus* unlikely. Initially, the authors thought that conjugation might still explain the results because certain so-called L-forms of bacteria, very small and lacking cell walls, might be able to squeeze through the bacterial filter in the U-tube and contribute their genes to cells on the other side. The later paper by Zinder and Lederberg in 1952 abandoned this possibility and showed that the filterable agent had the properties of a virus. The agent sedimented at high speeds in a centrifuge and was neutralized by an antibody to the temperate phage harbored by one of the two *S. typhimurium* strains. Moreover, the protein coat of the virus could protect the DNA inside from the action of DNAase.[13]

The full picture of transduction that emerged from these and later studies is quite complex for the nongeneticist, so I summarize it here. One "cross" of *S. typhimurium* yielding prototrophic recombinants involved two doubly auxotrophic strains. LT-2 (*met⁻ his⁻*) required methionine and histidine, and LT-22 (*trp⁻*) carried two mutations, one that blocks the synthesis of phenylalanine and tyrosine, and the other, fortuitously linked closely to the first, that blocks the synthesis of tryptophan. (Tryptophan alone can support the growth of the strain.)

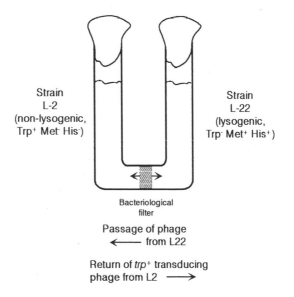

Strain
L-2
(non-lysogenic,
Trp⁺ Met⁻ His⁻)

Strain
L-22
(lysogenic,
Trp⁻ Met⁺ His⁺)

Bacteriological
filter

Passage of phage
⟵ from L22

Return of *trp⁺* transducing
phage from L2 ⟶

Fig. 9.1.   Diagram of the U-tube apparatus used to analyze the transduction process. Two populations of cells are separated by a filter preventing cell passage. Fluid is pumped between the compartments by applying air pressure alternately to the upper openings. See text for details of experiment.

Strain LT-22 is lysogenic for a temperate phage, now called P22, while LT-2 is not. The prototrophic recombinants were found in the arm of the U-tube containing LT-22 cells.

In Zinder's experiments, filtrates of the LT-2 culture or disrupted cells of LT-2 would not yield recombinants when added to LT-22 cultures. Instead, it appeared that strain LT-22 produced something that, when it passed into the LT-2 culture, rendered LT-2 capable of making a filterable agent that returned to the LT-22 culture. This agent, the transducing agent, rendered a few LT-22 cells prototrophic.

In this experiment, cultures of strain LT-22, the lysogen, produced phage P22 at a constant low level. These phages passed through the filter and infected some LT-2 cells. Many of the infected cells underwent the lytic response, as opposed to becoming lysogens. The phage produced by lysis of these cells then returned to the LT-22 culture and injected their DNA into the LT-22 cells. Because these cells were lysogens, they were immune to the phage and survived. So far, one would expect that no bacterial genes would have been transferred from LT-2 back to LT-22. The key to transduction, however, is that some of the phage particles liberated by lysis of LT-2 cells carried bacterial DNA instead of phage DNA. (In phage infections, assembly of progeny phage particles occasionally incorporates DNA fragments of the dying host cell rather than phage DNA into the phage heads.) Because the phage particle (head and tail assembly) can inject any DNA it carries into another cell, bacterial DNA from strain LT-2 was injected into the LT-22 cells. The bacterial DNA is innocuous, but it may be incorporated into the bacterial chromosome much like DNA introduced by transformation. In this way, the normal trp^+ (tryptophan) gene of LT-2 might find its way into the chromosome of the trp^- LT-22, making the latter independent of tryptophan (and, in this case, phenylalanine and tyrosine) supplementation.

It soon became clear that transducing phage particles derived from lysis of the sensitive strain (here, LT-2) could carry only a few contiguous genes because the capacity of the phage heads for DNA was limited to about 1% of the total length of bacterial DNA. Moroever, only bacterial DNA was carried in transducing particles; no phage DNA accompanied it. In this process, now called *generalized transduction*, any portion of the bacterial genome may be transferred. The transducing particles constituted only about 0.1% of the phage particles, and of these, only 1% carried any particular gene. In all cases studied by Zinder and Lederberg, only one character was transduced at a time, although later studies showed that many contiguous genes can be carried by a transducing phage.[14]

The impact of the discovery of transduction had an interesting intellectual dimension. Until then, phages were considered destructive pathogens with little *raison d'être* beyond their own reproduction. Upon the discovery of lysogeny and transduction, the temperate phages took their place as an important feature of bacterial biology. They represented not only vehicles of gene exchange, but also factors that could alter the physiology of the bacterial cell—in particular, their antigenic character. (The medical significance of the latter was not lost on Lederberg.) In addition, populations of lysogenic bacteria had a greater fitness if the low levels of phage they released could attack competing, sensitive bacteria.

Since 1951, the genetic systems of bacteria, phage, and related genetic elements such as plasmids have coalesced into a rich field of inquiry. Significantly, many of the leaders of this field have taken their skills and insight into investigations of animal cell viruses. In this new area, which suffered from confused concepts, a complex biological system, and experimental variability, those with elegant visions of simpler systems could formulate approaches to the more complex ones with precision, high intellectual standards, and, as it turns out, great effectiveness.

Three Model Organisms

The discovery of transduction led to *S. typhimurium* becoming a model for bacterial genetics. It would always be an accessory to *E. coli*, much as *A. nidulans* stands in relation to *N. crassa*. In both pairs of organisms, similar problems can be pursued, and many studies benefit from comparison. More important, each member of the pair offers unique and interesting questions to pursue, and this is enormously easier with the knowledge and technical resources developed in the other.

The discovery of transduction had a profound effect in solidifying the use of *S. typhimurium* in biochemical genetics. Perhaps its most important use stemmed from the fact that the *S. typhimurium*-P22 system was the first to offer a means of fine-structure genetic mapping of the bacterial chromosome. This potentiality lies in the fact that phage coats transfer only small amounts of DNA. Contrary to the initial impression, segments carrying about 200 contiguous genes can fit into a phage P22 head. When such a piece of DNA enters genetically marked cells, recombination between it and the homologous segment of DNA of the recipient chromosome can take place. The efficiency of recombination is high, so recipients can yield recombinants even if both the transducing DNA and the recipient DNA carry mutations close to one another. At the extreme, recombination can take place between mutations that affect adjacent nucleotides, well within a single gene (see chapter 11). The potential for studying genetic fine structure led to many studies of the organization of related genes in *S. typhimurium* and contributed to the knowledge of the organization of bacterial genes in general (see chapters 10 and 11). Among the first to exploit the system was Demerec, the former *Drosophila* geneticist who fostered microbial genetics at the Cold Spring Harbor Laboratory and who worked on mutation in *E. coli* at the time. He saw clearly that *S. typhimurium* and its transducing phage offered a better route to the fine structure of the gene than mutation did, and he quickly converted his laboratory to the study of *S. typhimurium*.[15]

In contrast to *S. typhimurium*, phage P22 did not become a widely adopted model for temperate phage but was a member of a group of phages that all contributed to the understanding of this category. Among the others was phage P1 and especially phage lambda (λ), which we take up next. Phages P22 and P1 would become workhorse technical resources for the genetics of *S. typhimurium* and *E. coli*, respectively, and in their way, models for generalized transducing phages.

Phage Lambda

In 1951 in the laboratory of J. Lederberg, Esther Lederberg discovered derivatives of *E. coli* K-12 that were sensitive to a phage for which the normal strain was lysogenic. Until sensitive strains were found, the phage had not been detected. J. Lederberg named the phage lambda (λ), using a Greek letter appropriate to his hypothesis that it was a nonchromosomal element like Sonneborn's κ (kappa) of *Paramecium* (see chapter 13).[16] The phage-sensitive cells arose as ultraviolet-resistant cells in mutation studies. Lwoff had found that UV induced massive lysis of lysogenic *B. megaterium* because, as later studies showed, UV led to the loss of the repressor protein in the majority of cells.[17] It turned out that the UV-resistant *E. coli* cells were those rare ones whose prophage had spontaneously excised themselves from the bacterial chromosome, but which had been lost before it could replicate. Such cells could therefore withstand much higher doses of UV.

The lysogenic and nonlysogenic *E. coli* strains, designated respectively K-12(λ⁺) and K-12(λ⁻), would become critical tools in studies of the mechanism of bacterial conjugation a few years later. The discovery further solidified the place of *E. coli*, the host of phage λ, as a model bacterium. Although the initial discovery of phage λ was coupled with the demonstration that it can transduce bacterial genes, the major importance of the phage was in the study of lysogeny and coordinated gene action.[18] Study of phage λ would yield the most detailed knowledge (for a time) of any replicating entity in the living world. After the discovery of phage λ, several related and overlapping phases of investigation developed. These phases were the nature of the association of the prophage with the bacterial chromosome; the immunity of lysogenic bacteria to superinfection by more phage λ; and specialized transduction.

Preparations of phage λ were easily made from lysogenic cultures by treating the cultures with ultraviolet light. The Lederbergs performed crosses among the various *E. coli* strains to define the inheritance of lysogenicity and thereby excluded the notion (which they favored previously) that it was a cytoplasmic rather than a chromosomal factor. The observations encouraging this view were that lysogenicity seemed to be linked genetically to the chromosomal *gal* gene.[19] At about the same time, Hfr strains of *E. coli* had been discovered, greatly facilitating genetic analysis.[20] In crosses, the DNA of the Hfr cell passes in a linear fashion into the F⁻ bacterium. The process can be interrupted by agitation in a Waring blender, allowing the experimenter to determine the time at which various genes first enter the F⁻ recipients in mass matings. A disrupted sample of cells is then plated each time on a medium both selective for F⁻ cells that have acquired genes from the Hfr and also toxic to the Hfr parent cells. With this technique, Wollman and colleagues confirmed, in a cross of lysogenic and nonlysogenic strains, Hfr(λ⁻) *gal⁺* × F⁻(λ⁺) *gal⁻*, that the prophage was indeed linked to the *gal* gene because recombinants lacking the prophage (λ⁻) appeared in interrupted matings just before those carrying the *gal⁺* allele.[21]

A complication arose in the reciprocal cross, Hfr(λ⁺) *gal⁺* × F⁻(λ⁻) *gal⁻*. In these crosses, the Hfr carries the λ prophage and the F⁻ does not. Few zygotes receiving the *gal⁺* allele were found. This was soon rationalized under the term *zygotic induction*, a process by which prophage DNA, as it enters the λ-sensitive F⁻ cell as part of the Hfr chromosome, enters a cytoplasm lacking the repressor protein. The prophage immediately leaves the DNA, multiplies, and kills the recipient, accounting for the lack of *gal⁺* recombinants. This observation was part of the evidence that lysogenic strains owed their phage λ immunity to a repressor protein.

In the Lederbergs' laboratory, Melvin Morse (b. 1921) discovered that phage λ can transduce the *gal* gene to other *E. coli* strains.[22] Phage λ accomplishes transduction rather differently than does P22. Unlike phage P22, the genes that phage λ can transduce are limited to the regions of the *gal* and *bio* genes, which lie on either side of the chromosomal site of the prophage. This strongly suggested once again that the λ prophage was part of the chromosome, possibly as a linear segment of DNA between *gal* and *bio*. The picture that formed at that time was that the phage DNA carried part of the bacterial chromosome with it into another host cell.

After much speculation, phage λ offered the first clear example of how a temperate phage adopts a prophage state. The idea was suggested by Alan Campbell (b. 1929) in a 1961 review of bacterial plasmids.[23] After the phage injects its DNA into the cell as a linear piece of DNA, the ends of the DNA are joined so that it becomes circular (fig. 9.2). In Campbell's model, a single recombinational event between the phage DNA circle and the circular bacterial chromosome renders the phage DNA a segment of the chromosome. (The F plasmid goes through the same general process as it integrates into the chromosome to form the Hfr bacterium; see fig. 7.3a.) The integration process is reversed during excision from the chromosome, normally using the same points of recombination, upon induction of the prophage.

The initial observations that phage λ could carry one or the other of the adjacent *gal* and *bio* genes in transduction could now be interpreted. Transducing phage λ DNAs form during prophage induction in an abnormal event of excision, in which the recombinational event takes place not in the segments involved in the insertion process, but at a position to one side or the other of the prophage DNA. In this way, part of the bacterial chromosome is included in the circular phage DNA as it leaves the chromosome. Because phage λ can carry only six or seven flanking genes to the right or the left of the prophage site, the phage is called a *specialized transducing phage*, and it therefore has little use in general genetic analysis of *E. coli*.

That role was reserved for the generalized transducing phage P1, discovered in 1950. Shortly before the introduction of *S. typhimurium* and P22, P1 was discovered by Giuseppe Bertani (b. 1923) in *E. coli*.[24] The P1 prophage is unusual in being maintained as a low-copy plasmid (an autonomously replicating circle of DNA) rather than as a segment of the bacterial chromosome.[25] Phage P1 serves well as a generalized transducing phage, with the same benefits for genetic fine-structure analysis in *E. coli* as P22 had in studies of *S. typhimurium*.[26] (P1 can also be used with *S. typhimurium* and other bacteria.) This development contributed to the continued wide use of *E. coli* and the

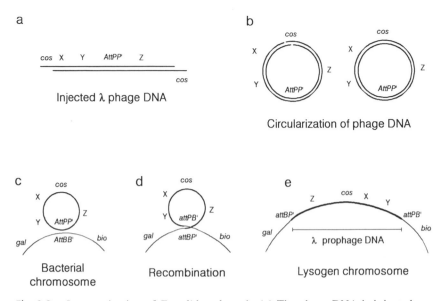

Fig. 9.2.   Lysogenization of *E. coli* by phage λ. (*a*) The phage DNA is injected as a linear piece of DNA, with complementary sequences on the two ends. (*b*) The two ends base pair to form the *cos* (cohesive) site, and the DNA becomes a continuous circle, closed by DNA ligase. (*c*) The *attP* site of phage DNA pairs with the *attB* site of the bacterial chromosome, which lies between the *gal* and *bio* genes. (*d, e*) Recombination between the *attP* and *attB* sites renders the phage DNA part of the bacterial chromosome. *X*, *Y*, and *Z* are phage genes to indicate the original and permuted order of the phage DNA.

maintenance of *S. typhimurium* as a secondary model organism in biochemical genetics and molecular biology.

The Supremacy of *E. coli* and Phage λ

Although phage λ has become the best known model temperate phage, it was not the only temperate phage studied at the time. Many temperate phages had been and would be discovered in the period 1953–65. Unlike members of the American phage group organized by Delbrück to study the virulent T phages, those studying temperate phages brought a variety of them to light, many representing new categories. Due to the explosion of knowledge about phage λ that occurred in the 1950s and early 1960s, phage λ kept its head above the din of information about other new phages. The simultaneous study of many types of phage, however, greatly expanded the knowledge of their genetic relationships, the variations in their biology, and their evolution. It would be some time before the phage bacterial hosts, *E. coli* and *S. typhimurium*, would gather a similar range of companions for comparative studies of molecular genetics.

The reason that phage λ held its own resulted from the intense interest in the problems outlined above: the nature of the prophage, its integration and excision from the bacterial chromosome, and the maintenance of its dormant state. The answers to these questions became applicable to one extent or another to all temperate phages. The ability of the prophage to remain dormant in the lysogenic cell was recognized as an activity of a phage gene that produced the repressor protein. This protein blocks the activity of almost all other phage genes and thus confers immunity. The phenomenon of repression was quickly generalized to include the control of bacterial enzyme-encoding genes by repressors of their own.

The genetics of bacteria proceeded simultaneously with investigations of metabolism, the latter being greatly accelerated by the former. Conjugation was used to map genes by the interrupted mating method, from which a coarse map of the genes was made. For work on the fine structure of genetic regions, of particular interest to those studying the structure of particular enzymes, transduction was used to determine relative positions of mutations in individual genes. Both conjugation and transduction could be used in *E. coli*, whereas conjugation in *S. typhimurium* became feasible only after 1960.[27] The latter development required strains of *S. typhimurium* harboring the *E. coli* F plasmid and lacking the factors interfering with the expression of such conjugative plasmids in most *S. typhimurium* stocks.[28] Therefore, experimentalists made an increasing commitment to *E. coli* and its transducing phage, P1. *S. typhimurium* and phage P22 developed mainly as important tools for work on specific projects, namely the antigenic character of the surface and the flagella of the bacterium, its pathogenicity, and the organization of genes of related function. Demerec was among the first to emphasize that genes encoding enzymes of a single biosynthetic pathway were often adjacent to one another on the bacterial chromosome and were expressed coordinately.[29] Such gene clusters were later called *operons* by Jacob (b. 1920) and Monod (1910–76), a finding that underlay their influential studies of gene expression in the early 1960s.

The reader will have noticed my invocation of DNA in explaining a number of phenomena in this chapter. Another book would have to be written to represent fairly the cognitive transition in molecular genetics between 1952 and 1954. In 1952, microbial genetics was a quite abstract pursuit, imputing more and more specific properties to the "genetic material." Within a few months of Watson and Crick's announcement of the structure of DNA in 1953, the thoughts of the more adventurous (and perhaps less critical) molecular geneticists had turned to the possible relationship between DNA on the one hand and genes, proteins, chromosomes, prophage, bacterial chromosomes, and genetic recombination on the other.

With the genetics of bacteria and phage well on its way by 1955, workers in these areas claimed even more forcefully that their studies could and perhaps did focus on fundamental aspects of life itself. This hubris is understandable in light of the rapid, extraordinary advances made in this and the next decade in the area of macromolecular synthesis. We turn now to how DNA changed the agenda and how microbial genetics became a backbone of molecular biology.

10

DNA

DNA: that registry of chance, that tone-deaf conservatory where the noise is preserved with the music.

—J. Monod

The Threshold of Molecular Genetics

In the previous chapters, we have seen many model microbes chosen and domesticated in just a few years. All were chosen before the structure of DNA was proposed by Watson and Crick in 1953, a proposal quickly accepted by the genetics community.[1] Before that moment, microbial geneticists sought to explore novel genetic systems and to infer the nature and function of genes strictly by genetic techniques and the study of the biochemical effects of mutation. Geneticists, like ships seeking harbor in a fog, had become adept at using phrases like "theoretical entity," "operational definition," and "unit of recombination"—none of them having a chemical referent. But after 1953, the agenda changed to the study of DNA and how its properties accounted for the information about the gene collected up to that time. This turn of events quickly heightened the model-organism status of *E. coli*, *S. typhimurium*, and their phages.

Neurospora, a genetically orthodox eukaryote, had been carefully chosen to pursue the metabolic role of genes, and genes had been characterized up to that time only in eukaryotes. *A. nidulans* had been chosen to explore new genetic systems and to increase the resolving power of genetic recombination experiments. *Neurospora* geneticists had accomplished their initial goal and eased comfortably into the task of demonstrating the benefits of the organism in the study of both metabolism and genetics. Their studies could be planned well into the future because the program was well formulated. Those studying

123

A. nidulans made good on their hopes both of finding a novel genetic system, parasexuality, and of performing fine-structure analysis on very small segments of the genetic material. They brought to a head the issue of pseudoallelism, which would soon be answered in principle by studies of phage T4. Yeast contributed little in this period to these areas; as the DNA era approached, the major contribution of yeast was the study of cytoplasmic inheritance, a subject I consider separately in chapter 14.

E. coli, S. typhimurium, the T-series of phages, and the transducing phages P22, P1, and λ all came into prominence as novel genetic systems, systems that began to bleed into one another. The most insular of the scientific groups in this area was the phage group—insular not intellectually, but by the demarcation of their model and their experimental aims. The Delbrück group wanted to infer laws of replication rather than biochemical mechanisms. For all the biochemistry of T-phage infections done in those days, the themes of the group were mathematical and formal modeling of replication and recombination. The rigor of the work was both creative on its own ground and therapeutic for others. The nature of the phage infection, which in a sense culminated in the Hershey–Chase experiment, was carefully worked out, excluding one hypothesis after another, but it never fully opened the black box to discern molecular details. This is unsurprising because, until 1953, the identity of the molecule of interest was still uncertain. The rigor of the phage group was thus not only admirable, but in some cases limiting.

Delbrück, in 1946, would insist that the program of the *Neurospora* group was deficient because it did not seek effectively the data that might falsify it. He casually described a model for stable cytoplasmic inheritance that eliminated the need for replicating entities—a model based on two mutually inhibitory metabolic sequences (see chapter 13). He initially resisted the idea of recombination in phage T2. Delbrück scorned the idea that lysogeny involved an inherited prophage until 1950. One can see that many microbial geneticists outside his own group were usefully challenged to do not merely adequate, but elegant experiments. But in his own arena, Delbrück's laboratory mainly wanted to study replication in the simple system he and his group had domesticated. Unfortunately for Delbrück, the discovery of DNA structure in 1953 accounted formally for life's most intriguing secret. This dissipated his hope of coming to an impasse that would have to be transcended with new physical laws.

In the initial discoveries of conjugation and of general and specialized transduction, Lederberg's laboratory was motivated by a desire to know whether and how bacteria exchanged genes—whether the phenotypic variety within a species reflected the existence of sexual mechanisms in nature. The discovery of bacterial conjugation ignited studies of the polarity of gene exchange, the discovery of Hfr strains, and elaborate mechanistic studies of chromosome transfer by the Parisians. The discovery of phage λ, after the discovery of P22-mediated transduction, led to a fusion of *E. coli* genetics with the study of lysogeny. As presaged by Pontecorvo's discovery of a mode of gene exchange without overt sexual display, bacterial gene exchange raised the interest of many biologists in genetics and new genetic systems to a high pitch.

Other tributaries had joined the mainstream of microbial genetics by 1953. The general nature of proteins became appreciated only in the late 1940s, and only by 1950 would Pauling and Robert Corey (1897–1971) propose the α-helix as a basic structural motif.[2] Their direct, model-building approach and the long efforts of the structural school of molecular biology, originating in X-ray crystallographic studies of proteins, gave a general picture of protein structure, but it was not yet a vivid and widely shared understanding. (Appendix 2 describes macromolecular structure and synthesis.) In fact, it would not be until 1952 that Frederick Sanger (b. 1918) would announce the amino acid sequence of insulin, using the new tools of controlled proteolysis and paper chromatography.[3] The proof that amino acid sequences were the responsibility of genes had to wait until 1956 with the demonstration of the amino acid substitution in the β polypeptide of hemoglobin imposed by the sickle-cell mutation.[4] But the same interests that drove the study of the structure of proteins branched into the study of DNA structure, with the laboratories of Pauling, Maurice Wilkins (b. 1916), and the Watson–Crick team competing in 1952 and 1953 for primacy in the formulation of a specific molecular model.[5] The motivations of these scientists varied; all were aware of Avery's work, of the common, albeit vague talk of "nucleoprotein" as the substance of the gene, and particularly in Watson's case, of the Hershey–Chase experiment that demonstrated that nucleic acid was the thread between generations of bacteriophage.

DNA Structure: A Watershed

Altogether too much has been written about the discovery of the structure of DNA. I have purposely discussed this subject rather little to this point because it did not emerge from the study of a model microbe. I will only summarize the main impact of the work on the choice and uses of model organisms. The event (and I speak now only of the announcement of DNA structure, not the work that led up to it) was similar to the discoveries of Mendel or the rediscovery of Mendel himself in its impact on genetics. Let us hear it from someone who was there, at the Cold Spring Harbor symposium in the summer of 1953:

> His [Watson's] manner more dazed than ever, his shirttails flying in the wind, his legs bare, his nose in the air, his eyes wide, underscoring the importance of his words, Jim gave a detailed explanation of the structure of the DNA molecule; breaking into his talk with short exclamations the construction of atomic models to which he had devoted himself at Cambridge with Francis Crick; the arguments based on X-ray crystallography and biochemical analysis; the double helix itself, with its physical and chemical characteristics; finally the consequences for biology, the mechanisms that underlay the recognized properties of the genetic material: the ability to replicate itself, to mutate, to determine the characters of the individual. For a moment, the room remained silent. There were a few ques-

tions. How, for example, during the replication of the double helix, could the two chains entwined around one another separate without breaking? But no criticism. No objections. This structure was of such simplicity, such perfection, such harmony, such beauty even, and biological advantages flowed from it with such rigor and clarity, that one could not believe it untrue. There might be details to modify, some further specifications to be made. But the principles, the two chains, the alignment of the bases, the complementarity of the two sequences, all this had the force of the necessary. All this could not be false. Even without understanding the details of the crystallographic analysis, even without an affinity for biochemistry, one could not but admire a structure that responded so well to the requirements of genetics. One of the oldest problems posed since antiquity by the living world, heredity, had just been resolved in the properties of a molecular species. The production of the same by the same, variation, the reassortment of characters in the thread of generations: all that flowed from the complementary distribution of some chemical radicals aligned along two chains. By all indications, it was a turning point in the study of living things. It heralded an exciting period in biology.[6]

This rhapsody, written by François Jacob (b. 1920) 35 years after the event, may well be an overstatement, brightened with hindsight. But overstatements were not uncommon even at the time. The agenda of genetics had changed with the startling explanatory potential of DNA, the idea that DNA, and DNA alone, could bear information in a form that cells could duplicate. Geneticists turned from the inferential work of the past to the confirmatory work of the future. How did the obdurately linear DNA specify the exquisite variety of three-dimensional proteins? How did DNA unwind and replicate? How did recombination resolve the long stretches of nucleotide pairs? How did mutations occur when cells were irradiated or treated with chemical mutagens? All of these were challenging chemical questions. All of them were of universal interest. All of them were questions having to do with the most fundamental activities of the cell. And the questions about the gene had a new twist, as Muller and others had hinted at long ago: how does one study the chemistry of a molecule of which there are only one or two copies in a cell?

These questions immediately solidified the choice of prokaryotes for the next phase of work in molecular genetics. As I noted in chapter 7, the rigorous distinction of bacteria, viruses and phages, and the eukaryotes would come only in 1962, informed in large part by the discoveries of molecular genetics and electron microscopy in the 1950s. The enteric bacteria *E. coli* and *S. typhimurium* had already yielded a multitude of mutations and a wealth of biochemical information. They were small and unicellular, and they displayed several technically useful systems of gene exchange. Lederberg had indeed brought genetics to the prokaryotes, showing that they shared with eukaryotes fundamental attributes of cell function and inheritance. Therefore the questions posed in the past could now be seen as universal, applying to all living forms. With some irony, we see the appropriation of fundamental biological research by students of life forms "bordering on the nonliving" 15 years before. The enteric bacteria and their phages provided geneticists, biochemists, and even physicists standardized media and growth techniques, methods for growing massive cultures,

synchronous growth, mutant selection, high-resolution genetics, and other tools
with which to pursue the burning questions. And with this came a codification
of new ways of thinking about biological experimentation.

The Divergence of Fungal and Prokaryotic Genetics

Many investigators working on fungi had two disadvantages: their origins and
their organisms. Trained in genetics or microbiology, they had burst blithely
into the study of biochemistry "without a license."[7] They had succeeded spectacularly in the years 1941–55, to the chagrin of some older biochemists. As
the latter had sought to make biochemistry a part of physical chemistry, geneticists had succeeded in making it part of biology. But while fungal geneticists thrived in the period, the biochemical knowledge they brought forth began
to require increasingly sophisticated training and refined techniques that many
of them lacked. For that reason, they had some difficulty legitimizing their findings in the eyes of classical biochemists.

Even interested, broad-minded biochemists would have had trouble with the
filamentous fungi: continued study of these organisms could not get to the heart
of the new questions. The growth of fungi was slow compared to bacteria. A
bacterial culture can be started and studied during a working day, whereas
cultures of even the fast-growing *Neurospora* must be started well before the
working day begins. The characteristics of filamentous growth, a tough cell
wall, cell heterogeneity, and the proteolytic enzymes that appeared in mass
culture made them less than optimal for biochemical experiments. Fungal cells
were 10-fold larger than those of bacteria, and their DNA was a much smaller
fraction of their biomass. Crosses of *Neurospora* took four weeks to mature;
bacterial and phage crosses were one-night affairs. Bacterial cells could be
plated directly and replica plated. In contrast, fungal conidia, used for plating,
were differentiated and dormant. Replica plating, used in *A. nidulans* for some
work, never became routine in *Neurospora*. These attributes stood in the way
of a rigorous approach to the analysis of DNA, its replication, and its function.
Overnight, the filamentous fungi became model eukaryotes rather than model
cells. And yeast was not quite ready for the big time.

An instructive example of the transition from fungi to bacteria and from biochemical genetics to sophisticated biochemistry is the career of Charles Yanofsky,
mentioned in chapter 3.[8] Unlike many working in biochemical genetics at the
time, he was trained as a chemist and began his studies of the tryptophan pathway of *Neurospora* under David Bonner at Yale. Yanofsky defined the intermediates of the tryptophan pathway in *Neurospora*, hoping to understand the role
of the different genes that specified the enzymes of the pathway.[9] He soon focused on tryptophan synthase, the last enzyme of the pathway. This reaction could
be resolved into two successive steps catalyzed by a single enzyme that retained
the intermediate compound, indole, as an enzyme-bound intermediate during the
overall reaction.[10] Yanofsky's methodology combined refined chemical techniques with the use of tryptophan synthase mutants and an increasingly detailed

study of the reaction, both in vivo and in vitro. With a highly purified preparation of tryptophan synthase, Yanofsky and his colleagues raised an antibody to it by injecting it into rabbits. The antibody allowed them to detect wholly inactive forms of tryptophan synthase (called "cross-reacting material," or CRM). Up to then, such defective gene products were invisible to the biochemist because enzyme activity was the only method of detecting enzyme protein prior to extensive purification.[11] The next step, in 1955, was to determine the amino acid sequence of the normal and mutant forms of the enzyme. This information would, first of all, identify amino acids critical to the catalytic activity or stability of the enzyme. More important, by comparing the position of amino acid substitutions in the polypeptide with the genetic map position of mutations within the gene that caused these substitutions, Yanofsky could test the expectation that the gene and protein were "colinear." And just at that time, Sanger had showed how to determine the amino acid sequences of polypeptides.

As we saw in chapter 3, a serious obstacle arose. Yanofsky's group could not purify tryptophan synthase of *Neurospora* easily, in sufficient amounts, and undamaged by proteolytic enzymes to continue the work in this organism. In the mid-1950s, the laboratory switched to the study of *E. coli* tryptophan (*trp*) mutants, developing the genetics of the cluster of tryptophan genes in parallel with similar studies of the *trp* genes of *S. typhimurium* by Demerec and Philip Hartman (b. 1926).[12] (This gene cluster would soon be recognized as an "operon" in the terminology of Jacob and Monod.) The Yanofsky laboratory developed further the genetics of tryptophan synthase in *E. coli*, using P1-mediated transduction to map mutations within the gene. They discovered that the two reactions catalyzed by the single *Neurospora* enzyme were catalyzed by an aggregate of two polypeptides in *E. coli*, products of adjacent genes, *trpA* and *trpB* of the five-gene *trp* operon.[13] The enzyme could be purified dependably both from normal and mutant cells (including those making only an enzymatically inactive CRM). Ultimately, by use of controlled proteolysis, paper-chromatographic "fingerprinting," and column chromatography, Yanofsky's laboratory determined amino acid sequences of parts of many mutant forms of the *trpA* polypeptide of the enzyme. In 1962, Donald Helinski (b. 1933) and Yanofsky demonstrated the colinear relationship of the amino acids altered in the polypeptide sequence to the order of corresponding mutations on the genetic map of the gene, determined by fine-structure transductional mapping.[14] Demonstration of this colinear relationship was a major advance in the field of molecular genetics. The sequence of events, from Yanofsky's early *Neurospora* work to the 1962 paper, illustrates as clearly as any other early example why bacteria captured many of the best minds in biochemical, and eventually molecular, genetics.

The New Agenda

The intellectual attraction of molecular biology to a younger generation lay both in the discoveries made with enteric bacteria and their viruses up to 1953 and in the clarity of experimental predictions possible after 1953. Indeed,

Watson and Crick's second paper of 1953 had specifically pointed to the informational character of nucleotide sequences, their possible mutability through errors of replication, and the possibility that the double helix could replicate by separation of its nucleotide chains and a template-directed synthesis of new complementary chains (see appendix 2).[15] In fact, the biological roles of the genetic material had played a large part in working out the structure itself. From that point on, explicit models of molecular behavior could be formulated easily and tested with rigorous experiments with the tools available.

The discoveries of geneticists and structural biologists up to 1955 had an important effect on the students attracted to the new molecular biology—a term increasingly heard about this time. Many of them had degrees in physical chemistry and biochemistry and, with the flexibility of youth, knew that they could bring chemical sensibilities to this new frontier of biology. Quite suddenly, molecular biologists working on prokaryotic systems could offer themselves as leaders in this new field. The new approach to the gene would be based more on the chemistry of DNA and protein than on genetic analysis. Genetics would pose the important questions and provide tools for proceeding; molecular approaches would answer them. While work continued in many academic institutions and in agriculture and industry on many organisms, molecular biologists would use only a well-developed few. The secrets of life seemed to lie with *E. coli* and its companions, which seemed a ripe garden of knowledge, tended and interpreted by the Merlins and Paganinis of the golden age.[16] This attitude proclaimed the triumph (the first and last) of a few universally relevant model organisms. After their heyday, no other organism could claim such broad biological significance. *E. coli,* its sister *S. typhimurium*, and their viruses were the crucibles of the molecular revolution.

11

Prokaryotes Take Center Stage

It is a truth universally acknowledged that there are only two kinds of bacteria. One is Escherichia coli *and the other is not.*
— J. Allan Downie and J. Peter W. Young

One would think that the discovery of the structure of DNA would render genetics obsolete. The previous 50 years had been devoted in large part to discovering the nature of the gene through studies of its function, mutation, and recombination. What was there for geneticists to learn after Watson and Crick's famous discovery? The answer lies in the transformation of genetics from its preoccupation with the identity of the gene to the use of genetics as an analytical tool. Genetic rationales would probe the properties of DNA in the living system and would continue to extend themselves into the analysis of many complex structures and processes. Genetics began to lose its identity as a separate field as it invaded almost every area of biological science. I devote this chapter to an exploration of the major trends of investigation with prokaryotes mainly in the years 1953–65, focusing on the use of model organisms we have considered so far in particular programs.

Phage T4: Clarification of Allelism

In 1953, geneticists had a messy problem on their hands: pseudoallelism. The word itself, as we saw in chapter 4, was coined to fudge the question of the divisibility of the gene. Few people except Muller had formulated the issues in straightforward language. E. B. Lewis, studying the *bithorax* region of *Drosophila*, thought that a series of genes, whose mutations were separable by recombination, might be organized on the chromosome in a way that facilitated

130

sequential reactions of their products, maintained at millimicromolar concentrations. Pontecorvo, accepting this view as valid, nevertheless offered another picture that might prevail in this and other instances. He believed, like Muller and others, that recombination might take place within genes. If that were so, recombination might not distinguish cases in which mutations lay within the same gene from those in which mutations lay in adjacent genes. In the cases at hand, both views found support in the critical observation that many closely linked mutations, which could recombine with one another at meiosis, nevertheless failed to complement well or at all in heterokaryons or diploids in the *trans* configuration.

In a valedictory article of 1954, Lewis Stadler (1896–1954) insisted on operational rather than on hypothetical definitions of the gene, pointing out that the *cis–trans* test had to do with the function of genes, while recombination and mutation had to do with their structure.[1] Thus the intactness of a gene required for function did not necessarily conflict with the observation that its structural components could be resolved by recombination and mutation. Profitable discussion was impossible if these experimental views became confused with one another. This admonition, echoing Muller's views over the years, was salutary because many geneticists had been unable to divest themselves of the idea that the gene was by definition indivisible on all three criteria.

Seymour Benzer, a physicist attracted to genetics by Schrödinger's *What Is Life?*, joined Delbrück's laboratory to work with phage T4. He soon took up the study of the *rII* region, one of the genetic regions (ultimately shown to consist of two adjacent genes) in which the *rapid lysis* mutations lay.[2] Such mutations, the first ones isolated by Hershey, could be detected visually because of the large plaques they formed on *E. coli* strain B. Benzer then discovered that these mutations, unlike the wild-type *rII*⁺ phage, would not form any plaques at all on *E. coli* K-12(λ). (The λ prophage, for an obscure reason, interferes with maturation of the *rII* mutant phage.) The K-12 strain, Benzer believed, would therefore serve to detect rare *rII*⁺ particles arising by recombination between different *rII* mutations during a mixed infection of strain B. Thus from a mixed infection of two mutants (e.g., *rII*ᵃ × *rII*ᵇ) in *E. coli* B, the progeny phage could be harvested and spotted *en masse* on a lawn of *E. coli* K-12. The spots could then be observed after incubation to see whether any plaques formed, indicative of the existence of a few *rII*⁺ particles.[3]

This test could detect as few as 1 wild-type particle in a population of 100 million phage (1 in 10^8), a population obtainable fairly easily with these tiny entities. The lowest rate of recombination that could reliably be detected was defined by the rate of back-mutation of the parents to wild type, the inevitable noise level of the test. In fact, Benzer found that the smallest non-zero frequency of recombinant particles between parents with very low back-mutation rates in his early studies was many fold higher, approximately 1 in 10^4. In an estimate of how far apart such mutations might be, Benzer assumed a random distribution of recombination events. He calculated that with a phage genome that had 1.66×10^5 nucleotide pairs, mutations recombining at the rate of 1 in 10^4 must be separated by no more than a dozen nucleotide pairs.[4] This was the first clear demonstration that recombination might resolve the genetic material into units smaller than the size of a functional gene. Delbrück said that Benzer's system could "run the genetic map into the ground."[5]

At the time, the word "gene" was the very thing that gave geneticists such difficulty. In the 1955 paper cited above, Benzer also demonstrated that groups of *rII* mutants could be placed in different "functional units," an operational term useful for the moment. Functional units were defined by the *cis–trans* test, performed by infecting the nonpermissive host, *E. coli* K-12, with two different *rII* mutant phages. Neither phage would be able to grow alone in this host, but two possible outcomes were observed in the mixed infection, depending on the mutants used. In the first, few or no plaques formed, indicating that both carried deficiencies for the same function. It was clear that the phages were injected into the host, but they could not cooperate to mount a successful infection. (Despite the multiplication and recombination that might occur after infection, the lack of the *rII*+ gene products at the very beginning leads to the abortion of the infection without liberation of mature phage.)[6] In the second possible outcome, a mixed infection of two other *rII* mutants immediately led to a successful infection, the products of which were largely mutant (accompanied by some recombinant) phage. The conclusion, well supported by later information, was that the deficient function of one of the mutant phages was carried out by the other phage, and vice versa. In other words, the two phages carried deficiencies in different functional units, and the mutant phages complemented one another. Complementing mutants fell into distinct parts of the genetic map of the *rII* region, adjacent to one another. Benzer named them *rIIA* and *rIIB*.

Infections in which each phage carries a different mutation is the *trans* part of the *cis–trans* test. The *cis* part of the test is simply to infect cells with a normal (a^+b^+) and a doubly mutant (a^-b^-) phage. Such infections are normally productive as a result of the activity of both genes of the normal phage. The formal necessity to compare the *cis* and *trans* arrangements lies in the need to establish that both mutations are recessive and thus that neither single-mutant phage would actively interfere with a productive infection in the *trans* arrangement.

Benzer's work gave us language to speak of genes without too much ambiguity. Indeed, he coined the terms *recon* (the smallest unit resolved by recombination), *muton* (the smallest unit affected by mutation), and *cistron* (the "functional unit" introduced above that must be free of mutations to function and that was later recognized as a region encoding a normal polypeptide chain). The last term, cistron, is now equivalent to the word gene. The other terms are obsolete; we now call mutons and recons nucleotide pairs. Benzer had codified the long-used, but inadequately formulated *cis–trans* test as the operational criterion for allelism (that is, whether two mutations were in the same or different genes). The gene as a functional entity thereby subtly replaced the gene defined previously as a unit indivisible by recombination. This change presaged a preoccupation with gene expression in the next decade.

Benzer's analysis of the *rII* region ultimately demonstrated that the two genes, *rIIA* and *rIIB*, could be mapped unambiguously as a continuous, linear, unbranched array of mutable sites, wholly consistent with the molecular topology of DNA. In the early studies, *point mutations*—revertible variants, most of which probably arose from nucleotide substitutions—were used to make the genetic map. The logic of the map was the same used by Sturtevant in the old days: the greater the recombinant frequency, the greater the genetic distance between the mutations. A fair ordering of the mutations used could be made with this technique, but for reasons that became clear only much later, the additivity of distances was rather poor.

The most rigorous map Benzer made took advantage of another type of mutation: deletions. Deletions of short or long segments of the *rII* region could be recognized by their failure to recombine with two or more (sometimes many) of the revertible mutants presumed to be consecutive on the early genetic map. Deletions were useful in the study of recombination because, entirely lacking certain segments of information, they could not back-mutate by base-pair changes to wild type and confuse tests for recombination. Crosses in which both parents carried deletions yielded wild type *rII⁺* phage only if the missing segments did not overlap; that is, only when the two parents, between them, carried all the information needed to reconstruct a normal gene (fig. 11.1). A large array of deletions was available, and crosses showed that they could be fitted into a continuous, overlapping series from one end of the *rII* region to the other. The largest deletions removed most or all of both *rII* genes. Others were limited to large or very small segments of one or the other gene. Still others overlapped the "right" end of the *rIIA* gene and the "left" end of the *rIIB* gene, thereby removing whatever might lie in the intergenic region between them. One of these deletions, deletion 1589, became an important tool in unraveling the nature of the genetic code, described in the next section of this chapter.

In Benzer's work, phage T4 made one of its most significant and universal contributions: the cognitive clarification of the nature of the gene by reconciling its classical genetic and newer molecular representations.

Phage T4: Formalities of the Genetic Code

The availability of the T4 *rII* mutants and the system that Benzer put in place opened the door to an extraordinary program of genetic inference, one directed at the formal nature of the genetic code. In 1961, Crick et al. published a study of the functional characteristics of an unusual class of *rII* mutants.[7] These mutations were unusual with respect to their location, back-mutation, and interaction. Benzer's laboratory had shown that the region at the left end of the *rIIB* gene (nearest the right end of the *rIIA* gene) was almost entirely dispensable for function. This conclusion was based on the deletion mutant 1589, mentioned above, carrying an *rII* region lacking the rightmost part of the *rIIA* gene and the left end of the *rIIB* gene (fig. 11.2). This deletion lacked activity of the *rIIA* gene, but complementation tests showed that, despite the loss of part of the *rIIB* gene, the deletion displayed *rIIB* function. In other words, in a mixed infection in which one phage carried deletion 1589 and the other mu-

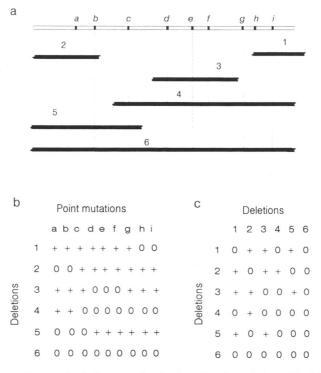

Fig. 11.1.   Rationale of deletion mapping in the *rII* region of phage T2. (*a*) The upper line represents the DNA of the *rII* region, with the positions of the point mutations *a–i* shown. Beneath this, deleted regions (black bars) in various deletion mutants are shown, with jagged ends signifying the indeterminate position of the ends with the information given. (Note that deletions shorten the DNA by the amount deleted; the remainder of the gene remains continuous.) (*b*) Crosses of deletions × point mutations, represented by the intersection of rows and columns. The outcome of each cross is shown as + (wild type recombinants recovered) or 0 (no wild type recombinants observed). The outcome depends on whether the point mutation lies within the DNA that is missing in the deletion parent or not; wild-type phage will form if, between them, the mutant parents contain all necessary information. (*c*) Crosses of deletions × deletions. The same logic applies as in *b*. Only non-overlapping deletions can yield wild-type phage when they are crossed.

tant having a mutation in the *rIIB* gene (but with an intact *rIIA* gene), mutant 1589 could supply *rIIB* activity.

The peculiar *rII* mutants isolated by Crick mapped in the *rIIB* region missing in deletion 1589. This was nominally a contradiction: how could any mutation block *rIIB* function if it lay in a dispensable region of the gene? These mutants had other curious properties: (i) they appeared in greatest frequency after mutagenesis of phage with the drug acridine yellow; (ii) most mutants were complete loss-of-function mutations; (iii) many of them could be induced to undergo back-mutation with acridine yellow; (iv) most of the back-mutants were not restored to the true wild-type condition, but were double mutants

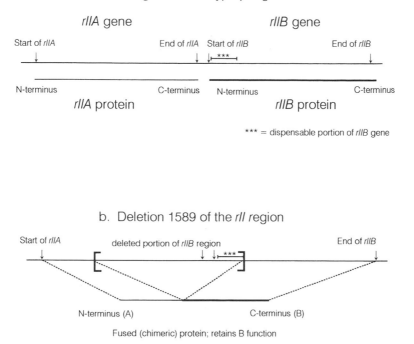

a. *rII* region of wild-type phage T4

rIIA gene   *rIIB* gene

Start of *rIIA*   End of *rIIA*   Start of *rIIB*   End of *rIIB*

N-terminus   C-terminus   N-terminus   C-terminus
rIIA protein   *rIIB* protein

*** = dispensable portion of *rIIB* gene

b. Deletion 1589 of the *rII* region

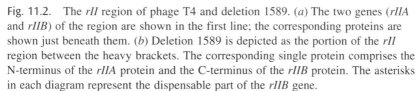

Start of *rIIA*   deleted portion of *rIIB* region   End of *rIIB*

N-terminus (A)   C-terminus (B)

Fused (chimeric) protein; retains B function

Fig. 11.2.   The *rII* region of phage T4 and deletion 1589. (*a*) The two genes (*rIIA* and *rIIB*) of the region are shown in the first line; the corresponding proteins are shown just beneath them. (*b*) Deletion 1589 is depicted as the portion of the *rII* region between the heavy brackets. The corresponding single protein comprises the N-terminus of the *rIIA* protein and the C-terminus of the *rIIB* protein. The asterisks in each diagram represent the dispensable part of the *rIIB* gene.

("pseudowilds") in which a second mutation had occurred near the first. (Geneticists call a mutation that reverses the effects of another mutation a *suppressor*.) Even more curiously, crosses performed to separate a suppressor mutation from the initial *rII* mutation revealed that the suppressor, as a single mutant, imparted a mutant *rII* phenotype! Thus the original and the second *rII* mutations were mutual suppressors, compensating for one another's defects (fig. 11.3).

The Crick group rationalized the results by suggesting that although the left end of the *rIIB* gene was dispensable, certain mutations might cause a polarized effect on the function of the rest of the gene. Specifically, they suggested, these mutations interfered with the proper translation of the gene, assuming that the process went from one end to the other. A hint in previous experiments suggested that acridine dyes intercalated between base pairs of DNA and that this might cause rare additions or deletions of a nucleotide pair from the DNA during subsequent replication. If this were so, then the polarized effect of such mutations could be seen in terms of reading the code for a protein from one

Wild type	CAC ACC GGA UCU UUG AAG GCA AAA GGG
	his thr gly ser leu lys ala lys gly...
-1 FS	CAC˙CCG GAU CUU UGA AGG CAA AAG GG
	his pro asp leu non
+1 FS	CAC ACC GGA UCU A˙UU GAA GGC AAA AGG G
	his thr gly ser ile glu gly lys arg .....
-1/+1 FS	CAC˙CCG GAU CUA˙ UUG AAG GCA AAA GGG
	his pro asp leu leu lys ala lys gly......

Fig. 11.3.   The effect of frameshift mutations. The wild-type mRNA nucleotide sequence and the corresponding amino acid sequence is shown at the top. (Note that spaces have been introduced between codons solely to facilitate reading.) A −1 frameshift (deletion of the A in position 4) leads to misreading of the subsequent nucleotides and the appearance of a nonsense (non) codon in the new reading frame. A +1 frameshift (introduction of an A in the wild-type leu codon) also leads to misreading of subsequent codons. With both −1 and +1 frameshift mutations, the proper reading frame is restored after the +1 mutation, and the nonsense codon caused by the first is not encountered as an in-frame codon. The amino acid alterations between the two frameshifts may or may not be functionally damaging. (The work done by Crick et al. on the *rII* region of phage T4 was localized to the region encoding the dispensable portion of the *rIIB* gene, and therefore only changes of reading frame extending beyond it would cause a mutant phenotype.)

end of a gene to the other, beginning at a fixed starting point. I will describe in more detail how this work illuminated the nature of the genetic code below.

Substantial thought had been devoted in the mid- to late 1950s to the nature of the genetic code, once the community became grounded in the structure of DNA. Initially, George Gamow (1904–68) proposed that the DNA of a gene was a structural template for amino acids.[8] The idea was that short sequences of bases along the DNA formed niches whose shapes corresponded to the shapes of amino acids. Once a series of amino acids was in place along the gene, Gamow believed, they could be chemically linked to form a polypeptide chain. However, no chemical affinities between sequential nucleotide combinations and amino acids could be divined, and the idea was quickly abandoned. Other theorists, Crick among them, retained the old idea that DNA carried the needed information (a code), but not the direct template, for protein synthesis. This provoked intriguing questions. These included whether units of the code overlapped one another; whether units of the code for individual amino acids corresponded to two, three, four, or variable numbers of nucleotides; and whether the code was read in a polarized manner from one end of a gene to the other. Did the code have punctuation between code "words"? Did the code have start and stop signals? These questions probed the formalities of coding and required little biochemical sophistication to pose them.

From their data on the induction and back-mutation of the acridine mutants, Crick et al. developed the following picture. If adjacent code words did not share nucleotides (i.e., if the code was non-overlapping), and if code words were read in a polarized manner, a single nucleotide deletion would cause all of the coding units that followed to be moved one nucleotide out of register. Such a picture (fig. 11.3) requires that for any long sequence of nucleotides, there is a biologically proper reading frame. The reading frame can be distorted by additions and deletions, particularly if the code is read in a polarized manner from one end of a gene. This idea gave the name "frame-shift" to the acridine mutants. A frame-shift mutation would cause all code words after the mutation to be different from the original. However, the effect of a nucleotide deletion could be reversed in large part by an addition nearby, such as might happen with a suppressor mutation. Except for the code words between the two sites of mutation, the proper reading frame would be restored. The study of primary mutants and their suppressors suggested and validated this hypothesis. Crick et al. deduced from the pattern of interactions of acridine mutations described above that the first mutation, if a "+" type, could be suppressed by a "−" mutant, but such a suppressor, which caused a mutant phenotype when it was alone, could not be suppressed by another "−" mutant. The investigators showed that almost all mutations could be assigned to two mutually suppressing classes. The interpretation readily explained why these mutations had such a drastic effect despite their location in a dispensable part of the gene. Reading was polarized, and a change of the reading frame at the beginning would make gibberish of the rest of the gene.

Crick's group then went on to show that three mutations of the same class, but not one, two, or four such mutations, yielded a pseudowild-type phenotype when they were all combined in a multiply mutant gene. This bore out the expectation that if +1 and −1 mutations compensated for one another, then addition or deletion of the exact number of nucleotide pairs in a complete coding unit should also restore function. According to the data, then, the code was a triplet code, in which each code word consisted of a sequence of three nucleotide pairs. In short order, the words "codon" and "triplet" were coined to designate a unit of the genetic code, encoding a single amino acid in a given polypeptide.

Crick et al. showed that frame-shift mutations in the *rIIA* gene had no effect on the expression of the *rIIB* gene. This demonstrated that each of the two genes had its own starting point and was read independently of the other. In an experiment of opportunistic brilliance, they used deletion 1589, which overlapped the end of the *rIIA* cistron and the dispensable part of the *rIIB* cistron. According to the rules developed so far, the fusion of the genes would cause the formation of a continuous, chimeric polypeptide, having *rIIA* amino acids at one end and *rIIB* amino acids at the other (fig. 11.2). The investigators constructed a double mutant that combined a frame-shift mutation in the remaining part of the *rIIA* gene with the 1589 deletion. Here, the *rIIA* mutation did inactivate *rIIB* function. This showed more clearly that the two genes in the wild type each had a fixed starting point for reading the code for its respective polypeptide. The deletion eliminated this starting point for *rIIB*. Venturing further to suggest another punctuation mark in the genetic code, Crick et al. hypothesized, with some evidence in their data, that "nonsense codons," encoding no amino acid, might be used to mark the normal stopping point for the reading of a gene. They would provide the periods at the ends of genetic messages, counterparts of fixed starting points, the capital letters at the beginning.

We see here the exploitation of phage T4 in the second of its important contributions: a detailed framework of the process by which DNA was translated into protein. The project preceded the discovery of messenger RNA (mRNA), the molecule used by the protein synthetic machinery, and an understanding of the RNA adaptor molecules to which amino acids attach before they are used in protein synthesis (see appendix 2).

Phage T4: Mutational Analysis of Phage Development

A third major contribution of phage T4 genetics concerns the type of mutations used to map the phage genome and the experimental use of these mutations in further work. The results of Horowitz and Leupold on temperature-sensitive mutants of *Neurospora* (see chapter 3) was forgotten until similar mutants were used by one of two groups working on phage T4.[9] Both of these groups wanted to make a genetic map of the phage, and to define, in the end, all the functions encoded by the phage genome.

The kinds of mutants used to study phage T4 biology were *ts,* for temperature-sensitive, and *am,* for *amber.* The *ts* mutants did not carry out a full cycle of infection of *E. coli* B at 37°C, but could do so at 30°C. The *am* mutants were of quite another type. As a class, they, too, could affect any gene in the phage DNA. But they could grow normally in *E. coli* K(λ) (with some exceptions), but not in certain other related strains nor in *E. coli* B. The *am* mutants were named fancifully after Harris Bernstein, a graduate student who had acquired the sobriquet *Immer Wieder Bernstein* ("Forever Amber") and who had helped isolate the mutants.[10] The *am* mutations were later shown to cause premature termination of translation of coding sequences in mRNA. They were, in fact, nonsense mutants of the type predicted by Crick et al. in the study described in the previous section. Nonsense mutations cause translation termination because a sense codon (encoding an amino acid) has been altered by mutation into a nonsense codon (which has no translation). If a phage DNA carries an *am* mutation, it cannot grow in certain *E. coli* strains. However, *E. coli* strain K12 carries a so-called nonsense-suppressor mutation. Such mutations are different from frame-shift suppressors and occur in special genes responsible for the translation process itself. Nonsense suppressor mutations overcome the effects of *am* mutations by changing the nucleotide sequence of genes encoding tRNAs, short RNA molecules that serve as adaptors for amino acids during polypeptide synthesis (see appendix 2). The tRNAs, when attached to amino acids, position the amino acids in the proper locations along mRNAs, the copy of the DNA sequence used in protein synthesis. They do so by having an anticodon capable of base-pairing with codons. Mutations in the anticodon of a number of tRNAs may change their anticodons so that they can pair with a nonsense codon. In doing so, they can bring an amino acid into that position during translation and thus can compensate for the chain-terminating effects of a nonsense (*am*) mutation. (Because *E. coli* has several copies of most tRNAs, the mutation of one such copy is not lethal.) An *E. coli* strain carrying a nonsense suppressor mutation supports the growth of certain *am* mutants; strains lacking the suppressor cannot do so. The *am* mutants are therefore conditional mutations in that sense.

Knowledge of the nature of *am* mutants came afterward, but it was clear upon their isolation that, like temperature-sensitive mutations, *am* mutations could affect any protein and could therefore lie in any gene. Because of the conditional nature of *ts* and *am* mutants—depending on temperature in one case and on the bacterial host in the other—all genes of the organism could be mutated and recognized in the form of viable variants.

The two groups working on this problem focused on the biology of phage T4. Therefore, the outcome of their work did not have the universal character of that of Benzer's and of Crick et al.'s contributions described earlier. However, their methods of using conditional lethals to explore the genetics of a complex sequence of events (and indeed, the structure of the phage) became universal tools for studying indispensable functions in all organisms subjected to genetic analysis. The two groups, led by Edouard Kellenberger (b. 1920) and Charles Epstein in Geneva and by Robert Edgar (b. 1930) at Caltech, summarized their initial work in a joint presentation at the 1963 Cold Spring Harbor symposium.[11] Their materials greatly expanded the repertory of mutants of the phage.

The *ts* mutants offered the opportunity to study the steps of replication and morphogenesis of phage T4 with mutants. (Comparable experiments were also done with *am* mutants.) These mutants can block these processes at any one of a number of developmental steps at the restrictive temperature. The methodology allowed the investigators to initiate an infection at the permissive temperature (30°C) or the nonpermissive temperature (37°C) and see how far the infection could proceed at the latter. Biochemical tests and electron micrography were used to visualize the state of the phage or its DNA in intact or artificially lysed host cells. More sophisticated experiments involved infection at one temperature and a shift, at different times during the infection, to the other. In this way, a clear sequence of discrete events dependent on particular genes (and thus proteins) could be discerned. The events spanned the earliest steps of replication to the final assembly of phage-coat proteins and DNA to make an infectious particle. Of some interest was the finding that the genes were clustered roughly on the genetic map according to their time of action in the phage infection.

Mapping the conditional mutants yielded a circular genetic map, which had to be reconciled with the knowledge that the phage DNA was linear, both during growth in the cell and in the phage particle. The resolution of this paradox came with the discovery that many T4 phage genomes are linked end to end in tandem during replication and recombination (fig. 11.4). As DNA is packed into phage heads, "headfuls" are cut successively from the long tandem array in lengths slightly longer than one complete set of genes. Thus phage DNA in mature phage particles all have two copies of some genes, one copy at each end. The mature DNA of a given particle is thus terminally redundant. How-

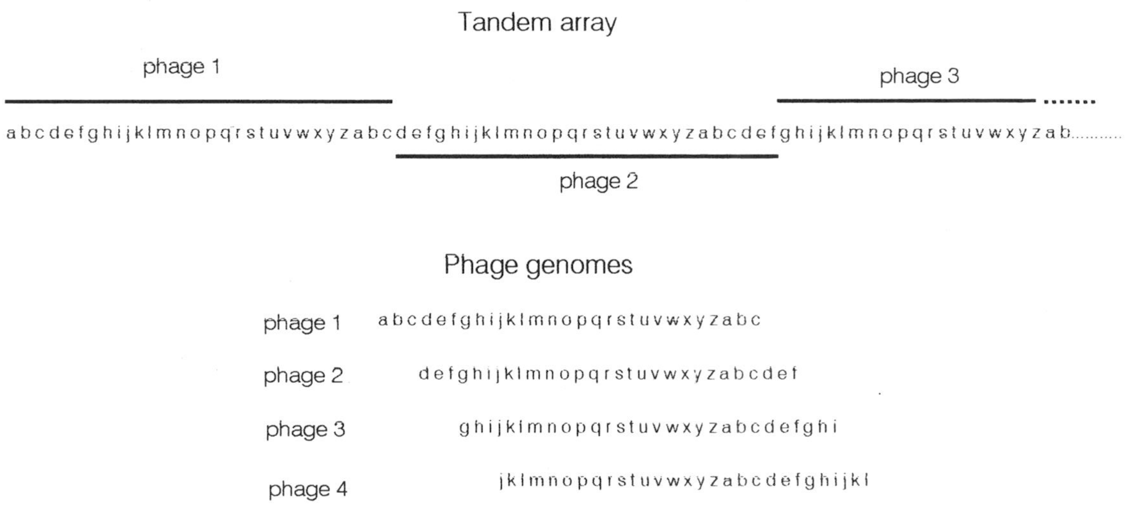

Tandem array

phage 1

phage 3

phage 2

a b c d e f g h i j k l m n o p q r s t u v w x y z a b c d e f g h i j k l m n o p q r s t u v w x y z a b c d e f g h i j k l m n o p q r s t u v w x y z a b........

Phage genomes

phage 1   a b c d e f g h i j k l m n o p q r s t u v w x y z a b c

phage 2   d e f g h i j k l m n o p q r s t u v w x y z a b c d e f

phage 3   g h i j k l m n o p q r s t u v w x y z a b c d e f g h i

phage 4   j k l m n o p q r s t u v w x y z a b c d e f g h i j k l

Fig. 11.4. Packaging the DNA of phage T4. The top array depicts a continuous DNA in which the genes *a–z* of phage T4 are repeated in tandem. Somewhat more than a complete set of genes DNA is packaged into each phage head, leading to a redundancy at the ends. When gene sequences of each genome are aligned, one can see that all genes are linked to one another; no ends are detected by recombinational studies. The genetic map is therefore circular.

ever, the population of DNAs in a crop of phage are also *circularly permuted*, meaning that the duplicate sequences at the ends of different phage DNA molecules are different. Therefore, any two adjacent genes remain next to one another in most particles, and thus no ends can be found on a recombination map; genetically, the map is circular.[12]

After these contributions, phage T4 investigations began to lose the universal relevance to biology that Delbrück had hoped for the phage in 1939. Phage T4 research made one more important, universally significant contribution in the discovery of messenger RNA (mRNA) in 1960. This was more a biochemical than a genetic investigation. The work started with the discovery of a transient species of RNA that appeared in low amounts in infected cells and appeared to have a short half-life (that is, once made, it was rapidly degraded). The nucleotide base composition of this RNA closely resembled phage DNA and not that of the bacterial host. This was, in fact, mRNA, rapidly made and rapidly destroyed, carrying the phage genetic message to the site of protein synthesis, the ribosomes (see appendix 2). The direct detection of mRNA resolved the paradox that the locations of ribosomes and DNA were somewhat distinct in bacterial cells. It also resolved the problem that the RNA found in ribosomes as part of their structure (rRNA) was not diverse enough to encode the many proteins of the cell.[13]

The T phages would continue to contribute to molecular biology, but they soon became organisms in their own right, with a fascinating biology, the study of which continues to this day. But the attention of many outside the T phage group had already become fixed on exciting findings arising from study of *E. coli* and its temperate phage, λ.

The Domestication of *E. coli*

We have seen *E. coli* acquire the status of a model organism with the demonstration that it displayed biochemical pathways in common with higher forms and that it had means of obtaining its nutrients and energy from simple sources. The bacterium gained respectability upon the discovery of sexual recombination between the F+ and F− mating types and became fascinating for many geneticists when the Hfr × F− crosses showed a novel, polarized donation of DNA from the former to the latter. With P1 transductional analysis and the discovery of the linkage of biochemically related genes in *E. coli* and *S. typhimurium*, the tools for thorough genetic analyses of many fundamental biochemical and molecular problems were at hand. By 1960, few molecular geneticists could resist the urge to thumb through the latest issues of *Science, Nature, Proceedings of the National Academy of Sciences (USA), Contes Rendus de l'Academie des Sciences*, and the new *Journal of Molecular Biology* as they arrived in the mail.

Before *E. coli* could be as useful as it became by 1960, its genetic system had to be domesticated for use by the many new entrants to microbial genetics. Élie Wollman (b. 1917; son of Eugène and Elisabeth) and François Jacob, at the Pasteur Institute in Paris, carried out this effort systematically in the years

1954–59, integrating their findings with those of many other investigators. They summarized their work in a major treatise, of which the English version of 1961 gives the most complete account.[14] The account embeds the study of λ lysogeny, their initial preoccupation, in the subject of bacterial sexuality. The integration of the two subjects led to an even greater focus on *E. coli* and λ as models and initiated what was to become, by the late 1980s, a grand vision of bacterial and phage genetics and biology. The book begins with a brief account of work prior to the discovery of Hfr strains. The next section develops the current picture of conjugation and chromosomal transfer during Hfr × F⁻ matings (fig. 7.4) and defines the states of the F factor in relation to mating. As we have seen in chapter 7, the F⁻ strain lacks this factor, which exists as a plasmid in F⁺ bacteria. In the formation of Hfr strains, the F DNA becomes a part of the *E. coli* chromosome by recombining with it. Studies of the transfer of the chromosome during Hfr × F⁻ matings included the development of the "interrupted mating" technique. This technique permits determination of the order of genes on a time-of-entry basis in synchronous mass matings. After mating, samples of the mating pairs are broken apart at different times in a Waring blender. Suitable genotypes of the male (Hfr) and female (F⁻) strains and selective platings allow the investigator to detect the entry of prototrophic markers into the multiply auxotrophic F⁻ cells. With this technique, one may draw a genetic map based on time (the minutes required before an Hfr marker enters F⁻ cells), rather than on the frequency of recombination between genes.

Two important findings that emerged from these studies deserve mention here. The first, which I discussed in chapter 9, is the phenomenon of zygotic induction, seen in crosses between Hfr(λ⁺) × F⁻(λ⁻) populations. In such crosses, the transfer of the λ prophage as part of the Hfr chromosome into the F⁻ cells is followed by the induction of the prophage because it is in a repressor-free cell. The induction is not seen if both parents carry the λ prophage. Therefore, in interrupted matings, the location of the prophage can be mapped to a particular site on the bacterial chromosome. The importance of zygotic induction, however, lies in the induction phenomenon. Induction gave the first clear evidence of a diffusible, cytoplasmic repressor of the prophage, encoded by the prophage in the lysogenic host, and absent from the nonlysogenic host, the recipient of the prophage in Hfr(λ⁺) × F⁻(λ⁻) matings. This conclusion soon became integrated into the view of negative control of the enzyme β-galactosidase by another repressor gene product, a story developed at the same time at the Pasteur Institute in studies by Jacques Monod. Further research on both regulatory systems would fully exploit the genetic techniques for *E. coli* developed by Wollman and Jacob.

The second finding of importance concerns the biology of *E. coli* and, by extension, almost all bacteria. Early observations of Hfr × F⁻ crosses showed that only portions of the bacterial chromosome entered the recipient cells. This reflected spontaneous breakage of the conjugation bridge between mating pairs after mating. Only a few pairs (ca. 0.1%) remain attached for the entire time Hfr cells require for the delivery of all of their genes to the F⁻ recipients. As more new Hfr strains were isolated, Jacob and Wollman found that different Hfr strains, isolated independently from F⁺ cultures, transferred different parts of their chromosome with high efficiency. Sometimes the order of transfer, by the interrupted mating technique, was the reverse of the transfer order by another Hfr. This indicated that the F DNA had integrated at different positions on the chromosome in different strains and perhaps in different orientations. With the interrupted mating technique, however, Jacob and Wollman found that the gene maps derived from various Hfr strains could all be arranged in a self-consistent circle (fig. 11.5). With this composite circular map, they could define the position and orientation of the F DNA in each one and suggest that the *E. coli* chromosome was a single, circular molecule of DNA. In 1963, John Cairns (b. 1924), using cells having DNA labeled with radioactive hydrogen (^{3}H, tritium), showed with gentle artificial lysis of cells and autoradiography of the liberated DNA that the chromosome was indeed circular.[15] The finding represents the form of the chromosome of most bacteria, and solidified circular DNA as a theme among prokaryotes, their plasmids, and many of their phages. More recently, linear chromosomes

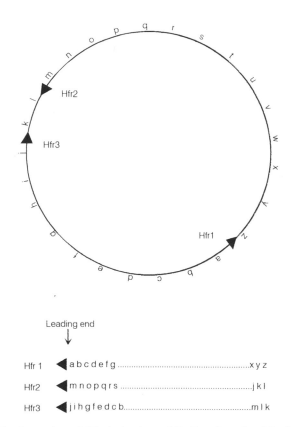

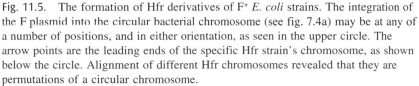

Fig. 11.5.   The formation of Hfr derivatives of F⁺ *E. coli* strains. The integration of the F plasmid into the circular bacterial chromosome (see fig. 7.4a) may be at any of a number of positions, and in either orientation, as seen in the upper circle. The arrow points are the leading ends of the specific Hfr strain's chromosome, as shown below the circle. Alignment of different Hfr chromosomes revealed that they are permutations of a circular chromosome.

have been discovered in a number of bacteria, with specialized means of replication and protecting the free ends of the DNA.[16]

E. coli and Semiconservative Replication of DNA

We turn now to one of the many contributions of *E. coli*, one that demonstrated a universal attribute of DNA originally suggested by Watson and Crick. They had proposed that the two strands of DNA could separate and, obeying the strict complementarity of the bases (A-T and G-C), each could specify the nucleotide order of a new complementary strand.[17] This notion, called *semiconservative replication*, was so attractive that few initially bothered to think of alternatives. On further reflection, however, one could not deny that the two strands of a DNA molecule might separate temporarily or locally and, together, specify the synthesis of an entirely new double-stranded molecule. That is, replication might be conservative. Moreover, no chemical constraints on a dispersive model of replication prevailed at the time: the original double-stranded molecule might replicate in small segments such that new and old fragments might become intermingled in the two daughter molecules. Much too little was known at the time to discard any of these hypotheses.

In 1958, data of Matthew Meselson (b. 1930) and Franklin Stahl (b. 1929) strongly supported the semiconservative model, by which daughter molecules consist of one parental and one new strand.[18] These data were obtained through a study of DNA molecules synthesized from a heavy isotope of nitrogen (^{15}N), which could be separated from normal DNA (in which the nitrogen, ^{14}N, was lighter), by high-speed centrifugation in a gradient of the rare alkali-metal salt, cesium chloride. Meselson and Stahl prelabeled the DNA of cells with heavy nitrogen by growing the cells for some time in the presence of ^{15}N-ammonium chloride. They then transferred the cells to a medium with [^{14}N]ammonium as the sole nitrogen source, which the cells would use in making new DNA. After one generation (one replication of DNA), all the DNA of the cells was intermediate in density, consistent with the picture that all molecules had one light and one heavy strand. After another replication, the DNA molecules consisted of equal amounts of intermediate and light forms. This would be expected if both the light and the heavy strands of hybrid molecules directed the synthesis of a light partner. The findings spectacularly demonstrated a method of studying DNA replication and confirmed a central prediction about how DNA might conserve information during successive replications. The findings found their way into textbooks within months. They form one of the key steps in the history of molecular biology, made by the use of *E. coli*, even by that time a supermodel.

E. coli: The Integration of Metabolism

Medical science and the fermentation industries had studied bacteria, including *E. coli*, for many years. In both fields, a great deal had become known about biochemical activities of bacteria. But much of it had developed in connection with studies that concentrated on the catabolism of various carbon sources by different species during growth and the energy yield from these substrates. In the late 1930s, comparative biochemistry and enzymology became better developed, and these areas underlay the interest of Beadle and Tatum (particularly Tatum) in biosynthetic reactions. As biochemical genetics developed in fungi and bacteria, information accumulated so rapidly that no picture of overall cell biochemistry could be complete for more than several weeks. In the 1950s, the integration of all of the biochemical reactions of the cell was well beyond the understanding of biochemists or biochemical geneticists and promised to remain so forever. During these years, contemporary with vigorous studies on *Neurospora* metabolism, a group of physicists, biologists, and chemists at the Department of Terrestrial Magnetism at the Carnegie Institute of Washington in Washington, D.C. embarked on an ambitious program with *E. coli*. These individuals, Richard Roberts (1910–80), Philip Abelson (b. 1913), Dean Cowie (1913–77), Ellis Bolton (b. 1922), and Roy Britten (b. 1919), set out to measure the intake and the ultimate disposition of the nutritional sources of carbon, sulfur, phosphorus, and nitrogen during exponential, steady-state growth. Their tools were new: radioactive isotopes of the first three elements had become available after World War II, and ^{15}N had been available for some time. These labeled elements provided a sensitive means by which the fates of labeled precursors could be studied quantitatively in metabolism as they entered the cell and became incorporated into protein, carbohydrates, nucleic acids, and lipids.

The work of this group was published as *Studies of Biosynthesis in* Escherichia coli, first in 1955, reprinted with additions in 1957 and again in 1963.[19] The first publication includes a chapter on amino acid synthesis in *Torulopsis utilis* (a yeast), and similar studies of *Neurospora* were added to the 1957 version. Studies of yeast and *Neurospora* illustrated the degree to which findings in the bacterium could be generalized to eukaryotic microorganisms. The experiments represent an early example of an engineering approach to a living system. The authors emphasized the balance of synthesis and utilization of intermediates and suggested that control mechanisms might maintain that balance and respond to changes in the environment. They speculate about compartmentation or the formation of variant chemical derivatives of metabolic intermediates that would assure their proper destiny in the cell. The authors noted in particular the way in which cells stopped making an amino acid, for instance, when the same amino acid was provided in the medium. Such cybernetic feedback controls are vital attributes of well-engineered dynamic systems, properties explored both theoretically and in practice during World War II. Indeed, the science of cybernetics owes a great deal to the development of artillery-control systems. Some quotations from the last chapter of the Carnegie publication (1963 reprint) give a flavor of the state of the field in 1957:

> The composition of the cell does not change markedly whether glucose, glutamic acid, or alanine is provided as the sole carbon source. It is, therefore, evident

that the amino acids are not used as fast as they are provided, but rather that they are provided as fast as they are needed.

Such results seem to indicate that there is a definite coupling between protein synthesis and nucleic acid synthesis. . . . Though this type of coupling may well be due to an intimate relation between protein and nucleic acid synthesis, there is also a possibility of a coupling through the supply of intermediates.

Such a crude mechanism could not be expected to achieve the delicate balance which is actually achieved in growing cells. . . . The process of enzyme formation must be exceedingly responsive to the concentration of the products of the enzyme. A study of the ways in which the synthetic systems of the cells can be unbalanced and the ways in which balance is restored should lead toward the comprehension of the regulatory mechanisms which are of such vital importance throughout living systems.[20]

It is no surprise that several of the authors of this treatise had taken the phage course at Cold Spring Harbor. Given the opportunity presented by isotope technology, their background in physics, and the excitement surrounding microbial genetics and physiology, the authors would naturally want to determine whether a free-living organism could be treated as a system, as a whole, in terms of the flow of mass. The work of this group was highly biochemical, involving little direct use of genetics.

Another effort began in the same years. It concerned the balanced synthesis of DNA, RNA, and protein and of ribosomes achieved by *E. coli* in exponential growth, studied with isotopes and direct bulk measurements of these components. This, too, was a physical and mathematical approach, by which growth and the increase in these components could be modeled according to parameters implying internal control mechanisms.[21] The later demonstration of the coordination of DNA synthesis and the cell cycle at different growth rates made this picture more intriguing, and both of these programs carried out with *E. coli* would be a model for similar work on yeast that began shortly thereafter.[22]

E. coli: Enzyme Regulatory Mechanisms

The work of the Carnegie group took little technical advantage of genetics. Most of it was done on *E. coli* B, which had not been developed as a genetic system. The mechanisms that might assure the fine-tuning of metabolic supply and demand were entirely unknown. Two forms of enzyme regulation were recognized soon thereafter: feedback inhibition of the activity of initial enzymes of a biosynthetic pathway, and induction and repression of gene expression, governing the synthesis of catabolic and biosynthetic enzymes. One can appreciate the role of *E. coli* as a model for all of living things simply by noting that most of the language and diagrammatic imagery of regulatory biology emerged from the studies that followed.

Feedback inhibition was an important discovery in the field of biochemistry. In this form of regulation, the catalytic activity of the first enzyme of a biosynthetic pathways is frequently inhibited by the end product of the same pathway. Several properties of this phenomenon soon became clear. The end product need not resemble the substrate of the enzyme, and thus the end product and the substrate did not compete for the active site on the enzyme. Indeed, the inhibition was exerted by the action of the end product on a separate site, called an allosteric site. Finally, feedback inhibition was virtually instantaneous because it depended on the ambient concentration of the end product within the cell. Therefore, the mechanism was capable of fluent fine-tuning of the flux of intermediates through the pathway. Feedback inhibition was clearly a highly adaptive, Darwinian feature of biochemical systems, and its widespread distribution in many forms of life was established soon after its discovery in the late 1950s in *S. typhimurium* and *E. coli* in the laboratories of Harold Erwin (Ed) Umbarger (b. 1921), Arthur B. Pardee (b. 1921), and many others following them.[23]

The other control mechanisms had to do with enzyme synthesis. Already, in the late 1940s, "adaptive enzymes" had been recognized as a conspicuous feature of microbial life. However, virtually all examples had to do with the enzymes induced by nonstandard carbon and nitrogen sources. The prevailing idea of their origin was that these nutrients had a direct role in physically shaping the appropriate catabolic enzyme. (This notion arose from a proposal, since disproved, that antigens have an instructive role in shaping antibodies.) The Carnegie group's work prepared the way for analysis of regulatory systems governing adaptive enzymes. We turn, therefore, to the operon model and the closely allied picture of the control of gene expression. In the development of both, genetics takes a leading role.

The theories about the operon and gene control arose from a collaboration between Monod and Jacob at the Pasteur Institute. In his early work, Monod had studied the physiology of protozoa, but switched before World War II to bacteria and took up the study of adaptive enzymes. The enzyme he chose, β-galactosidase, hydrolyzes the disaccharide lactose to the 6-carbon monosaccharides, glucose and galactose. The enzyme also hydrolyzes artificial β-galactosides, one of which releases a colored product upon hydrolysis. This analogue greatly simplified measurement of enzyme activity in later analytical work. Presentation of lactose to cells induces them to synthesize high levels of the enzyme, but only if glucose, a more readily assimilated carbon source, is absent. In cultures grown in mixtures of glucose and lactose, all the glucose is consumed before lactose is used, and a pause in growth takes place during which cells synthesize β-galactosidase. The enzyme enables growth to resume at the expense of glucose and galactose derived from lactose. (Monod named the two-phase growth pattern *diauxie*.) Considerable confusion prevailed well into the late 1950s about the basis of this phenomenon.[24] I will focus only on the induction of β-galactosidase by lactose and its chemical analogs.

As I noted previously, even in the late 1940s many biochemists thought that the catabolic substrates of enzymes physically shaped the enzymes that would catalyze their breakdown. Soon thereafter, it became clear that substrates simply induced the mature enzyme, which the cell already had the capacity to synthesize. Monod's work on β-galactosidase induction by lactose took advantage of work done by the Lederbergs on the same enzyme, work that provided a number of chemical analogs of lactose. (The genes governing β-galactosidase activity are distinct and are located at a different place on the *E. coli* chromosome from the *gal* genes adjacent to the λ prophage site.) Some of the chemical analogs of lactose could induce β-galactosidase without being broken down by it, and some analogs could be catabolized even though they could not themselves induce the enzyme. These findings demonstrated two things: (i) induction and catabolism were separate processes, and (ii) the structural information for the enzyme lay wholly within the cell. The latter point was more clearly demonstrated by the isolation of *constitutive* mutants—mutants that did not require addition of the inducer for the production of the enzyme. But even here, Monod felt that constitutive cells might nevertheless have an "internal inducer," a compound for which he sought evidence for some time.

During the time that Monod was struggling with this problem, biochemical geneticists in America, Henry Vogel (b. 1920), Werner Maas (b. 1921), and Pardee separately discovered that the synthesis of certain biosynthetic enzymes was also a regulated process. This phenomenon added another control mechanism to the cell's repertory, distinct from feedback inhibition. In the latter, an end product acts on an initial enzyme of the path to block its catalytic activity. In the new one, called *repression*, a small-molecule end product blocks or reduces the rate of synthesis of many or all enzymes of a biosynthetic pathway. In retrospect, both mechanisms clearly prevailed in the experiments of the Carnegie group, who had studied only the synthesis of small-molecule intermediates, rather than the biosynthetic enzymes themselves. Pardee's group was able to distinguish these mechanisms in his work on the pyrimidine pathway. But Vogel, who earlier had clearly enunciated the possibility of repression of arginine-biosynthetic enzymes by arginine, saw that the phenomena of induction of a catabolic enzyme and repression of a biosynthetic enzyme might have common features.[25]

In 1959, Jacob, working on the genetics of *E. coli*, Monod, working with β-galactosidase, and Pardee, who had discovered repression of pyrimidine biosynthetic enzymes, all worked at the Pasteur Institute. They did an experiment (the famous "PaJaMo" experiment) that would touch off an explosion of knowledge about the control of the *lac* operon, a set of three contiguous genes.[26] These genes encoded the enzyme β-galactosidase; a *permease,* a protein in the cell membrane that transports β-galactosides into the cell; and β-galactoside transacetylase, a third protein without a clear biological function. The genes are now called *lacZ, lacY,* and *lacA,* respectively. Jacob and Wollman by that time had refined the interrupted mating technique and had explored the kinetics of mating. They had also demonstrated the phenomenon of zygotic induction, the release of λ prophage from repression as it entered a nonlysogenic F⁻ cell from an Hfr cell. They also determined, using mutants lacking β-galactosidase (*lacZ⁻*), the time after mixing suitable parents at which the *lac* operon entered F⁻ cells in their Hfr × F⁻ matings. Monod had by that time obtained constitutive mutants (ultimately named *lacI⁻*) that synthesized β-galactosidase without addition of an inducer. Happily, the *lacI* gene, although distinct from the *lac* operon, was adjacent to it. Pardee was drawn into an experiment that would test for the existence of Monod's hypothetical internal inducer in constitutive (*lacI⁻*) mutants. At that time, Monod thought that the *lacI⁻* mutants actively made such an inducer as a product of the mutant *lacI⁻* allele.

The experiment, although not complex, required a clever strategy. The investigators used Hfr cells and F⁻ cells, each of them of two types, *I⁻Z⁻* and *I⁺Z⁺*, so that reciprocal crosses could be done. Without an inducer in the medium, neither of these genotypes can make β-galactosidase: the first because the *lacZ* gene is mutant; the second because the gene is uninduced. If the mating were to deliver the *I⁻Z⁻* genetic segment from an Hfr donor to a F⁻ *I⁺Z⁺* recipient, the supposed internal inducer made by the *I⁻* gene should immediately provoke synthesis of the enzyme from the resident *Z⁺* gene. The experiment failed: no enzyme was formed unless an external inducer was added. This observation, in addition to contradicting expectation, showed that the resident *I⁺* gene was dominant over the entering *I⁻* allele because the F⁻ cells remained inducible (rather than becoming constitutive) even after they had received an *I⁻* allele. To a geneticist, a dominant allele is normally the active one, the mutant being an impaired or inactive variant. Finding that *I⁺*, rather than *I⁻*, was dominant was a surprise to Monod.

When the reciprocal cross was done, in which the *I⁺Z⁺* segment of an Hfr entered an *I⁻Z⁻* recipient, enzyme formed immediately, without any induction at first, but it soon stopped accumulating. In fact, the enzyme quickly became inducible and would continue to form in the recipient only if inducer were added. This confirmed the conclusion that the active, or dominant, form of the *I* gene was the *I⁺* allele. Its product (now called the *lac* repressor) was absent from the F⁻ (recipient) cell at first and seemed to accumulate soon after transfer of the *I⁺* gene from the Hfr cell. As it did so, it inhibited (or, as we say now, "repressed") the further formation of β-galactosidase, the product of the *lacZ⁺* gene that entered with the *I⁺* gene. (A complexity of this experiment was that the donor, in the second experiment, had to be killed with streptomycin so that it would not respond to inducer. The recipient was streptomycin resistant, and therefore only the F⁻ exconjugants could make the enzyme when inducer was added.)

The PaJaMo experiment, in short, came out exactly opposite to expectation. The information about enzyme induction could now be interpreted as follows. The enzyme in uninduced, wild-type (*I⁺Z⁺*) cells was under negative control (repressed) by a product of the *I⁺* gene. The action of the external inducer was to interfere with this negative control, allowing the *Z⁺* gene to express itself. The need for an internal inducer disappeared. Jacob could see that the induction of the prophage λ during zygotic induction could be understood in exactly the same way. The prophage, in lysogenic cells, was under repression by a product of one of its own genes. When the prophage entered a nonlysogenic F⁻ cell in a mating, it would thereby escape repression; phage development and lysis would ensue. The general picture drawn by Jacob and Monod (fig. 11.6) filled in extremely rapidly after the PaJaMo experiment. It is a seminal point in molecular biology and genetics because it initiated modern thinking about

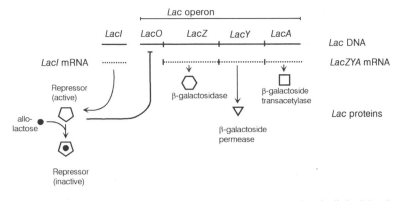

Fig. 11.6.   The *Lac* operon. The DNA of the operon and the closely linked *LacI* (repressor) gene is shown at the top, with the corresponding mRNAs beneath the DNA. The geometric shapes signify the individual proteins. The repressor, only when it is not associated with the inducer (allolactose), binds to the operator site (*LacO*) and prevents the transcription of the operon mRNA (flat-end line).

the control of gene activity. The first full picture of the regulatory system was detailed, although eventually it did not live up to its authors' hopes of representing a universally applicable mechanism of gene control.[27] But it served as a highly typological model with which many other investigators began studies of other systems, in both prokaryotes and eukaryotes.

The experiments that led to this model of gene regulation had another component, the operon model of the *lac* region.[28] The region comprised a contiguous set of three genes, as indicated above, but all three were controlled by the inducer, lactose, mediated by the *I* gene nearby. The gene cluster was another example of the clusters encoding enzymes of various pathways in *E. coli* and *S. typhimurium*. Jacob and Monod believed that if the product of the *I* gene, which they now referred to as a regulatory gene, was to control the activity of the cluster of *lac* genes, which they termed structural genes, there must be a target on the DNA for the *I* gene product. Indeed, they defined such a target with constitutive mutations of another sort, distinct in position from the nearby *I* mutants, and lying between the *I* gene and the *lac* structural gene cluster. They were within the operon proper and were called O^c mutants for "operator constitutives." According to the developing picture, these mutants had a defective target for the *I* gene product and therefore could not be repressed. This target acted as an on–off switch and was called the operator for that reason. A distinguishing attribute of O^c mutations and their normal allele (O^+) was their ability to control only those genes physically attached to them. In partial diploids made with altered F plasmids carrying a copy of the entire *lac* region, they showed that an O^c mutation could cause constitutive behavior only of the genes contiguous to it—that is, in a *cis* relation to the operon. This was the case even if an O^+ allele were present in the same cell. This confirmed that the *lacO* gene did not make a diffusible product: it was a *cis*-acting site on the DNA, and its behavior was denoted by the term "*cis*-dominant."

The operon model grew from the realization that the *lac* structural genes, *lacZ*, *lacY*, and *lacA*, were adjacent to one another and were coordinately regulated, behaving similarly in response to added inducer and to the *lacI* and *lacO* mutations. Jacob and Monod proposed that the cell expressed all the structural genes as a unit, hypothesizing that the gene cluster was first transcribed into a single, long RNA molecule, which was then translated into the three proteins encoded by the messenger. (Messenger RNA had just been detected biochemically as a defined species of RNA in 1960.)[29] Because of the unitary transcript (RNA), the repressor (*lacI* gene product) and operator (*lacO* site) can coordinately control the expression of all three structural genes. The model was generalized by reference to the many gene clusters known in *E. coli* and *S. typhimurium*. The possibility that it applied to genes encoding biosynthetic enzymes required only that the end products of biosynthetic pathways would enable the repressor of the cluster, rather than, as in the *lac* system, disable it.

The model of regulation emerged with many biochemical and genetic findings that put the "Central Dogma" at center stage (see appendix 2). This slogan, formulated by Crick, conveyed the idea that DNA not only specifies its own replication, but that it is transcribed into RNA, a molecular species that is then translated into the form of polypeptide chains.[30] One implication of the conception, which Gunther Stent (b. 1924) compared in importance to the theory of natural selection, is that the flow of information is unidirectional.[31] Thus information in proteins, the products of gene expression, cannot be compelled by the environment to return along the same path and change the genes from which they came—a clear bar to Lamarckian inheritance.

Phage λ: Two Seminal Contributions

Work on phage λ proceeded from the early studies of lysogeny well into the 1980s and became the most thoroughly studied of any replicating entity for a time during that period. Soon after the nature of the lysogenic state of *E. coli* K-12(λ) and the physical state of the prophage became known, Meselson and Jean Wiegle (1901–68) and, later, Meselson, explored the genetic phenomenon of recombination at the molecular level.[32] They infected *E. coli*, grown in ordinary medium, with genetically different λ phages. They labeled the DNA of both phages with heavy isotopes of nitrogen (^{15}N) and carbon (^{13}C). The isotopes allowed them to separate and thus to detect phages with DNA of different densities by density-gradient centrifugation. In these infections, they detected recombinant genotypes among phage particles whose DNA was made almost entirely of ^{15}N and ^{13}C. This proved that recombination could occur by breakage and reunion of parental nucleotide strands, without the synthesis of appreciable new DNA. (The caveat in this description refers to the allusion by Meselson, in the second article, that perhaps DNA synthesis, up to about 10%, might accompany the formation of recombinants.) Later analysis of phage λ recombination has revealed that several recombination systems, derived from both bacterial and phage genes, prevail in normal infections, and knowledge of the recombination process has now become quite complex and sophisticated. However, the findings demonstrated at the time that recombinant molecules could form from parental material and threw some doubt on the then popular copy-choice scheme. A thorough discussion by Stahl of the initial experiments and later studies, many from his own laboratory, exemplifies how the early, relatively simple models of a fundamental process become elaborated in later studies to the point that the phenomena are beyond the reach of the lay biologist.[33]

The second major area of study with phage λ concerns the control of transcription, both in connection with the integration and excision of the phage DNA into and out of the bacterial chromosome, and with the growth cycle of the phage during the lytic cycle. Overall, these events comprise highly programmed steps that assure either the attainment of the prophage state or the replication of phage DNA and the maturation of a crop of phage particles. In the end, the delicate choice of whether a phage λ DNA entering a cell would progress to lysogeny or lysis preoccupied many investigators, with the choice being seen as a "genetic switch." Many of the events are now fully understood in terms of the control of transcription (mRNA synthesis) of phage or prophage genes in a temporally prescribed manner. Thus, after one mRNA is translated into protein, the protein might then regulate the transcription of another gene. In some of these cases, the proteins promote transcription. In others, the proteins inhibit transcription. In still others a protein causes termination of transcription at a particular stopping point, but in the end, another protein blocks this termination function and allows transcription to continue unabated through much of the phage DNA as the infection becomes committed to the lytic cycle. At the time, the attraction of this biological system was, first, the simplicity of the organism, most of whose genes were by that time known, and second, the complexity of gene interaction leading from a naked segment of DNA to a burst of mature phage particles. The gene interaction here represents a developmental sequence, a process far more complicated than the formation of phage heads, or the self-assembly of ribosomes from preformed components, or the induction of the *lac* operon. Phage λ became, in short, a potential model for developmental processes in higher organisms—processes that had remained a complete mystery at the molecular level up to that time.

The phage λ community developed an extraordinarily detailed knowledge of the molecular events in the lysogenic and lytic phases of λ infections. The genetic techniques became increasingly sophisticated, and the mutations available for studying this or that process became increasingly numerous. To the outsider, the knowledge was baroque in the depth and subtlety of its application, and yet it imparted confidence, precisely because it was complicated, that this phage did indeed have important things to say about the biology of higher organisms.

By the mid-1960s, the question whether repressors were proteins or RNA was settled in favor of the former. In 1966 and 1967 the laboratories of Walter Gilbert (b. 1932) and Mark Ptashne (b. 1940) isolated the repressor proteins of the *lac* operon (the *lacI* gene product) and of phage λ (the *cI* gene product), respectively.[34] Ptashne continued his study of the λ system, focusing on the questions of how the repressor worked and of its role in the lysis–lysogeny decision as a phage infects a nonlysogenic cell. This "decision" is not entirely that of the phage because environmental conditions or the physiological state of the *E. coli* host can lead one way or another. Ptashne showed elegantly that the decision relies on the competitive interaction of two proteins.[35] One is the repressor, the other is the product of the λ *cro* gene, both of which begin to be made at the beginning of an infection. If the *cro* protein is made in sufficient abundance, it gets the upper hand and promotes the transcription of all genes of the phage DNA required to sustain a lytic infection. The repressor protein, if it is sufficiently abundant, has two actions. The first is to block the transcription of the *cro* gene. The second is to promote the transcription of the *cI* gene, which encodes the repressor protein itself. Therefore, if the repressor protein gets the upper hand, the phage DNA will, through the agency of certain other genes of the phage, recombine with the host chromosome and become a prophage. The prophage DNA makes only a very few proteins thereafter, one of them the repressor (and another that interferes with infections by *rII* mutants of phage T4). The lysis–lysogeny decision is therefore the resolution of a labile initial state. The induction of the prophage is another story, but involves, in the end, the destruction of the repressor protein in conditions that threaten the integrity of the host.

The entire DNA sequence of phage λ is known, and work on the details of λ biology continues. However, many of the workers on this phage have moved on to even more complex biological systems, and λ is no longer a mainstay of discovery in molecular biology. Nevertheless, phage λ investigations, some described above, have sharpened biologists' appreciation of the temporal integration of gene activity that might formally represent the events of embryogenesis. An understanding of the roles of genes during the development of higher animals was the grail sought by many geneticists since the time of Morgan. Phage λ provided them a taste of success.

12

Prokaryotes

Later Contributions

The outstanding accomplishment of the romantic period was therefore not so much the deep insights it provided but rather the introduction of previously unknown standards of experimental design, deductive logic, and data evaluation into the study of the genetics of bacteria and their viruses.

—G. S. Stent

Continuing Investigations

E. coli could with little exaggeration claim primacy in forging our picture of the fundamentals of life—the physical and chemical underpinnings of gene structure, replication, and expression. While many think this statement can be justified only by remaining blind to much else, the prokaryotes gave us the language and much of the knowledge of molecular biology relevant to all organisms to this day. The imitative character of much contemporaneous work on other organisms demonstrates the influence of studies with *E. coli*. Nevertheless, investigations underlying continued progress even with *E. coli* and its phages were made with other organisms. For instance, work with animal materials led to the appreciation of Chargaff's rule (the equivalence of adenine and thymine and the equivalence of guanine and cytosine in DNA from all sources), the demonstration of the importance of RNA in protein synthesis, and the first steps toward the synthesis of protein in the test tube.[1] I explore the role of biochemistry more fully in chapter 17.

After the genetic code had been defined fully (by 1963), some sighed that it was all over. This feeling signified only that the goals of many up to that time had been so narrowly conceived that their attainment left nothing to aim for. For those who thought this way, continued study of these phenomena would consign them to gleaning the stubble of a freshly harvested terrain—the cleanup job of ordinary, journeyman scientists. For others, the Central Dogma and the explication of the genetic code were simply open doors to the real, newly appreciated, unique biochemical mechanisms of protein and nucleic acid synthesis, the broad, formal outline of these life processes now behind them. For still others, observation and imagination would define problems previously undreamt of. Fortunately, tastes vary, and work continued in earnest, in many ways, and on many fronts. Let us now look, if only briefly, at the sweeping developments after 1963.[2]

DNA Synthesis

The study of DNA transactions with biochemical and genetic techniques gave surprisingly detailed information about the enzymology of replication, spearheaded by the work of Arthur Kornberg (b. 1918). He had won a Nobel prize in 1959 for his discovery and description of DNA polymerase I of E. coli and the invention of many techniques for studying replication and DNA chemistry. Soon thereafter, Cairns showed that DNA polymerase I was dispensable to the survival (if not the good health) of E. coli: a mutation that eliminated all but a tiny fraction of its activity still allowed growth. This observation provoked a search for the real DNA polymerase (DNA polymerase III), which was rewarded some years later with the discovery and characterization of the multicomponent replication complex postulated by others. This has led to the recognition of still other DNA polymerases, mainly concerned with the repair of DNA damage, of which DNA polymerase I is just one. This work led to later, strictly biochemical work, still in its development, of eukaryotic DNA polymerases in animal and yeast cells.[3]

Mutagenesis and Repair

The application of ultraviolet irradiation or mutagens to cells induces an array of DNA repair enzymes with overlapping functions. Mutations that cause sensitivity to mutagens, particularly ultraviolet light, frequently impair one or more of these repair activities. Such mutations presented many obstacles to analysis, and some time was needed to sort out the individual action of each one. Among the complications is that many repair systems are redundant; loss of one can be tolerated because of the activity of another. Another complication is that genes used in DNA repair have in many cases similar roles in DNA recombination, which also proceeds by several redundant pathways. Eventually, this area became clarified to the point that we have detailed knowledge of mechanisms by which DNA damage may be fatal, or can be reversed, or becomes immortalized as mutation. The work revealed an aspect of DNA transactions largely unforeseen in the 1950s; namely, that the stability of DNA as a bearer of information owes as much to the efficiency of repair as it does to the accuracy of replication.

Indeed, we now appreciate the role of mutation in higher organisms far better as a result of the groundwork laid in *E. coli*. With the studies of higher organisms that correlated the action of carcinogens and mutagens, biologists sought means to measure environmental factors for their mutagenic potential. Prior investigations of mutagenesis in *E. coli* and *S. typhimurium* provided Bruce Ames (b. 1928) with a number of bacterial mutants lacking one or more DNA repair mechanisms. These strains were used to measure the frequency of mutations caused by UV light, environmental pollutants, industrial chemicals, pharmaceuticals, food additives, components of cigarette smoke, and biochemicals found in natural foodstuffs. Ames introduced revertible, point mutations of known character (e.g., base substitutions of all kinds and frameshifts) into the repair-deficient strains. The mutations were for the most part auxotrophs as a result. Application of potential mutagens to a panel of such mutant strains led to reversion to prototrophy of particular auxotrophs. The nature of the DNA alteration of the revertant was identifiable as the reverse of the known, original, auxotrophic mutation. The "Ames test" has been steadily refined since its initial uses and became the origin of many studies of the etiology of cancer in experimental animals and humans.[4] Indeed, we now know that many cancers derive from mutations of repair systems. The findings have forced us to appreciate the major role of DNA repair in the remarkable stability of the genome— a property emphasized by Schrödinger and Muller, among others, long before.

Recombination Mechanisms

The study of DNA synthesis and recombination at the DNA level paralleled the study of mutagenesis and repair. This reflects the overlapping responsibilities of various genes and enzymes in all three areas. Many systems devoted to recombination (general, site specific, or nonspecific) emerged from combined genetic and biochemical analysis.[5] These have been incorporated into models of recombination in higher organisms, developed in the interim on the basis of strictly genetic evidence in the fungi (see chapters 15 and 16).

Regulation of Gene Expression

Regulatory mechanisms became the most intensely studied area in the molecular biology of *E. coli* and its phages for some time. Many biologists, both near and more remote from the front lines, adopted the initial Jacob–Monod picture of gene expression almost by default, much as they had adopted the first picture of DNA structure and its genetic implications. The uncritical acceptance of the operon model, particularly the aspect of negative control by a repressor, reflected the elegance and detail of its presentation. The acceptance was abetted by many confirmatory studies of the *lac* operon and other gene clusters at the molecular and biochemical levels and by the detailed enzymological studies of RNA polymerase in vitro with defined DNA sequences. The *E.coli lac* system provided us with many excellent tools for such studies. Phage λ soon entered the picture because the RNAs made after infection or during its emergence from the prophage state could be measured and eventually put into a coherent biological scheme.

Another factor in the acceptance of the Jacob–Monod model of regulation was its vigorous defense by the two investigators and the isolation of the *lac* and phage λ repressor proteins in 1966 and 1967 (see chapter 11). Some unnecessarily detailed predictions of the model failed to be proved, however, or to have somewhat less elegant real-world counterparts. The most serious lapse of the model was the exclusion of an alternative to negative control, one which Jacob and Monod might have suggested at the outset. The alternative was, simply, positive control, in which a regulatory protein might activate rather than repress gene expression by binding to an operator segment of DNA.[6] The discovery of such a mechanism, which Ellis Engelsberg (b. 1921) presented and defended for at least four years against severe opposition led by Monod, came out of his study of the arabinose-utilizing (*ara*) operon of *E. coli*.[7] Somewhat embarrassingly for Monod, evidence emerged soon thereafter that a positive control protein (in addition to the *lacI* repressor) was required for normal expression of the *lac* operon itself. Positive control in response to low-molecular weight internal metabolites and external inducers is now recognized as much more the rule than the exception in eukaryotes.

Ribosome Assembly

An interesting question arose in the 1960s regarding the origin of complex structures that incorporated a number of different polypeptides and perhaps other components. The question was whether the components assembled under the direction of a pre-existing template, or whether their own structures were sufficient to guide a process of self-assembly. Biochemists soon favored the latter mechanism following the discovery that certain denatured proteins could refold to their proper conformation and the demonstration of certain steps of the assembly of the phage T4 coat protein in vitro. A spectacular example of self-assembly was provided by the work of Masayasu Nomura (b. 1923), working on the *E. coli* ribosome. The ribosome, the particle on which polypeptide synthesis takes place, consists of two large, separable subunits, and contains 54 polypeptides and three rRNA molecules. Purification of all of the macromolecules eventually permitted an ordered assembly of the entire ribosome. This allowed Nomura to state (in relation to the small subunit) "the entire information for the correct assembly of the ribosomal particles is contained in the structures of their molecular components, and not in some other non-ribosomal factors."[8]

Transposable Elements

A mutational phenomenon, described in 1969 by James Shapiro (b. 1943) and Peter Starlinger (b. 1931), led to the detection of transposable elements. This class of DNA elements was entirely new to those working with *E. coli*, and credit properly goes to those who defined them at the molecular level.[9] However, the formal genetic behavior of such elements had been described in corn by Barbara McClintock (1902–1992) in the 1940s, work that earned her a Nobel prize once it was "legitimized" by the findings in *E. coli*.[10] In both organisms, segments of DNA appeared to move from one place to the other in the genome, often replicating in the process. Investigation of the phenomenon in *E. coli* quickly demonstrated a variety of such elements, each one a segment of DNA inserted into the bacterial chromosome. Their replication, independent of chromosomal replication, took place by an unusual means, by which the copy could be transferred into another piece of DNA. These "jumping genes" therefore have the means of increasing the number of copies in the genome and of causing mutations in the recipient DNA as they do so. Transposable elements can also depart their location in DNA, causing deletions of the flanking DNA in the process. *E. coli* thereby became a model for the behavior of transposable elements, a variety of which are found in all classes of higher organisms. Indeed, the origin of much of the large amount of DNA lacking any obvious function in higher organisms can be traced to the introduction and multiplication of these "selfish DNAs."

Restriction and Modification

Luria, in his studies of T4 and phage λ, had discovered a curious inability of progeny phage grown on one strain of *E. coli* to multiply efficiently in another strain. After growth in one host, only 0.1% or so of the progeny phage infecting a second host would form plaques. When this phenomenon was more fully analyzed (more as a curiosity for some time), it was found to be due to an enzyme in the second host that cleaved the infecting phage DNA into innocuous pieces before it could replicate. However, the few successful phage infections in the second host yielded phage that could infect the same host with high efficiency. Luria termed these phenomena "restriction" (of the initial phage crop by the new host) and "modification" (of the phage that survived).[11]

Werner Arber (b. 1929) described this system in detail in the 1960s, proposing the existence of bacterial enzymes that degraded phage DNA.[12] He also suggested that the DNA that escaped destruction became modified (and thus protected) by methylation of the nucleotide bases at the target sites. Several investigators then sought enzyme systems that would destroy or modify DNA. Hamilton Smith (b. 1931) later isolated a "restriction enzyme" from the bacterium *Hemophilus influenzae* that cleaved DNA at a specific nucleotide sequence.[13] Most bacteria seemed to have one or more restriction enzymes with a specific target. In each case, the restriction enzyme had a second enzyme activity or was accompanied by a second enzyme that could methylate and thus protect the bacterium's own DNA at these sequences. This explained the survival of the few lucky phage genomes in the restricting infections: their DNAs were methylated before they were cut. Vast numbers of other restriction enzymes have now been isolated and purified from other bacterial species. Because these restriction enzymes can cut DNA at specific 4–6 base restriction sites of known nucleotide sequence, they have become a major tool of science and industry in characterizing and manipulating DNA. The fragments of DNA resulting from restriction enzyme action can be joined together by other enzymes in any desired way. These techniques, together with those used to determine the exact nucleotide sequence of DNA or RNA of any length, opened a wide door to DNA cloning and genetic engineering (see appendix 3).

Plasmids

The first DNA plasmid to be discovered was the F (fertility) plasmid of *E. coli*. Many other plasmids became extremely well characterized in the years 1955–75. The large conjugative F plasmid was an example of a number of other large plasmids of medical importance. These appeared in the medical literature as *resistance transfer factors* (RTFs), plasmids that carried as part of their genomes up to 10 different antibiotic resistance genes. Because of their F-like ability to promote conjugal transfer from one cell to another, or even from one species to another, resistant strains of many bacteria arose and thrived in hospital environments. The discovery of these strains in nature has led to an appreciation of novel means of gene exchange in bacteria.

At the same time, small plasmids, common in natural populations of bacteria, were discovered and characterized. The smaller plasmids were extremely useful in later DNA cloning work. In contrast to the 100-kb F and other conjugative plasmids, which are maintained at about one copy per cell, the small plasmids (3–10 kb) may be represented more than 100 times per cell. Many of these have been artificially modified ("engineered") such that they contain little more than an antibiotic-resistance marker and an origin of replication that can be recognized by the *E. coli* replication machinery. These attributes allow antibiotic selection of cells they transform and enable their replication in *E. coli*, respectively. When techniques for the transformation of *E. coli* became widely used, the introduction of foreign DNA into *E. coli* as part of a plasmid became possible. To do so, investigators simply cut circular plasmids with a restriction enzyme and add linear segments of foreign DNA, followed by an enzyme (DNA ligase) that recircularizes the plasmid. Some plasmids incorporate the foreign DNA into the restored circle in the process. Transformation of *E. coli* with these chimeric plasmids and their replication thereafter is the essence of cloning genes. These techniques, together with DNA transformation, have revolutionized the study of molecular biology in all organisms (appendix 3).

Protein Secretion

The genetic study of the ability of cells to translocate proteins through membranes began in *E. coli* in the 1970s. The picture of this process had taken form during extensive biochemical studies of these phenomena in higher organisms in the previous decade. The genetic and molecular work in *E. coli* would lead quite directly to similar studies in yeast. Together, they have given us a detailed picture of the process of targeting proteins to organelles, the movement of proteins through membranes, the amino acid sequences that facilitate these processes, and the steps in the unfolding and refolding of proteins.

Behavior

It might seem fanciful to think of E. coli as a subject for the study of behavior. Yet studies of the reaction of E. coli to chemical attractants and repellents have given us one of the most elegant analyses of the connection between environmental stimuli and cell movement. The set of genes active in this process include cell-surface receptors that recognize specific kinds of chemicals and the so-called che (chemical) genes that transmit the signal from the receptors to the flagellar apparatus. There, other che genes activate the motion of flagella clockwise (leading to random tumbling of the cell) or counterclockwise (leading to coherent rotation of all flagella, effecting forward motion). The biochemical changes in the proteins encoded in the che genes include phosphorylation and methylation, and the system as a whole displays adaptation—that is, a tendency to detect concentration gradients at different absolute levels of a chemical stimulus. The work on E. coli chemotaxis represents a simple model for sensory systems and intracellular signaling pathways in multicellular organisms. (Inevitably, the relevance of the findings to higher forms was overemphasized in scientific papers, grant applications, and in the press at the time.)

The End of the Beginning

Monod said that what is true of E. coli is true of an elephant. This adage suggests the hubris of molecular biologists at the beginning of the 1960s. E. coli, with its phages, had been the model organism with which basic findings in metabolism, replication, recombination, mutation, regulation, and many other areas had been made. The coordination of so many strands of discovery—the connectivity of many levels of organization—in one organism painted much of the landscape of molecular genetics and molecular biology. But starting in the 1980s, the study of the enteric bacteria started to become a more esoteric and less populated field. While there was still much to learn, bacterial molecular biologists could no longer proclaim E. coli as a model for many remaining, quasi-universal questions. In fact, one might say, no organism could ever be a model for all forms of life again, as E. coli and its phages were for a brief time.

In the 1980s, those wanting to make findings on the scale of relevance I have described had to do one of two things: change organisms or temper their ambitions. The latter was, of course, unnecessary if they did the former. For those infected with the enthusiasm of the 1960s, however, the large world of eukaryotes lay ahead. Investigators of the then-primitive molecular biology of eukaryotes during the long hegemony of prokaryotes needed help, and the National Institutes of Health had money to spend on science that might benefit medicine more directly than would further studies of E. coli and phage λ.

In many accounts of the early days of molecular biology, including this one, authors ignore developments in other fields of biology and with other organisms. The field of microbiology, begun in Leeuwenhoek's time, had a rich history and would remain exceedingly broad during the latter half of the 20th century. Microbiological investigation and the use of bacteria were staples of medicine, agriculture, academic institutions, and industry. Electron microscopy had much to reveal. Biochemists worked with a large number of species; taxonomists struggled to maintain order in the kingdom; and ecologists explored the enormous range of bacterial habitats for a variety of reasons.

But even molecular biology adopted many species beyond *E. coli*, *Salmonella*, and the phages discussed earlier. As I have noted previously, bacterial molecular biologists departed strongly from Delbrück's admonition to stay with a small, well-chosen group of related organisms. Starting with *S. typhimurium*, other bacteria were brought into play by microbial geneticists. These included *Bacillus*, *Caulobacter*, *Agrobacterium*, *Azotobacter*, *Haemophilus*, *Klebsiella*, *Myxobacterium*, and *Rhizobium*, all of which had attributes of great biological interest, and for all of which genetic techniques were developed. Among the phage, a huge variety came to light early on, many of which contributed strongly to molecular genetics. These included the single-stranded DNA phage ΦX174, the RNA viruses f2 and Qβ, and some RNA viruses with a DNA intermediate in their life cycle. The many phages opened a window on their previously unsuspected variety, their mechanisms of growth, the relation to their hosts, and their evolution. Thus the habits of molecular biology soon pervaded the field of prokaryotic microbiology, and, in return, molecular biologists not only became aware of the diversity of microbial life but began to embrace it. The discovery (or better, the definition) of the new domain Archaea among the prokaryotes gave new life to the study of the origins of life (see chapter 18).

We can now appreciate the change of focus in this field. The bacterial molecular geneticists that remained with prokaryotes would initiate important new areas of investigation, and these would influence the study of prokaryotes more than investigations of higher organisms. In the process, these later studies used, appropriately enough, *E. coli* and its phages as models or points of reference for the studies of the domains Eubacteria and Archaea. As the studies have progressed, these biological models begin to be less and less useful, often confining, in the newer lines of investigation. As I argue in chapter 18, the academic preoccupation with model organisms has now given way to an appreciation of ecology, evolution, and comparative biology of microbes, sustained more recently by the sequencing of a number of microbial genomes, 30 of them in the year 2000, and more than 80 in 2002.[14] This information is now increasingly used to explore microbial relationships, gene function, and gene interaction. While this has solidified the molecular-genetic language in which modern studies of prokaryotes are carried out, genomic work has caused *E. coli* as an organism to occupy a smaller scientific niche.

This is not to say that the study of *E. coli* has lost its internal vigor. Indeed, due to the enormous amount that was already known about it, *E. coli* continues to be used, but for studies of new kinds. In a review of current *E. coli* studies, Schaechter et al. discuss the contents of the *E. coli* genome, now fully sequenced, the multifunctional character of proteins, redundancy of gene function, coordinated regulation of large groups of genes, internal cell structure in relation to function, sensing of the environment, evolution of the global *E. coli* population, and the ecological adaptations of *E. coli* as it cycles between soil and gut.[15] Notice that these are integrative, rather than reductionist, themes— an embrace rather than a snub of complexity. In the authors' words, "within the limits of our comprehension, we may be in sight of the holy grail of molecular biology, i.e., to understand the integrated genetic and metabolic patterns of an organism. That organism will probably be *E. coli.*"[16] Whether the understanding achieved here, if it ever is, will precede similar understanding of simple eukaryotic cells, and whether it will apply other than broadly to higher cell types, we cannot yet know. But *E. coli* remains a compelling model for living things in general to some who work with it because, as ever, it shares much with the rest of the living world.

We may draw from the history of studies with the prokaryotic models an old lesson, to be repeated hereafter in the case of other organisms: after its contributions to the study of life, a model organism opens the study of other members of its kingdom or taxon, and may in some cases be buried by its relatives.

Molecular Biologists Move On

Many of the founders of the field of molecular biology departed from the prokaryotes once the main lines of the structure, activity, and regulation of genes were laid out. Many moved on to the study of animal cells, animal viruses, and the development of multicellular organisms. They did so with the intellectual standards and visions of efficient experimentation characteristic of work with *E. coli,* phage λ, and the T phages. While the animal systems were a good deal messier, and in some cases intractable, the molecular analysis of animal and higher plant systems was greatly accelerated by the entry of prokaryotic biologists and their students. These workers had dealt with the simplest organisms they could find. They developed elegant, sophisticated models of many molecular processes and integrated them into a picture of the properties of the living cell. With confident minds, prepared with detailed paradigms and expectations, they could now approach the molecular biology of eukaryotes and their viruses in the areas of cell biology, physiology, development, and gene expression, even in organisms not suitable for genetic work. Benzer entered (even invented) the field of insect behavioral genetics in *Drosophila*. Stent attempted to develop the neurobiology of the leech. Brenner started the domestication and analysis of the development of the nematode, *Caenorhabditis elegans*, from scratch with spectacular results.[17] George Streisinger (1927– 84), a key student of phage T4, began similar developmental studies of the transparent zebrafish. Ephrussi and Pontecorvo both developed parasexual techniques in cultured, differentiated animal cells. In chapter 15, I introduce a younger generation, trained in prokaryotic molecular genetics, that would contribute to the systematic development of yeast molecular biology in the 1970s and thereafter.[18] Jacob illustrates a feeling of such transitions as he describes his own move to the study of mice:

> At the end of the 1960s, it was clear that the center of gravity in biology was shifting. Although the study of bacteria and viruses still had much to teach us, it was slipping to second place. If we didn't want to stand around rehashing the same questions, we needed the courage to abandon old lines of research and old models, to turn to new problems and study them with more suitable organisms.
>
> The word 'courage' is not too strong. The daily interaction over years with a living organism, however humble, entails a certain familiarity. You could almost say that you acquire a certain tenderness for it. After fifteen years of working with a particular colon bacillus, I had accumulated hundreds of mutants. In each of these mutants, one or another of the cellular functions, many of them indispensable to the life and reproduction of the bacteria, had been altered. To abandon this work and all that it offered; to renounce the kind of intimacy that comes with the knowledge of little unwritten quirks, the folklore that surrounds the work on any one organism; to start again from zero with another, unknown organism whose idiosyncrasies I would have to discover—all this was a considerable sacrifice. It was a little like leaving a loved one. But, at the same time, the new project was an exciting one. It would mean entering an unknown world, beginning a new life, becoming young again. . . . [ellipsis in the original][19]

Historians and philosophers of science critical of the reductionist themes of *E. coli* and phage molecular biology often ignore the concurrent theme of cellular integration. This theme appeared quite early with the overall balance of accounts of central metabolism and macromolecular synthesis, culminating in the discovery of feedback inhibition and the regulation of gene activity. These activities were never sidelined; they remained central in the field. Even when remarked upon by historians, the findings and the conceptual descriptions are often called an "engineering" view of life—a view that fails to appreciate both the holistic attributes of living things and their variety. Perhaps one difference between the researchers and some of those who denigrate scientists' analytical ambitions is that the first seek discovery and detailed understanding, and the second prefer wonder—wonder threatened by further, more detailed investigation.

Yet Darwin believed that understanding the natural world should increase our awe at the blind process and fortuitous products of natural selection. Monod, in a remark of striking import, said, "A *totally* blind process can by definition lead to anything; it can even lead to vision itself."[20] Although the scientist's duty is to render mysteries uninteresting, new mysteries thereby and inevitably appear.

13

Cytoplasmic Inheritance

The Ciliates

Treasure your exceptions.
—W. Bateson

The Cytoplasm in Development

I return now to a major preoccupation of biology in the early part of the 20th century. This is the role of the cytoplasm, particularly in development. In chapter 2, we saw that most sophisticated biologists at one time considered inheritance, embryogenesis, and evolution parts of the same field, not easily dissociated. But while Morgan was an embryologist first and last, he was a pure geneticist in between. He consciously ignored embryology and evolution as he embarked on his experimental work on *Drosophila*, determined to study with few preconceptions the transmission and assortment of characters or their determinants from parents to offspring. In fact, for his entire professional life, he considered the gene a theoretical entity that stood in for whatever lay at the basis of the phenotype and for whatever was passed from generation to generation. When he turned again to embryonic development in the 1930s, he was no closer to uniting this field with genetics than he was before. After Ephrussi had read Morgan's book, *Embryology and Genetics* (1934), Ephrussi told Morgan that he seemed not to have integrated the two fields adequately. Morgan replied, "You think the title is misleading! What is the title?" "*Embryology and Genetics*," Ephrussi said. "Well, is not there some embryology and some genetics?"[1]

This chapter, coming as it does after the review of gene regulation, is subject to the bias of hindsight. We must therefore force ourselves to imagine the intellectual issues regarding the role of the cytoplasm in development and heredity as they were argued at the time and to withhold immediate judgment of their legitimacy. The *Zeitgeist* described in the earliest part of this chapter was swept away by a gale of later discoveries and therefore forgotten even by most biologists. A thorough and readable account of this matter, in the context of cytoplasmic inheritance, is J. Sapp's 1987 book, *Beyond the Gene: Cytoplasmic Inheritance and the Struggle for Authority in Genetics*, to which I owe much in the following account.

Embryonic development presented contradictions to biologists after the rediscovery of Mendel's work. In the view of embryologists in the first decade of the 1900s, the ground plan of development, the properties of the fertilized egg that destined it after cell division and differentiation to become a characteristic member of its species, lay in the cytoplasm. The view arose from the conviction, appropriate even now, that the chromosomes, and by inference the nuclear genes, did not seem to differ as cells of the embryo began to differentiate and the embryo took shape. Yet the differences among the various cell types seemed to be an inherited property because more cells of particular kinds multiplied in lineages leading to bone, blood, muscle, nerves, digestive organs, and skin. Embryologists had to contend in addition with the observation that many developmental mutations studied by geneticists lay in the nucleus. The paradox resulted in a consensus among many embryologists. They continued to believe that the cytoplasm of the egg established the basic body plan, attributes of polarity, left–right bilaterality, the sequestration of germ cells, the three embryonic germ layers (ectoderm, mesoderm, and endoderm), and so forth. The discrete mutations of the geneticists, they believed, were much less significant, affecting only the last stages of development.

Morgan championed the more rigorous nucleocentric view that almost all inherited attributes, whether identified by mutations or not, reflected the action of the nuclear genes. He thus demoted the cytoplasm to the role of a medium in which genetic instructions were carried out. He was too pragmatic, too careful, to spell out how the nucleus could exert its effects. Instead, his argument rested on the rarity of cases in which the hereditary determinant of any variant phenotype was inherited through the cytoplasm of the egg. He did acknowledge the maternal inheritance of plastids, a problem explored in detail in Germany, and about which controversial opinions prevailed for a time. But many purported cases of cytoplasmic inheritance seemed either to be vague in character or simply an extension of the mother's genotype, a delayed effect that would disappear in the next generation. A classic example of this is the difference in shell coiling in snails: some coil to the right, some to the left. The direction of coiling of a snail shell is established by the genotype of the mother, who imposes it by way of the disposition of cleavage planes in the early divisions of the fertilized egg. Morgan would also agree that various symbionts might be transmitted through the egg cytoplasm and rarely, if ever, through the cytoplasm-poor pollen or sperm.

Nevertheless, a large body of contrary opinion persisted, mainly in Europe, with a number of adherents in the United States. According to this contrary view, hereditary alterations of fundamental determinants of embryogenesis in the cytoplasm were rarely seen because derangement of such determinants would not be viable or, if the determinants were plentiful, their redundancy would obscure the effect of mutating any one of them. Correns had described variegation of four-o'clocks, a variegation that reflected the segregation of different kinds of plastids at cell division during the growth of shoots and leaves. He also showed that the determinants (later known to be the plastids) were inherited through the ovule. The findings also suggested that the importance of plastids might make their complete absence lethal and that their numbers in each cell might make many mutations undetectable. Similar studies showed that the expression of some of these maternally inherited plastid determinants were dependent on nuclear genes, a point that reemerges in studies of *Paramecium*.

In another, later study, carried out in the plant *Epilobium* over many years by Peter Michaelis (1900–75), F_1 hybrids of two species (*E. luteum* and *E. hirsutum*) differed considerably depending on which one served as the female parent. The F_1 derived from the *E. luteum* female was healthy and fertile but that from the *E. hirsutum* female was stunted and male-sterile. This suggested a non-Mendelian phenomenon. Michaelis proceeded to perform successive backcrosses starting with ovules of the second, stunted F_1, crossed with the parental *E. luteum* pollen. The progeny of each cross were backcrossed, pollinated again by *E. luteum*. By this means, the cytoplasm of the stunted hybrid, derived from *E. hirsutum*, would be maintained, while the nuclear genes were gradually replaced with those of *E. luteum*. Even after 25 backcrosses, the stunted phenotype persisted. Michaelis could with good reason point to the relative autonomy of the cytoplasm (whose inherited determinants he called the "Plasmon"). He argued that this was not simply a matter of plastid inheritance, since so many characteristics besides plastids were altered.[2] No particulate basis for the differences was adduced; the word Plasmon was used to refer to a collection of particles, a structure, or a state of the egg cytoplasm. The theory was holistic, and the experiments had a different character from those of the fast-moving Morgan school, which used discrete mutations in pedigrees lasting only two or three generations. Despite the thoroughness of the backcrossing experiments, few American geneticists before World War II believed that the case for cytoplasmic inheritance as a general phenomenon could be sustained. In 1939, Sturtevant and Beadle, Morganists to the core, would ask in a general textbook in which they discussed a number of other cases of maternal inheritance:

> Is the specificity of the cytoplasm to be referred back to the specificity of the genes, or is it a permanent property that reproduces itself regardless of the genes present? That is, do the genes modify the nature of the cytoplasm, and, given time, mould it into a specificity determined by their own properties? At present the most probable answer to these questions is: Only the chloroplasts are known to have permanent properties independent of the genes in the chromosomes.[3]

The upshot of the controversy was an estrangement of the Morgan school and many European biologists (particularly in France), trained in a biology that failed to dissociate embryology from genetics and therefore allowed the cytoplasm a role in heredity. The latter group displayed neo-Lamarckian tendencies because they thought the cytoplasm (or Plasmon) might be susceptible to persistent, hereditary changes induced by the environment. Indeed, certain reports of such changes emerged from studies of ciliates, and the ciliates thereby became a focus of many of the arguments about cytoplasmic inheritance after World War II.

Jennings and Jollos

The field of protozoology began with the invention of the microscope by Leeuwenhoek. The protozoa include amoebae, trypanosomes, a host of flagellates, the ciliates, and sporozoans and thus include the agents of sleeping sickness, Chaga's disease, malaria, and other human maladies. The infectious protozoa had been well studied by pathologists, medical microbiologists, and immunologists for some years since the beginning of the 20th century. Amoebae are found in the wild in many aqueous environments, and they became a metaphor for an undifferentiated state of life since their discovery. The ciliates can also be found in ponds at any place on the globe and have fascinated both schoolchildren and biologists by their motility, complex morphology, and ubiquity.

Early 20th-century studies of *Paramecium* and other ciliates by American protozoologists such as Herbert S. Jennings (1868–1947) and European biologists, among them André Lwoff, revealed the general cytology, life cycles, and behavior of these organisms. Jennings's interest in genetics led him to establish pure lines of *Paramecium* by continuous propagation, both asexually and sexually. His work focused on the interaction of organism and environment in the elaboration of the phenotype, and, although not disputing Mendelism, he inevitably considered the possibility of the inheritance of acquired characteristics, particularly if such characteristics were inherited through the cytoplasm. His work could not proceed with the protocols of Morgan because controlled matings between different strains of *Paramecium* failed. More important, Jennings's interests concerned holistic aspects of the cell, interests that led him to the study of other organisms to detect Lamarckian phenomena. During this time, Tracy Sonneborn (fig. 13.1) became his graduate student. Jennings and Sonneborn each described several cases of the persistence of environmentally induced alterations in other organisms. One, described by Sonneborn, was the induction of "double monsters" of flatworms by lead ions, a morphology maintained during further propagation of animals by asexual growth. Jennings described an interesting inheritance of shell pattern in a shelled amoeba, in which the pattern of a new shell, built of bits of sand on a template of the previous shell, would reproduce damages to the latter inflicted by the investigator.[4]

During the same years, investigations by Victor Jollos (1887–1941) in Germany, initiated at about the time of World War I and continuing into the 1930s, showed that environmental insults to *Paramecium*, especially temperature

Fig. 13.1.   Tracy Sonneborn. Photo courtesy of John R. Preer, Jr., Indiana University.

shocks, led to changes in behavior and the rate of growth persisting over hundreds of generations. These changes, though they lasted many asexual generations, eventually faded out. Significantly, they disappeared immediately upon sexual conjugation. These changes, called *Dauermodifikationen*, attracted considerable attention, especially in Europe, but also in the laboratory of Jennings, focused on just such phenomena.[5] At the time, there appeared to be a number of similar cases in bacteria and other microbes. Some of the changes, such as the acquisition of resistance to antibodies and antimicrobial agents, seemed to be permanent, although the actual basis of such changes was poorly understood at the time, as discussed in chapter 7.

Indeed, in the late 1930s the distinction between mutation and selection by antibacterial agents was poorly appreciated, to say nothing of the understanding of genes and sexuality in bacteria. Even the differences between protozoa and bacteria, both primitive forms of life, had not been formalized. The experiments of Jollos in *Paramecium* and similar studies using other eukaryotic organisms gave promise of a tool to study the role of the cytoplasm in heredity. At the same time, such work might yield information about the role of the cytoplasm in embryogenesis. In this context, *Paramecium* was to become a model organism with relevance to the biology of higher animals, followed by *Tetrahymena*, another ciliate more suitable for modern work.

Finally, another research program would contribute to the development of interest in the role of the cytoplasm in heredity. In an extensive study, Lwóff and his mentor, Edouard Chatton (1883–1947), maintained that the regular rows of cilia that characterize the ciliates was a hereditary property of the cytoplasm.[6] Their observations on a group of ciliates called the apostomes, which live in association with marine crustaceans, lent themselves to speculations about the source of information necessary to maintain cell morphology during cell divisions. *Paramecium*, with a similar ciliary apparatus, was thus poised to become an organism amenable to studies of the inheritance of cell structure and thus a possible model of early embryonic development in higher animals.

The *Paramecium* Life Cycle

When serious genetic studies of ciliates began, the biological structure of the group was not appreciated. The form once called "*Paramecium aurelia*" is actually a large group of species separated from one another by fertility barriers. Initially, these species were called "varieties" (1, 2, 3, . . . etc.), then "syngens," and finally species. The varieties of the *Paramecium aurelia* group were then given systematic names, *P. primaurelia, P. biaurelia, P. triaurelia, P. tetraurelia*, and so forth. The same problem arose in the genetics of "*Tetrahymena pyriformis*" when studies of this group began. In both cases, species differed in details of their life cycles, but shared enough biological traits that useful correspondences and differences for comparative analysis could be appreciated at the outset.[7] In this discussion I use, where appropriate, the current names of the species used.

The structure and life cycle of *P. primaurelia*, the first to be studied systematically, are both complex. The organism, like all ciliates, is nominally unicellular, but so complex morphologically that some protozoologists call it a highly differentiated, "acellular" organism. The cell measures up to 1000 times the length of a bacterium and can be seen under low magnification. It has a slipper shape (fig. 13.2), with a groove serving as a mouth and gullet, an anal

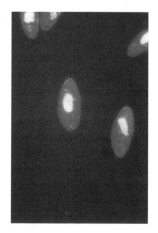

Fig. 13.2.   *Paramecium tetraurelia*, stained with a fluorescent stain that binds to DNA. The conspicuous macronucleus in the center of the cell is flanked by two micronuclei. Micrograph courtesy of John R. Preer, Jr., Indiana University.

pore, and several contractile vacuoles. The entire surface of the cell is covered with longitudinal rows of cilia that beat in a coordinated fashion to propel the cell through aqueous environments. The cell is excitable and moves easily, changing direction appropriately in response to noxious or attractive stimuli. Lwoff had developed much comparative information about the apostomatous ciliates in the 1930s, focusing on the cell cortex and the regular rows of cilia, which he called "kineties." The cortex is a cylinder of firm, organized cytoplasm, patterned by the rows of ciliary units.[8] The cilia are underlain by, and are generated from, basal bodies that appear to divide in an asymmetric fashion as the cell divides. This pattern is autonomous, to the point that naturally occurring or microsurgical reversals of patches of cortex are reproduced almost indefinitely in cell division in the new orientation.

Paramecia, until the late 1970s, could be cultured practically only in media with bacteria as the food source.[9] This unfortunate limitation prevented them, despite their being among the first microorganisms to be domesticated by geneticists (in 1937), from contributing to the revolution in biochemical genetics and many other research programs that required a synthetic medium. For this reason, the ciliate *Tetrahymena* would supplant *Paramecium* as a model ciliate when simpler media for *Tetrahymena* were devised in 1951 and after its mating types were discovered in 1953.[10]

P. primaurelia has two micronuclei and one large macronucleus (fig. 13.2). The cell divides by fission, with the micronuclei dividing mitotically, one daughter of each going to each daughter cell. The macronucleus, derived from micronuclei after conjugation, has much more DNA than the micronuclei and divides amitotically, with one portion going to each daughter cell. The term *amitotic* designates a nuclear division in which chromosomes and a mitotic spindle do not become visible prior to division; the dividing nucleus apportions its DNA in a less disciplined way to the daughter nuclei. (The significance of this process became clear only later.) Before the cell divides, the ciliary rows lengthen by interpolation of additional basal bodies. This anticipates the transverse division of the cell and assures maintenance of the original pattern in each daughter cell.

Mating takes place by conjugation (fig. 13.3), and here the difficulties in considering *P. primaurelia* (and, later, *P. tetraurelia*) as a model organism become most apparent. The micronuclei are each diploid (as Sonneborn demonstrated in his seminal studies). Two cells join side-by-side at the outset. Both micronuclei of each cell undergo meiosis, thereby forming eight haploid nuclei in each partner. Seven of these degenerate, as does the large macronucleus. At this point, the remaining haploid nucleus enters a paroral cone that juts into the partner. The nucleus divides once, and one of the daughter nuclei from each partner moves into the other partner. This reciprocal exchange of nuclei leaves each exconjugant with one resident and one donated nucleus, and thus the same genetic constitution. The two nuclei fuse to form a diploid nucleus, and this then divides twice. Two of the resulting four nuclei become micronuclei, and the other two become individual macronuclei. The two exconjugants then undergo fission to form two cells each. As each exconjugant does so, the two

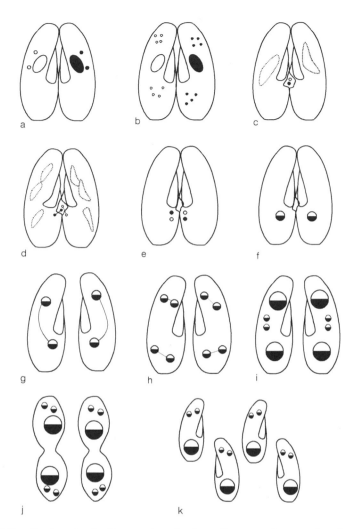

Fig. 13.3. Conjugation in *Paramecium*. In these diagrams, the macro- and micronuclei of the two parents (*a*) are distinguished by filled or open circles. Upon contact (*b*), both micronuclei of the two parents undergo meiosis. Only one haploid product survives in each cell. This nucleus migrates to the paroral cone (*c*). Meanwhile, the macronucleus begins to degenerate (*d*). The haploid nuclei divide once, and the cells exchange daughter nuclei (*e*). The donated and resident nuclei fuse to form a diploid nucleus (heterozygous if the parents differ genetically) in each cell (*f*). The diploid nuclei divide twice (*g*, *h*), with one daughter nucleus of the second division enlarging by endoreduplication of DNA to form macronuclei (*i*). During the subsequent cell division, both micronuclei divide, but the two macronuclei do not, thereby restoring the nuclear condition of the original parents (*j*, *k*). The descendants of the four cells at stage *k* will be *caryonides*. Based on diagrams of Beale (1954).

micronuclei divide, but the two macronuclei do not. Thus as fission is completed, each daughter cell contains two micronuclei and one macronucleus. The latter grows by a process of DNA duplication (endoreduplication), some details of which make it far more complex than suspected in the early years.

A second process, autogamy (originally called "endomixis"), takes place without conjugation. Most of the events of conjugation take place here in a single cell, with meiotic divisions of the two micronuclei yielding eight haploid nuclei. During this time, the macronucleus degenerates and so do seven of the haploid nuclei. The remaining nucleus divides once, and the products fuse, this time forming a homozygous diploid nucleus. This in turn divides twice to form the new micro- and macronuclei. Most of these details were described by Sonneborn and his followers, and controlled matings between different strains could be accomplished only after Jennings's time.

Sonneborn and the Mating-Type System

Sonneborn, continuing in Jennings's laboratory as a postdoctoral fellow in 1930, took up the *Paramecium* mating problem in earnest. Controlled matings were essential if work on this organism was to be credible to the Mendel-Morgan school. If the phenomena suggesting cytoplasmic inheritance were to be believed, Mendelian inheritance had to be demonstrated for other characters, proving that the phenomena have an exceptional genetic basis. In 1937 Sonneborn was finally able to report the conditions necessary for mating and at the same time to describe a system of mating types, two for most species.[11] (Originally called mating types I and II, the two mating types of all species are now designated odd [O] and even [E].) Sonneborn also showed that conjugation may permit cytoplasmic exchange between conjugating pairs. Conjugating pairs normally remain attached for a short time, with little or no exchange of cytoplasm. Occasionally they remain attached for a long time, and extensive cytoplasmic exchange may take place. This gave Sonneborn a useful tool to manipulate the cytoplasmic transactions separately from the nuclear ones, particularly because in certain situations he could prevent nuclear exchange in the process.

Even in the analysis of mating types of *P. primaurelia*, several peculiarities of *Paramecium* genetics asserted themselves. The two mating types are not allelic variants of one another, and thus even a homozygous animal may express either mating type. The original mating-type phenotype was maintained during continuous asexual growth uninterrupted by conjugation or autogamy. However, in *P. primaurelia*, the mating types of parents did not determine the mating types of the exconjugants. The two diploid exconjugants of a mating form a total of four first-division fission products (fig. 13.3). Sonneborn showed in this work that the descendants of the four, despite their identical genotypes, might yield clones of mating type O, mating type E, or a mixture of O and E. The rule governing this phenomenon was that the mating types expressed by the four macronuclei of the exconjugant pairs were determined, independently and randomly, when the macronuclei form. These findings led to the designation of the asexual clones derived from the four daughters of the two exconjugants as caryonides (fig. 13.3). The term *caryonide* denotes a clone in which all macronuclei are derived by the amitotic process characteristic of asexual fission.

When the caryonides derived from an O × E mating of *P. primaurelia* become sexually mature (in about a week), they become capable of mating with cells of the other mating type, or they may undergo autogamy. In the latter process, the same randomness of mating-type determination, at the point of formation of macronuclei, prevails. Note that in the case of autogamy, the genotypes of the first-division products are not only identical, but homozygous as well, yet they may yield both mating types and will do so during successive autogamies thereafter. This demonstrates that the mating types do not segregate as alleles; any homozygous animal has the information for both. But the maintenance of the different phenotypes despite the identical genotypes made the system of considerable interest to Sonneborn, even if it provoked skepticism on the part of many orthodox geneticists. The lack of any demonstrated Mendelian variant in *Paramecium* at the time, in fact, reinforced this skepticism.

Curiously, the first Mendelian factor was discovered only in 1947 with the demonstration that the ability of cells to express mating type E is controlled by a gene (mt^+, the wild-type allele). The mutant allele of this gene (mt^O) restricts all descendants of the caryonide to mating type O.[12] Whether the *mt* gene controls the interconversion of O and E or whether it controls the expression of alternative, nonallelic genes is not known with certainty; relevant literature on several *Paramecium* species is cited in a recent review.[13]

The mysteries of the macronucleus underlie most investigations because the macronucleus determines the phenotype of the organism. The nature of the duplication and processing of macronuclear DNA as it formed remained unknown for many years, and even now it is obscure in many details. Moreover, the macronucleus was not always derived from the first diploid nuclei formed after meiosis. Instead, macronuclei can regenerate from fragments of the old one as the latter degenerates. Sonneborn's expermental skill played a great part in the knowledge of these peculiarities, and he used them opportunistically to render, for example, the macronuclear genotype different from the micronuclear genotype. Sonneborn and his group wrestled with increasingly sophisticated models of how the phenotype was determined, and the obscurity of the arguments to many nonprotozoological geneticists rendered the case for cytoplasmic inheritance, plasmagenes, and other models tentative in their minds at best. Nevertheless, Sonneborn was an extraordinarily capable scientist, and his findings and arguments were respected by those who mattered.

The discovery of mating types set the scene for developing the genetics of an animal-like eukaryotic unicell, although the genetics of filamentous fungi was already well under way. Characteristic of Sonneborn's confidence is the final sentence of his 1937 paper: "It may perhaps be said that with the present work the genetics of *Paramecium* enters the quantitative and predictable stage, with tools and methods of analysis which should lead rapidly into a systematic, coherent body of knowledge in close touch with the rest of genetic science."[14]

The inheritance of mating type became a major theme of *Paramecium* studies, and the variants of the system in other *Paramecium* species preoccupy investigators to this day. The obscure genetics of mating type shares curious attributes of other studies for which *Paramecium* and other ciliates became known. These include the gradual differentiation over many generations of certain phenotypes controlled by macronuclei (unlike the immediate determination of mating type); the mutual exclusion of certain alternative physiological states; the inheritance of cortical patterns of cilia; and the existence of bacterial or algal symbionts. The elaborate oddities of *Paramecium* made it and other ciliates one of the most exciting genetic systems explored up to that time. But these very oddities soon rendered *Paramecium* and other ciliates out of bounds as models for anything but themselves.

Kappa

The findings that propelled paramecia to the status of a model organism were Sonneborn's discovery of "killer" strains and the inheritance of surface antigens.[15] Sonneborn observed that when certain stocks of *P. tetraurelia* were mixed, one (called sensitive) would die, but the killers would survive. Cell contact was not required for this to happen in most species in which killer strains were found. Sonneborn's group named the presumed toxic material(s) "paramecin," which later studies showed was particulate and sedimentable by centrifugation. A single particle of this material could kill a single animal, as statistical studies demonstrated, and it could be removed from suspension by adsorption on sensitive animals. This unusual phenomenon was found in several species of the *P. aurelia* group, but not all, and the behavior of animals as they died varied considerably from one species to another and from one killer factor to another. In some cases, animals simply became immobilized and died, others underwent spinning or reverse spinning motion, and others vacuolated and died soon afterward. Chemical investigations in 1947 showed that paramecin was sensitive to proteases, suggesting that protein was an important component of the agent.

The killer trait was heritable in sexual or asexual growth, even though some varieties lost the trait if the division rate of well-fed animals was rapid. The loss of the trait brought with it a sensitivity to paramecin, as though the two traits went hand in hand. The genetic determination of these traits was most unusual. It happens that matings of killers and sensitives can be made by environmental manipulations that mute the killing reaction, and in fact in certain varieties, killing does not take place during conjugation, provided that the process begins immediately and that the animals are separated soon thereafter. Sonneborn established in the early 1940s that normal matings between killers and sensitives led to the maintenance of the two different phenotypes in the two animals after they separated. Therefore, the phenotypes did not follow the pattern corresponding to the nuclear exchange and the formation of new macronuclei. However, if the mating was allowed to proceed for more than 30 minutes, cytoplasmic exchange often took place as well, and in many cases this led to a transformation of the sensitive conjugant into a killer exconjugant. This indicated that a cytoplasmic entity determined the killer trait, which Sonneborn named "kappa" for the *P. tetraurelia* factor. (This established the convention of using Greek letters to designate cytoplasmic entities.) And, bearing out the correlation noted above, the new killer had lost its sensitivity to the paramecin of this variety.

As noted above, rapid division of the killer stocks of other varieties often led to the loss of kappa. This could be interpreted as a loss of an entity that could replicate, but whose division could be outpaced by that of the host itself. Sonneborn showed this also happened in *P. tetraurelia* with certain manipulations of temperature. Kappa thus became a spectacular case of cytoplasmic inheritance, an example of what Sonneborn and others called a "plasmagene." This example was promoted by Sonneborn for some time, the kappa story being embellished by the finding that a simple Mendelian gene of *Paramecium* (*K* and its allele *k*) determined whether kappa would be maintained in the host.[16] This example of nucleo-cytoplasmic dependence echoed other such cases, such as plastid inheritance in plants.

The kappa story began to lose its luster in the late 1940s and early 1950s. Radiation inactivation studies by John Preer (b. 1918) demonstrated that the target size of kappa was a good deal larger than that of paramecin.[17] Therefore, the temptation to think of kappa and paramecin as the same thing (similar to a bacteriophage, perhaps) was unjustified. In these studies, high-energy X-rays are applied to an active, presumably particulate entity; the larger the entity, the lower the dose needed to hit (inactivate) it. Fortunately, the large host, the *Paramecium* cell, is unusually resistant to radiation, possibly because of the many copies of each essential gene in the macronucleus. Preer, realizing that the X-ray inactivation data yielded a target size for kappa that was similar to the size of a bacterium, sought and actually saw small particles the size of bacteria in killer strains. Under the light microscope, the particles stained with the Feulgen technique, diagnostic of DNA. Significantly, Preer found no such particles in nonkillers. For a short time, these particles, and others like them (pi, mu) in other species seemed to be the embodiment of plasmagenes because they did not appear to have DNA localized in a central area like bacteria do. But soon it became clear that the cytoplasmic particles were indeed bacterial symbionts, and therefore hardly generalizable as models for the cytoplasmic determinants sought so long by embryologists. Geoffrey Beale (b. 1913), a student of Sonneborn, worked in the laboratory in the late 1940s and then took a position in Edinburgh, Scotland. He corresponded with Sonneborn often thereafter, but developed some skepticism about the plasmagene hypothesis. In 1954, he published a book on the major themes of *Paramecium* genetics, with a rather dispassionate opinion of kappa:

If plasmagenes of the type postulated by Darlington, i.e., cytoplasmically located determinants of hereditary characters, were of widespread occurrence, kappa particles would be excellent plasmagenes. Kappa particles are determinants of a hereditary character (i.e. the killer character), and are transmitted through the cytoplasm. Unfortunately, however, kappa particles probably constitute the best example of such plasmagenes. And we have to face the fact that most paramecia do not contain kappa particles at all. Hence the kappa plasmagenes could not be of great importance in the heredity of most paramecia. One could counter such an argument by assuming that all paramecia contain plasmagenes, killers containing one kind and sensitives another. But since it has been shown that killers contain visible particles, which are absent from most sensitives, in order to maintain this argument any longer it would be necessary to assume that sensitive paramecia contain plasmagenes which are both inert and invisible.[18]

The demise of kappa as a plasmagene did not render it uninteresting; indeed, it became, much later, one of the best supporting arguments that other cytoplasmic entities with genetic continuity, mitochondria and chloroplasts, might have evolved from an association similar to that of kappa and *Paramecium*. The kappa system has been studied in more detail; in the early 1950s, Preer observed kappa particles as they seemed to divide ("doubles"), and observed refractile inclusions, or "R-bodies" in the kappa particles. Later work showed that these inclusions are the paramecin released by the symbionts and encoded by plasmids and viruslike genomes carried by the symbiont.[19] Even now, however, the process of secretion of paramecins from their hosts, the immunity of killers to these toxins, and the action of the various paramecins on sensitive paramecia are unclear.[20]

The Paris Meeting of 1948 and the *Paramecium* Antigens

Before the nature of kappa became clear, many seemingly related phenomena came to light. In yeast, Spiegelman and Lindegren promoted the cytogene model, in which a copy of a gene might enter the cytoplasm and replicate as long as conditions requiring it prevailed (see chapter 5). Avery's group showed that *Pneumococcus* could be transformed permanently with added DNA. Indeed, Sonneborn's group demonstrated that with low efficiency, kappa could be transferred to sensitive cells by growing them in the presence of ground-up killers. A case of carbon dioxide sensitivity, a maternally inherited trait in *Drosophila* called sigma, had been described by Philippe L'Héritier (b. 1906) and Georges Teissier (1900–72). Several groups had studied adaptive enzymes as possible environmentally induced, hereditary changes, with models somewhat different from that of Lindegren and Spiegelman. In yeast, Ephrussi had discovered cytoplasmically inherited respiratory deficiencies. All of this took place well before the identification of DNA as the genetic material, at a time when developmental biologists cried out for cytoplasmic determinants of cell differentiation. The behavior of the surface antigens of *Paramecium,* described hereafter, was still another case of apparent cytoplasmic inheritance. These widespread exceptions to Mendelian genetics excited the genetics community sufficiently that Lwoff organized a meeting in Paris in 1948 at which the principal investigators working on them described their investigations.[21]

Lwoff, still a protozoologist in part, presented the first talk of the meeting on kinetosomes of the apostomatous ciliates, mentioned previously.[22] This is a subject on which he had worked since the 1930s, with a fascination for the highly organized nature of the cortex of the ciliate cell. The cortex, Lwoff believed, was the origin of or the milieu for cytoplasmic, hereditary determination of much of the cell's organization. He emphasized the genetic continuity of the kineties, in which the basal bodies of the cilia (kinetosomes) would divide, with the descendent kineties still arranged in their proper pattern after cell division. He showed that kineties would adopt specific features characteristic of their position and that the basal bodies could generate trichocysts (stinging organelles) rather than cilia. To Lwoff, this suggested that cytoplasmic factors could call forth different potentialities of these versatile organelles.

Lwoff believed that the ciliates were suitable models of cells in general and that his findings had relevance to embryogenesis in multicellular animals. He could list, among organisms in general, perhaps 20 cases of elements "endowed with genetic continuity," and suggested that the number would be greater if mutations of such elements were not lethal. Here, and in a small book devoted to this question written at the same time, he elaborates the argument and denotes the kinetosomes as "plasmagenes."[23] His definition of plasmagenes was borrowed from C. D. Darlington (1903–81), who called them "the undefined residues of heredity, not associated with any visible body; this is the 'cytoplasmic' or 'molecular' system whose constituents are generally known as 'plasmagenes.'"[24] Notice that the word "plasmagenes," as opposed to "Plasmon," suggests particulate entities. Many geneticists at the time struggled unsuccessfully to keep an open mind in this matter. It was hard to do because Lwoff's 20 examples included manifestly particulate entities ("unités").

In the second talk at this meeting, Beale reported on the inheritance of surface antigens of *Paramecium*, then a new topic of interest in Sonneborn's laboratory.[25] Surface antigens are easily detected with antibodies raised in rabbits by injecting them with various stocks of *Paramecia*. Each stock can express a variety of antigens, but only one at a time; they appeared to be mutually exclusive. The antigens (A, B, C, etc.) are inherited in asexual reproduction true to type under constant conditions, but different antigens are expressed at different temperatures. Beale described work with two strains (nos. 51 and 29) of *P. tetraurelia*. The strains had four antigens in common (A, B, C, E), but only strain 51 had antigen D and only strain 29 had antigens F and H. Crosses between the strains showed that the ability to display antigen F was inherited in a strict Mendelian fashion. In fact, for each of several antigens one can define minor variants in different strains, and these minor differences are also inherited in a strict Mendelian manner. These findings gave credence to the exceptional inheritance patterns that emerged from further study.

The curious feature of the antigens lay in their behavior in crosses of strains, each displaying a different antigen of the set common to both strains. In crosses of a strain displaying antigen A with one displaying antigen B, one could follow the phenotype of the exconjugants in further divisions. One must remember that the conjugation process normally leads to no exchange of cytoplasm but that the exconjugants are nominally genetically identical, at least at first. Each one contributes one of its identical haploid nuclei to the other, and each forms diploid nuclei and macronuclei after fusion of the donated and resident nuclei. In the case of antigen phenotypes, however, the exconjugants each retained the antigenic type they displayed before conjugation. If these F_1 progeny were crossed to form the F_2 generation, no segregation of antigenic types took place. To Sonneborn and Beale, it seemed that cytoplasmic determinants caused the maintenance of antigenic types—an example of plasmagene inheritance. In their interpretation, a copy of the gene product could indeed replicate indefinitely in the cytoplasm. The observations were strengthened by the exceptional cases in which cytoplasm was also exchanged during conjugation. In that case, both exconjugants then displayed antigen A in asexual growth and in the F_2 with the stability seen in the previous case.

Another curious observation was that temporary treatment of animals with antibody against the expressed, or primary, antigen caused their transformation, after they had recovered mobility, to one of the secondary antigenic types. This phenotype was maintained in further growth in defined conditions, even if conjugation or autogamy intervened. The "transformations" were not due to selection of a few variants, since all members of the treated set were immobilized, and substantial numbers of transformed animals appeared. Sonneborn and Beale interpreted this to mean that the antibody to the primary antigen led to interference with the replication of its plasmagene, with another antigen-determining plasmagene becoming predominant through lack of competition with the original. The transformations to different alternate types could also be controlled by abrupt temperature shocks, they could be delayed by lack of nutrition, and they could be effected without antibody by continuous cultivation at different temperatures. The facts appeared consistent with a competitive set of replicating, cytoplasmic entities, derived from the genes. Reinforced by the discovery of kappa, the data supported a good case for the existence of plasmagenes. In contrast to kappa, antigens, as proteins, seemed to be much closer to gene products. As part of the routine phenotype of *Paramecia,* they could be used as a model for any characteristic displayed by other organisms. Sonneborn and Beale suggested that differentiation of cells during embryogenesis might have the same basis as the changes in antigenic types. That is, constellations of plasmagenes, derived from the genes, order themselves into different concentrations in different tissues according to the local environment in the embryo and remain a semiautonomous feature of the cytoplasm thereafter.

Delbrück's Challenge

Delbrück also attended the Paris meeting, at which Beale presented the work on antigens just summarized. In the discussion following Beale's presentation,

Delbrück called attention to the following principle: "many steady-state systems are capable of several alternative steady states. They may switch from one to another steady state under the influence of transient perturbation."[26] (The French version uses the term *"équilbre de flux"* for "steady-state," and this in turn gave his model the awkward but distinctive English epithet, the "flux equilibrium" model.) Delbrück presented a simple graphic model (fig. 13.4), showing two pathways of gene action, each having an intermediate that inhibits the other pathway. As one of these paths gains ascendancy, the other is shut down. By temporarily blocking the ascendant one (and thus the inhibitor made by it), a metastable state ensues that may allow the other pathway to get the upper hand. This seductive model was not particularly novel to chemists, and indeed, it had the potential, endorsed by Delbrück, to explain multiple equilibria in more complex systems.[27] Like many good ideas with few clear examples in actual biological systems, it gave considerable license to those who would oppose the plasmagene hypothesis. Such challenges, however, would lead to excellent experiments thereafter by the plasmagene proponents, who had the burden of the proof. Some succeeded in demonstrating replicating, cytoplasmic units. Some failed to do so and became marooned in investigations that bogged down until the advent of advanced molecular techniques and the understanding of nuclear gene regulation.

The Model Status of *Paramecium*

The *P. aurelia* group took its place as a collective model organism that promised an understanding of the role of the cytoplasm in inheritance and embryogenesis. Sonneborn campaigned relentlessly for the significance of his findings on mating type, kappa, the antigens, and several other problems pursued later in another, more easily cultured ciliate, *Tetrahymena*. With his observational skills, Sonneborn recognized unorthodox phenomena that seemed to challenge established genetic principles. His experimental ability and his description of

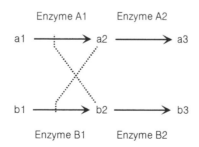

Fig. 13.4. Delbrück's flux-equilibrium model. In this open, steady-state system, two parallel metabolic sequences may proceed, but not simultaneously: intermediate a2 inhibits enzyme B1; intermediate b2 inhibits enzyme A1. A switch from one sequence to the other may be accomplished by depriving the predominating sequence of substrate (a1 or b1).

experiments rendered these phenomena understandable to sophisticated scientists worldwide. His rigor led him to test his own hypotheses according to standards of the field and to acknowledge negative or inconclusive outcomes of critical experiments. But throughout, he sought to test the weaknesses of generally accepted thinking, and in doing so he revealed an extraordinary range of interesting phenomena in ciliates. In many cases, however, these phenomena seemed to have had relevance to the ciliates alone, or the problems had simpler counterparts in other organisms—in particular the discovery of genetic continuity of mitochondria and chloroplast DNA (chapter 14).

The demise of *Paramecium* as a model came with the failure of kappa and kinetosomes as generalizable phenomena, with the complexity of antigen expression, and with a later study during the 1960s of "metagons," a purported self-replicating mRNA required to maintain the mu symbiont. The gradual realization that metagons did not exist and the recognition that peculiarities of the macronucleus, found only in the ciliates, were at the root of many unusual phenomena hastened the disappearance of *Paramecium* from the scene.[28] Investigations of other organisms showed that patterns of inheritance that might have a cytoplasmic basis could be explained by differential gene expression under the control of specific extra- or intracellular conditions. The Jacob–Monod model of regulation eclipsed arguments based on cloudier phenomenology from the ciliates, and soon the ciliates became a forgotten empire, especially with the death of Sonneborn and the movement of his students to other ciliates or other organisms altogether.[29] Even Lwoff, originally a ciliate biologist, was captured by the question of lysogeny in 1950. Obscure as lysogenic bacteria were, even displaying similarities to *Paramecium* carrying the infective/heritable kappa, they were much simpler organisms.

In a 1997 article, Preer asks, "whatever happened to *Paramecium* genetics?"[30] He lists eight possible reasons they seemed to disappear from the scene, of which he rejects a number. Are the paramecia just too hard to culture, too complicated, too queer, irrelevant, unable to attract funds, or useless in a reductionist scientific field like molecular biology? No, Preer insists; none of these reasons suffice, although he acknowledges that Sonneborn was no longer there to maintain the visibility of *Paramecium*. In addition, the introduction of other ciliates diffused the visibility of any one of them. But Preer evades the actual question that might be asked: why were the *Paramecia* (and, by extension, other ciliates) ever considered model organisms? From what we have seen here, *Paramecium* had promise as a model that was not fulfilled. That promising phase did not last long, due in part to many of the arguments Preer rejects. It went from ascendancy to neglect in short order, a fate already met or in store, over a longer time, for many of the other organisms discussed in this book.

Tetrahymena

A renaissance in ciliate studies after about 1980 enriched the knowledge of this group and its diversity. The studies have not only explored ciliate biology, but have identified ciliates as the first or even the best model systems to study two particular, universal, biological phenomena, discussed later. A major step in this renaissance was the development of defined media, mating techniques, and genetic rationales for ciliates known initially under the single species name *Tetrahymena pyriformis*.[31] Like the *P. aurelia* group, this taxon had become familiar to protozoologists before the 19th century, especially as the compound microscope became a favorite observational tool. After 1950, attempts to corral the variants of the organism and to give them individual names led geneticists to focus on a single species, *T. thermophila*, and to develop a limited set of genetic stocks of the organism derived from only four individual progenitors.[32] Papers on *Tetrahymena* now outnumber those published on any other ciliate, and the work proceeds without the ideological trappings of the debates on the role of the cytoplasm in heredity.[33] Current work uses molecular techniques to ensure progress on many of the old problems and to define new ones.

   T. thermophila cells are smaller than those of *Paramecium* spp. They grow quickly and display many of the peculiarities of the *P. aurelia* group. Bacteria are not essential in the (albeit complex) media on which *T. thermophila* grows well; even a defined medium is available but has too many components to make it easy to use. Many investigations echo those on the *P. aurelia* group, which continue in parallel. The *T. thermophila* macronucleus is relatively small, having about 50–100 copies of its genes, in contrast to the 1000–2000 copies in *Paramecium*. The fewer copies of DNA allowed Sally Allen (b. 1926) and David Nanney (b. 1925) to detect segregation of two alleles of a single gene (originating in the diploid nucleus that gives rise to the macronucleus) during the random assortment process of chromosomal units during macronuclear divisions.[34] This observation led in turn to the recognition of the programmed breakup, during macronuclear growth, of the chromosomes into much smaller units of DNA, all of which replicate independently during cell division.[35]

   The fragmentation of macronuclear DNA, peculiar as it is, has culminated in one of the most important recent discoveries about chromosome structure, relevant to all eukaryotes. This is the structure and synthesis of telomeres, the DNA sequences found at the ends of linear chromosomes. The 3' end of linear DNA nucleotide chains, templates for the synthesis of the complementary strand, cannot be copied by the DNA polymerase complex; the complex must bind the end of this template and begin its copying at an internal position. Telomeres represent an extension of 3' ends of templates, maintained by regular additions, that allows the 3' ends of the template to be copied without the erosion of important genes internal to them. These ends are added by *telomerase*, a polymerase enzyme containing an RNA molecule that includes a short sequence of about six nucleotides. These nucleotides serve as a template for DNA synthesis, the DNA being added, in many tandem copies, to chromosome ends. Due to the large number of chromosome ends after the fragmentation of macronuclear chromosomes, *T. thermophila* has huge numbers of telomeres and is a rich source of telomerase. Accordingly, the first basic picture of telomeres emerged from this organism.[36] It is of interest that more modern work on the subject now uses yeast as the experimental organism, one of many examples of the appropriation of important research programs to an organism more amenable to molecular and genetic experimentation.[37]

Another important phenomenon was uncovered in *T. thermophila*, the self-splicing ability of certain nucleotide sequences in RNA molecules, an ability recognized as a catalytic activity of RNA. Until that time, biochemists thought that only proteins, among biological macromolecules, could catalyze biochemical reactions, a view contradicted in 1981 by the discovery of "ribozymes." It had been known for several years that genes in eukaryotic organisms contain *introns*, nucleotide sequences in DNA and RNA that interrupt coding sequences. After the primary RNA transcript of the gene is made, introns are removed from it as the primary transcript is processed into the mRNA used as a template for protein synthesis. In many organisms, the same is true of some ribosomal RNAs and tRNAs as well as mRNA. The single ribosomal RNA gene of *T. thermophila* contains an intron. Because of macronuclear DNA amplification, the gene exists in about 10,000 copies per cell and generates a large amount of its transcript. With this advantange, *T. thermophila* became an ideal organism in which to study the removal of the intron from the primary transcript of the rRNA gene. Thomas Cech (b. 1947) and his associates disovered that the RNA could remove its own intron in a complex self-splicing mechanism.[38] Many other reactions now known to be catalyzed by various RNAs are found in all eukaryotic organisms.[39] Once again, the studies of this phenomenon continue in yeast and other organisms, depending on the advantages presented by each for specific problems.

In concluding this chapter, I call attention to an unusual aspect of the ciliates and their use as model organisms. Protozoologists and geneticists were unaware of the cryptic diversity of the paramecia and tetrahymenids. At the outset, Sonneborn and the early *Tetrahymena* workers explored phenomena that seemed to vary in detail from one species (then, "variety") to the next. Some species had two, some one, and some no micronuclei. Strains differed in whether macronuclear lineages became uniform or varied in the course of successive cell divisions. Strains, especially among tetrahymenids, displayed senescence, an inability to grow well and finally to mate. The number of macronuclear DNA copies varied. This diversity was a partial deterrent to understanding any of the epigenetic phenomena uncovered. As the rules of the various games changed, less and less faith could be put into generalizing any of these phenomena within the ciliates, much less to other organisms. At the same time, the diversity of ciliates was appreciated from the beginning. Biologists know that one cannot appreciate the value of observations on one organism until related forms are studied. And so at the outset ciliate geneticists were granted a vision of these organisms not as typological models, but as a stew of evolutionary diversity. This vision approximates what workers with other model organisms find much later as they convert their interests from universal phenomena to an exploration of variation among related taxa.

14

Organelle Genetics

Yeast and Chlamydomonas

*A bright mind that does not apply itself to useful endeavor is like a firefly,
which scintillates in the dark, emits no heat, and leaves nothing but a
fleeting memory in its wake.*

—Father Baltasar Gracian

The Origins of Yeast Mitochondrial Work

To most geneticists after 1950, cytoplasmic inheritance did not threaten the
basic principles established by Mendel and Morgan. Mendelian genetics had
proven so general by that time that it was the students of cytoplasmic inheritance that were threatened. The spotty reports of cytoplasmic inheritance, from
plastid inheritance to *petites colonies* of yeast; kineties to kappa; and sigma to
adaptive enzymes, all intrigued a small, but growing coterie of investigators.
They were obliged to grant the general applicability of the chromosome theory.
But they believed that the unusual phenomena would eventually extend the
picture of cell heredity, whether or not they reflected an underlying Mendelian basis. The respect of the workers on cytoplasmic inheritance for Mendelian genetics led them to bend over backward to falsify Mendelian explanations.
This might seem to represent stolid resistance to Mendelism, but this was not
the case. Their strategy simply recognized that that a Mendelian basis was
conceptually the most easily falsifiable hypothesis among candidate explanations. Discussion of these matters, originating in their modern form with the
Paris meeting of 1948 (chapter 13), became quite respectable, and onlookers
could enjoy the progress of both fields without having to take sides. The exceptional phenomena held much more excitement than many later writers would

acknowledge because the world was as yet uncontaminated by the explanatory allure of DNA structure. Something unusual was out there, a group of mysteries with an allure all of its own.

We may usefully continue the story of cytoplasmic inheritance with a precocious look at an elegant monograph by Ephrussi, drawn from three lectures given in 1952 in Birmingham, England.[1] We must remember that although Ephrussi's scientific interests were cell differentiation and cell heredity, he had fully assimilated the findings and viewpoint of Morgan at Caltech when he worked with Beadle on the biochemical genetics of *Drosophila*. In the introduction to his book, Ephrussi repeats the paradox of development: cells differentiate into many types from germ cells, while the genetic content of their nuclei (presumably) remains unchanged. His objectivity about the paradox emerges in the following words:

> Obviously, what is required is more than deductions from the behaviour of germ cells; what is needed is direct genetic analysis of somatic cells, for the assumed functional equivalence of irreversibly differentiated somatic cells, however plausible, is only an hypothesis. Crosses between such cells being impossible, only nuclear transplantation from one somatic cell to another, or grafting of fragments of cytoplasm, could provide the required information; such experiments, however, will have to await the development of adequate technical devices. In the meantime, the closest approximation to the evidence we would like to have is provided by the study of lower forms which propagate by vegetative reproduction and possess no isolated germ line.[2]

Ephrussi noted that one could not at that time argue definitively for the near-exclusive predominance of nuclear genes in cell heredity. The indispensability or the redundancy of cytoplasmic determinants might deprive us of variants with which they could be studied. In the appendix to his book, Ephrussi documents both possibilities with examples, one from his own work on the cytoplasmic inheritance of mitochondria in yeast. This program is discussed in detail in the first part of his book and represents a modern approach to organelle heredity.

The work on cytoplasmic inheritance, mentioned in chapter 5, began in 1946. Ephrussi gave a major summary of it at the Paris meeting of 1948 on cytoplasmic inheritance.[3] He indicated that his initial interest was the chemical induction of mutations by acriflavine, a chemical known then as a potent bacteriocidal agent. He was aware of the ability of acriflavine to complex with nucleic acids, and he justifies the choice of this drug by pointing to the possibility that nucleic acids are a constant constituent of "toutes les unités biologiques douées de continuité génétique" [all biological elements endowed with genetic continuity].[4] (This assertion may seem bold for the time, but Ephrussi's wife, Harriet Ephrussi-Taylor [1918–68], also a speaker at the conference, had worked in Avery's laboratory soon after he had demonstrated a genetic role for DNA in *Pneumococcus*.) The nucleic acids should be susceptible to chemical alterations, possibly mutations, induced by acriflavine. Ephrussi's initial results, however, did not fulfill this promise. Instead, they yielded a surprising and coherent body of information on the petite-colonie variants (chapter 5) that he then describes.

These respiratory-deficient variants appeared in platings of vegetative yeast at low frequencies but failed to segregate as a genetic character in crosses to the normal strain.

The induction of petites by acriflavine was most unusual. By all tests Ephrussi could administer, petites seemed to be induced, rather than merely selected for by the toxic drug. The induction contrasted strongly with expectations based on the then-recent Luria-Delbrück fluctuation test, used to prove the preexistence of spontaneous mutations of *E. coli* to phage resistance. Not only that, but acriflavine, at a high concentration, could induce the irreversible transformation of normal yeast to petite in 100% of the population treated. This striking fact makes Ephrussi's failure to remark on the narrow range, the specificity of mutational types induced by acriflavine, which is uncharacteristic of the action of other mutagens, understandable. Ephrussi showed that the appearance of petite yeast was an occasional (ca. 0.2%) but irreversible event upon ordinary platings of normal yeast. The phenotype of petites showed that they had a serious respiratory impairment and that they lacked certain cytochromes characteristic of normal yeast grown in the presence of air. Cytochromes are required for aerobic respiration, and if either air or the cytochromes are missing, glucose, the usual carbon source, can only be fermented—a much less efficient metabolic process. Thus normal yeast grows as slowly on glucose without air as petites grow on the same medium in the presence of air. In addition, petites fail to grow at all if they are presented with a sole carbon source, such as glycerol, that cannot be fermented.

Matings of most haploid petites, whether spontaneous or induced, to normal cells yielded normal diploids, and few or no petites appeared among the meiotic products of the latter. Diploid yeast yielded spontaneous petites at about the same frequency as haploids, making it improbable that the phenotype reflected the occurrence of a recessive mutation. Tests for a multigenic basis for the altered phenotype by backcrossing the meiotic products many times to the petites were negative. Ephrussi, in his careful account at the 1948 meeting, concluded that petites suffered from the loss of a cytoplasmic particle, of which there would have to be relatively few (perhaps 10–20) per cell. He acknowledged (and, in many experiments, tested) counter-interpretations, with a discussion added to the published version that considered whether the flux equilibrium model, presented by Delbrück days before at the same meeting (see chapter 13), might be applicable. Ephrussi believed that the irreversibility of the petite phenotype, among other things, indicated that the Delbrück hypothesis was unlikely.

Ephrussi was familiar with a report by Winge and Laustsen on the cytoplasmic inheritance of a variant of yeast affecting germination efficiency.[5] They found that genetically identical strains might differ in the number or type of "chondriosomes." Winge and Laustsen had speculated that chondriosomes had genetic continuity, a point with a precedent in the apparent genetic continuity of plastids. In his Birmingham lectures, Ephrussi asserted that the phenotypic attributes of the petites suggested a mitochondrial impairment or loss—a tentative conclusion because the nature of mitochondria was not entirely clear at the time.[6] Ephrussi made the point that yeast, unlike almost any higher organism, "possesses a remarkably efficient alternative energy-yielding metabolic pathway [fermentation]. It is to this circumstance, hardly realized with similar perfection in many other sexually breeding organisms, that we owe the possibility of detailed study."[7] In other words, in yeast, the loss of aerobic respiratory capacity was not lethal, as it is in most organisms. With these words, Ephrussi supported the argument stated earlier that cytoplasmic inheritance might actually be widespread but experimentally invisible in most organisms.

The petite yeasts, then, were a solid case of cytoplasmic inheritance, far more accessible to experimentation than the cases described in *Paramecium* or higher plants. Ephrussi was well known as a developmental biologist, as a pioneer in the biochemical genetics of *Drosophila*, and as one sympathetic in his immediate postwar investigations to the role of the cytoplasm in heredity. He was also the first professor of genetics in France and the director of one of the CNRS institutes for genetic research established in 1946, just after World War II. Ephrussi therefore had credentials in the burgeoning community of microbial geneticists in Europe and America and authority against the prevailing neo-Lamarckian, anti-Morganist bias of French genetics in the areas of heredity and evolution. He had collaborated with Lwoff, Monod, and L'Héritier for some years before the war, and he and Lwoff and Monod had a major role in developing microbial genetics in France after the war.[8] Their success was aided by the fact that this field did not directly confront the establishment biologists working with higher organisms.

Further study of petites by Ephrussi and his student Piotr Slonimski (b. 1922) revealed petites with other properties (fig. 14.1). The first were those arising from mutations in the nucleus, and which behaved in a perfectly Mendelian manner. These were of interest because they implied that both nucleus and cytoplasm collaborated in the formation of normal respiratory capacity. The study of such mutants, isolated later in large numbers by Alexander Tzagoloff (b. 1937), would yield much information about mitochondrial biogenesis. In the second unusual category of petites, the respiratory impairment could, like an infection, gradually dominate the phenotype of a diploid strain formed by mating such a petite and a normal haploid cell. That is, as the strain proliferated, the percentage of respiratory-deficient cells increased. These "suppressive" petites seemed to be a form of the postulated particle (by then named rho) that could take over the phenotype. Rho resembled the kappa particle as it gained a foothold in *Paramecium* after cytoplasmic exchange in matings. Possibly in the case of suppressive petites the takeover reflected a replicative advantage of the variant entity over its normal counterpart.

Although the path of carbon compounds in the mitochondrial Krebs cycle had been determined in the 1930s, and the central role of ATP was clear in the early 1940s, mitochondrial research did not begin in earnest until the late 1940s when techniques for isolation of intact mitochondria from mammalian liver became common. Therefore, by 1954 the role of the mitochondrion as the seat of energy metabolism had been clear for just a few years. Slonimski and his

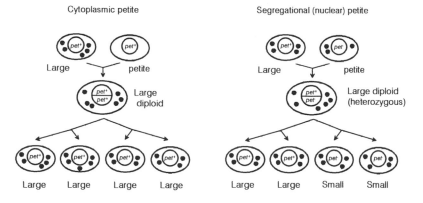

Fig. 14.1. Inheritance of cytoplasmic and nuclear petite phenotypes in *S. cerevisiae*. Crosses proceed by fusion of haploid parents to form the diploid, and meiosis follows to form four haploid meiotic products in both sequences. In these diagrams, mitochondrial DNA is signified by small, dark circles, and the condition of the nuclear *pet* gene is signified by a + or – superscript. (Note: yeast cells lacking mitochondrial DNA nevertheless have promitochondria, membrane-bound organelles, formed from nuclear gene products, that are indispensable for viability.)

colleagues established that the petite phenotype reflected a mitochondrial derangement by that time. The isolation of yeast mitochondria suitable for intense study was delayed for some years due to the difficulties of breaking the yeast cell wall without destroying the organelles within (see fig. 5.3). The use of mechanical breakage of cells with glass beads or enzymatic digestion of the cell wall overcame this problem by the mid-1960s.[9]

These promising studies reached something of an impasse in the late 1950s with more and more descriptive work, but no clear mechanistic understanding of the origin and basis of the petite phenotype. Progress resumed after the discovery of mitochondrial DNA. This discovery was an extended and continuous process that culminated only in the early 1960s, well after the elucidation of DNA structure in 1953. It started with electron microscopic observations of fibrils sensitive to DNAase in mitochondria and plastids of plants and of mitochondria of animals. These led to isotopic studies in which these fibrils in mitochondria became labeled with radioactive nucleic acid precursors and finally to the isolation of mitochondrial DNA. The latter step was complicated by the problem of contamination by nuclear DNA during cell breakage and cell fractionation. Indeed, proof of mitochondrial origin of the mitochondrial DNA required characterization of its distinct bouyant density, determined by equlibrium sedimentation in cesium chloride, in the mid-1960s.[10] In 1964, Schatz et al. definitively demonstrated yeast mitochondrial DNA.[11] Soon thereafter, petites were shown to have either no such DNA or DNA with severe deletions that presumably enabled it to replicate faster and to replace the normal organellar genome in the suppressive variety of petites.[12]

At about this time, the genetics of the mitochondrial genome (as opposed to its molecular characterization) resumed with the discovery in 1967 of a variety of antibiotic resistance mutations that resided in mitochondrial DNA.[13] This opened the way for experiments leading to genetic mapping of point mutations, escaping the limitations of working with gross alterations of mitochondrial DNA with a common phenotype (respiratory deficiency) as seen in petites. The remarkable ability to perform crosses of organellar genomes reflected a striking advantage of yeast over multicellular organisms. This was the ability of both parents of a mating to contribute organelles to the zygote, an attribute virtually excluded by the anisogamous nature (unequal size) of gametes of higher organisms, in which the maternal parent contributes virtually all of the cytoplasm. Curiously, this very advantage had also arisen in studies of chloroplast inheritance that began in the early 1950s with *Chlamydomonas reinhardtii* and that would continue in parallel with work in yeast. *Chlamydomonas* is our last model organism, to which we now turn.

Chlamydomonas: A False Start

Few geneticists know or remember that the unicellular alga *Chlamydomonas* was the first genus in which tetrad analysis was performed, before 1920.[14] This would seem to have positioned it for a future as a prime model organism in microbial genetics. However, this study (which some think might be fraudulent) did not catch fire, and it was left to Dodge and Lindegren to introduce the first microbe, *Neurospora*, in which formal genetics could be pursued easily with tetrad analysis.[15] This development has tended to obscure the story of Franz Moewus (1908–59), working on *Chlamydomonas eugametos* in Germany in the 1930s and 1940s. Moewus's major work, begun in the laboratory of Max Hartmann (1876–1962), was devoted to testing and extending Hartmann's theory of relative sexuality. The theory held that sexual unions would take place only between cells (or organisms) that differed sufficiently in their sexuality to allow it. The "gametes" of the alga (vegetative cells conditioned for mating), like the haploid cells of yeast, did not differ morphologically, but the organism was nevertheless heterothallic. Any given mating required that the gametes differ, and no matings within a clone would occur. This led to the designation of male and female sexes, which we know now simply as plus (mt^+) and minus (mt^-) mating types. Fusion of mt^+ and mt^- gametes form a zygote which, after suitable incubation, undergoes meiosis with the formation of four meiotic products. These can be dissected and used for tetrad analysis. In Moewus's hands, mating types segregated 2:2, like ordinary Mendelian characters.

The major interest in Hartmann's laboratory was the nature of the chemical difference between the mating types. Relative sexuality, according to Hartmann, was a complication of the simple two-type system just described. Mating potential resided, he believed, in the strength of the chemical difference, which might vary from extreme male to extreme female. The theory predicted successful matings of two cells that were classified as male with respect to a particular female culture if they themselves differed sufficiently from one another (weak male vs. strong male). Moewus provided evidence for diffusible substances, which we would now call pheromones, produced by sexually competent cultures that could induce a copulatory potential in other cultures. These substances, and the genetic control of their formation, were the subjects of his further work.

Moewus moved to the laboratory of the chemist Richard Kuhn (1900–67) in 1937 to collaborate with his group in determining the chemical nature of the pheromones. Moewus continued his investigations, reporting environmental manipulations that induced conversion of pheromone precursors to active substances and mutants that affected these steps. The scheme he developed purported to show that the sexual substances, each with its particular effect, arose from the carotenoid derivative protocrocin. Sapp, describing this history, shows that Moewus had anticipated Beadle and Tatum's use of a microorganism in the analysis of biochemical function by three or four years.[16] Moewus's intense genetic work led to many publications that described an elaborate genetic map, together with a rational, formal picture of the pathway of pheromone production.

This work attracted an enormous amount of attention and was championed vigorously by Sonneborn in America.[17] But even with this endorsement, the work invited substantial skepticism on the basis of the astonishing numbers of tetrads analyzed and the rather too-perfect correspondence of theory and results in the physiological experiments. Thus as Sonneborn became an increasingly lonely defender of Moewus's general contributions, skepticism deepened on the part of other investigators as they failed to reproduce (or at least believe) Moewus's results. In something of a showdown, Moewus himself failed to reproduce his results when he was put to the test in 1954. This "trial" took place at the Woods Hole Marine Biological Laboratory in Massachusetts, to which Moewus had been invited to demonstrate his interesting system. Under close observation by Francis Ryan, Bernard Davis, Boris Ephrussi, and Tracy Sonneborn, Moewus was unable to repeat crucial findings and was evasive about certain steps in the experiments. He could not repeat most of his important findings in a more thorough test, lasting 16 months, in Ryan's laboratory in 1954–55.[18]

Moewus is now forgotten as a contributor to the main themes of microbial genetics, and his data survive mainly as a spectacular case of scientific fraud. As Sapp has shown, this harsh characterization was constructed in the scientific community slowly and without definitive proof of intentional fraud during the 1940s and 1950s. Sapp reports an interview with Moewus's widow, who collaborated with her husband in most of the experiments, revealing her to be the sole remaining defender of their findings.[19] Even Sonneborn felt forced to spurn Moewus in the end, feeling perhaps that he had given undue credit to ideas and too little attention to their credible verification. We are left, in Father Gracian's words, with nothing but a fleeting memory of this episode. Moewus's only major, lasting contribution was the demonstration that *Chlamydomonas* could be used easily for experimental genetics.

The Modern Domestication of *Chlamydomonas*

After this false start, work that would render *Chlamydomonas* a model genetic organism began in the early 1950s, with little dependence on the prior efforts of Moewus. Two investigators initated work with *Chlamydomonas*, though with different aims and with different species. One was Ralph Lewin (b. 1921), who had done his Ph.D. work in 1950 on *C. moewusii* (possibly the same as *C. eugametos* of Moewus) and had broad interests in the biology and genetics of algae. Lewin was extremely skeptical of Moewus's results and quickly abandoned any attempt to repeat them or to base his own investigations on them. The other investigator, Ruth Sager (1918–97; fig. 14.2), focused on *C. reinhardtii* to study the genetics of the chloroplast, starting at about the same time as Lewin. Both investigators obtained their strains from the eminent American phycologist Gilbert M. Smith (1885–1959), who had tried without success to repeat some of Moewus's work. The strain of *C. reinhardtii* used by Sager and the one later adopted for work by Wilbur Ebersold and R. Paul Levine (b. 1926) in the mid-1950s generated the major stocks for almost all American genetic investigations thereafter. The strains differ in their ability to use nitrate as a nitrogen source, suggesting either a mutational divergence or some confusion about their exact ancestry.[20]

Sager, at Woods Hole in 1954, would become a key player in the Moewus affair. She was almost the only one there with experience in the genetics of *Chlamydomonas*, and, though incidentally and only recently familiar with Moewus's past work on *C. eugametos*, was quite skeptical of it. She contributed some of the expertise required to make informed observations during the demonstration experiments and on the behavior of one of Moewus's mutants.[21]

Sager's Ph.D. work on corn (1948) was done at Indiana University with Marcus Rhoades (1903–91), an eminent geneticist who had defined one of the clearest cases of cytoplasmic inheritance involving chloroplasts. (He, too, was at the Paris meeting of 1948.) Given the interest in cytoplasmic inheritance

Fig. 14.2.   Ruth Sager. Photo courtesy of Kristin Lacey, Dana Farber Cancer Institute.

emerging both in America and in Europe, and given the widespread attention to the use of microbes in genetic research, Sager sought to continue studies of chloroplast inheritance in a microbial system. G. M. Smith had done enough to show that the relative sexuality model of Moewus could not be confirmed with the isolates that he had. Indeed, he simply confirmed Moewus's findings that any successful mating gave excellent 2:2 segregations in tetrads for the mt^+ and mt^- mating types. To Sager, this confirmed the suitability of *Chlamydomonas* for Mendelian genetic work, an essential feature of any system in which cytoplasmic inheritance was also to be pursued. She began her work at the Rockefeller Institute, much of the time in the laboratory of Sam Granick (1909–77). Granick was interested in chloroplast biogenesis, and, with Sager, hoped that the use of a green unicell with the potential for genetic analysis would be profitable.[22] Later, Sager described her choice of *Chlamydomonas* as a model as follows:

> Following the precept that the best organism for solving a particular problem is the simplest one, I cast about for a suitable eukaryotic microorganism. Discussions with C. B. van Niel [(1897–1985), an eminent bacterial physiologist] and G. M. Smith led to the choice of *Chlamydomonas reinhardi* for a number of reasons. In the first place, it is a unicellular sexual microorganism with a simple sexual life cycle, easy to control under laboratory conditions. Secondly, the organism grows well on a simple defined medium, can be grown in mass culture, and can be handled by standard bacteriological methods. Thirdly, of particular importance for our purposes, the organism contains one chloroplast and several mitochondria per cell, making available both of these organelle systems for investigations in the same organism.[23]

This recollection has much in common with Beadle's recollections of his and Tatum's choice of *Neurospora*. *Chlamydomonas* falls in the category of an organism chosen deliberately for a particular program of research. The choice did not depend (at least as this description would have it) on some surprising result or unusual phenomenon such as those that brought *Paramecium* or phage λ to the forefront of genetic research. Nor was it as casual as those that brought the T phages and *E. coli* into the field.

Chlamydomonas (and I will speak hereafter only of *C. reinhardtii*)[24] has many features that suited it to Sager's ambitions (fig. 14.3). As she mentions, it is unicellular, but more important, cells can be plated on agar medium, and the colonies that form may be replica-plated with filter paper disks. When *Chlamydomonas* cells are transferred from agar to liquid culture, they quickly develop two flagella and become motile. The cells have a single, large chloroplast, several mitochondria, and a light-sensitive eyespot. The medium for *autotrophic* growth (without an organic carbon source, in the light; not to be confused with nutritionally deficient auxotrophs) contains inorganic salts of nitrogen (ammonium or nitrate), sulfate, phosphate, calcium, and various trace elements. The carbon required is derived from atmospheric CO_2. A culture grown in the dark (heterotrophic culture) requires only acetate in addition to the inorganic salts as a source of carbon and energy. The dispensability of photosynthesis made feasible an extensive mutational analysis of chloroplast

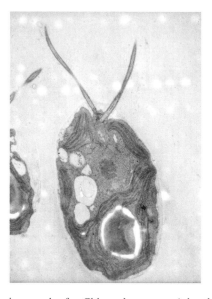

Fig. 14.3. Electron micrograph of a *Chlamydomonas reinhardtii* cell, showing the two anterior flagella, the multilamellar chloroplast at the base, extending (in three dimensions as a cup) around the margins of the cell. The chloroplast encloses a large pyrenoid, the site of CO_2 fixation, at its base, and electron-light starch granules at center left. The nucleus lies, poorly defined, in the center, with a slightly darker nucleolus. Electron micrograph courtesy of David Mitchell, SUNY Upstate Medical University.

function, to which both nuclear and organellar genes contribute. This is the same advantage enjoyed by yeast in its dispensability of mitochondrial function.

Haploid mt^+ and mt^- cells mate after suspension of agar-grown colonies in water in the light, which leads to starvation for nitrogen. In these conditions, they rapidly form two flagella per cell. After populations of the two mating types are mixed, cells clump, and then mating type pairs form head-to-head, as it were, at the flagellar ends. The pairs shed their cell walls and fuse to form diploid zygotes, in which the chloroplasts also fuse. The zygotes form a new cell wall. The zygotes, plated on agar, go through meiosis ("mature") in about five days in the light. They stick to the agar, making it easy to sweep unmated gametes away with a stream of water or a razor blade. Zygotes can be collected in a pile with a needle and then induced to germinate by spreading or replating them on fresh medium in the light.[25] Individual zygotes may be isolated from one another and from remaining, unmated gametic cells with a glass loop under a dissection microscope. Germination of zygotes begins with meiosis and culminates in the production of four (or, after one division, eight) haploid meiotic products. These may be pulled out into a line, thus arraying the resulting colonies for phenotypic analysis. For this, the colonies may be replica-plated to various media, much as is done with yeast. Because of the simplicity of these plant unicells and the ease of genetic analysis, *Chlamydomonas* is often called "the green yeast."

With these techniques, the Mendelian genetics of *C. reinhardtii* was well developed by the laboratories of Sager and Paul Levine by the late 1950s.[26] Before going into chloroplast heredity pursued by Sager, I describe the work by the Levine laboratory at Harvard between 1955 and 1975. Levine's group was equally influential in establishing *Chlamydomonas* as a model organism. Levine developed the nuclear genetics of the organism thoroughly and began his work on the genetics of photosynthesis thereafter. Originally interested in the mechanisms of recombination, Levine and Ebersold refined the techniques of tetrad analysis in the organism. They hoped to pursue genetic anomalies that turned up in other laboratories and to study the effect of environmental manipulations on the process of recombination. This was done at a time when the structure of chromosomes was still unclear; in particular, the number of strands of DNA was unknown, and the idea of "linkers" between discontinuous DNA segments (genes?) remained a real possibility. The work on recombination was ill fated, but the genetics of the alga nevertheless continued.

The search for more genetic markers became complicated by a severe lack of auxotrophs, largely because the uptake systems for amino acids in *Chlamydomonas* seemed to be absent or inefficient.[27] Only a few mutants, those with arginine and vitamin deficiencies, could be found. But the mutant hunts uncovered a variety of heterotrophic mutants—those that could not live on light and atmospheric CO_2 but required acetate as a carbon and energy source, even in the light. Such mutants, called *ac* (acetate-requiring), might have been expected, since wild-type cells could grow in the dark if acetate were provided, and one might predict a large number of mutable chloroplast functions. The Levine laboratory then turned its focus from recombination to these presumed photosynthetic mutants, especially because postdoctoral and other associates with suitable backgrounds in both photosynthesis and genetics were at hand.

According to members of the laboratory, *Chlamydomonas* "had to earn its citizenship in the community of photosynthetic research model organisms."[28] At that time, a number of other microbial systems (e.g., the algae *Chlorella* and *Scenedesmus* and the prokaryotic blue-green alga *Anacystis*), all lacking developed genetic techniques, had been used efficiently for studies of photosynthesis in vivo. Spinach had served as the classical source of chloroplasts for work in vitro. These largely biochemical and biophysical investigations had defined the path of electron transport and the assimilation of CO_2 at a high level of sophistication, although the number of gene products was not known, and confusion prevailed on some important steps in the process. The work with the *Chlamydomonas ac* mutants would soon yield a satisfying model of the genetic control of photosynthesis, including the resolution of some problems remaining from other organisms.[29] As time went on, the work branched into the structure of the chloroplast in relation to its role in photosynthetic reactions. The great preponderance of the photosynthesis-defective mutants were nuclear, and soon a large number of students and postdoctoral fellows from the Levine laboratory were committed to *Chlamydomonas* as a research organism.

Nicholas Gillham, a graduate student in the Levine laboratory between 1958 and 1962, would participate not only in this work but would soon focus on non-

Mendelian inheritance of some *ac* mutants as well as a number of drug-resistant mutants. His efforts, described later, proceeded largely independently of the Sager laboratory, to which I now return.

Organelle Genetics of *Chlamydomonas*

Sager's first major findings in 1954 set the stage for modern genetic analysis of chloroplast genomes.[30] She had isolated two streptomycin-resistant variants, one resistant to low concentrations of the drug, another resistant to five times that level. The first behaved as a simple Mendelian variant in crosses with the sensitive strain, with resistant and sensitive alleles segregating 2:2 in tetrads. The other variant, however, when mated with the sensitive strain, displayed two unusual characteristics. First, the trait did not segregate at all. The descendants of meiotic products were either all resistant or all sensitive, and they remained so in further vegetative growth. Second, the outcome depended on whether resistance was carried by the *mt+* or the *mt−* parent. Specifically, this uniparental pattern of transmission showed that for this character, the progeny displayed the phenotype of the *mt+* parent (fig. 14.4). Several significant exceptions appeared, however. In a few tetrads, all four meiotic products had the resistance phenotype of the *mt−* parent instead of the *mt+* parent. Several years of work would reveal still another, biparental, pattern, in which the nonchromosomal traits of both parents appeared among the meiotic products.[31] Application of ultraviolet light to the *mt+* parent (the one that normally contributes chloroplast genes to the progeny) greatly increased the frequency of such bi-

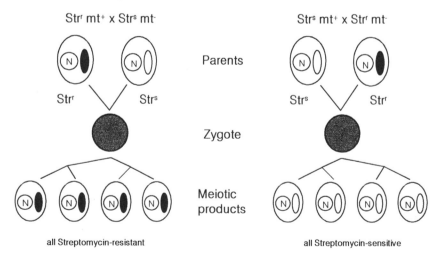

Fig. 14.4.   Uniparental inheritance of the streptomycin-resistance phenotype (*Strʳ*/ *Strˢ*) of *C. reinhardtii*, determined by a chloroplast gene. Note that the meiotic products carry only the chloroplast determinants of the *mt+* parent. Ovals: chloroplasts; circles: nuclei (N).

parental tetrads. Later work showed that these patterns correlated well with the fate of chloroplast DNA.[32] The significance of the biparental inheritance pattern was the possibility that recombination between the organellar genomes might take place and that recombinants might be detected among the mitotic descendants of the meiotic products. Before that could be done, more nonchromosomal mutants were needed.

Sager had described experiments suggesting that streptomycin was a mutagen for nonchromosomal but not for chromosomal genes.[33] Gillham was skeptical of this and began to work on nonchromosomal genes, showing that the evidence for streptomycin mutagenesis was not secure.[34] Nevertheless, the two laboratories managed to isolate of a variety of nonchromosomal mutations, which included some heterotrophic (ac) strains. A large class of mutations comprised those resistant to herbicides or to bacterial antibiotics other than streptomycin. Thus a stable of mutants for tests of recombination became available.

The genetics of the organelle genome of Chlamydomonas began with the analysis of the mitotic pedigree of individual meiotic products of the biparental tetrads. In the descendants of a given haploid meiotic product, several recombinant nonchromosomal genotypes were found.[35] To make this clear, realize that when zygotes form, the parental chloroplasts fuse. If, in zygotes yielding biparental inheritance, the chloroplast DNA of both haploid gametes survives, there is an opportunity for recombination. The multiple copies of chloroplast DNA in the diploid zygote, among which recombinant molecules might arise, are randomly apportioned to the meiotic products. Upon mitotic division, each meiotic product will give rise to descendants in which the various DNAs acquired from the zygotic cell are sorted out into pure lines. This is a quasi-random process, but there are few enough clusters of DNA molecules, or nucleoids (perhaps five or six) to assure such segregation within several divisions.[36] The recombinational analysis of the chloroplast genes could be done with colonies arising from either individual meiotic products or entire zygotes of the biparental class. The technique is simply to allow a colony to grow and then to determine the phenotypes of individual descendants after plating the cells of a colony on agar. From biparental zygotes, a variety of recombinants could be found, and certain linkages could be discerned. This technique did not yield a full, formal genetic map because of the large size of the DNA (192,000 nucleotide pairs) and the high frequency of recombination, which together obscure physical linkages. The current map is based on the actual DNA sequence, with the position of many of the classical mutations known precisely. Nevertheless, the recombinational genetics of an organellar genome had been pioneered with Chlamydomonas.

More recently, the inheritance of mitochondrial DNA in Chlamydomonas was shown to be uniparental, but the inheritance is the reverse of that of chloroplast DNA: the mt⁻ parent normally contributes mitochondrial DNA to the meiotic progeny.[37] The uniparental patterns of inheritance of chloroplast and mitochondrial DNAs reflect the selective destruction of the noninherited genome: rapidly for chloroplast DNA after the formation of zygotes, slowly for mitochondrial DNA. In the case of the former, this is arguably correlated with some modifications of the inherited DNA that may protect it from the action of a known nuclease.[38] The latter appears to be a nuclear gene product.

Through the efforts of Sager, Levine, and Gillham, *Chlamydomonas* became a model organism. Although organelle genetics was the leading phenomenon that brought it to serious attention after Moewus's day, the mutational dissection of photosynthesis, the structure of the chloroplast, mitosis, the sexual cycle, light responses, and the genetics of flagellar biogenesis also contribute to its prominence.[39] Much of the work on the last two subjects began much earlier with intense biochemical study in other organisms, but the availability of a model system for their genetic analysis has facilitated many major advances thereafter. Once well established, *Chlamydomonas* stock collections, a *Chlamydomonas* source book, and meetings bound the community further.[40] The study of *Chlamydomonas* biology continues with this momentum as a high-connectivity model organism for the study of a variety of biological problems.

The Parallel Fates of *Chlamydomonas* and Yeast

The early work on the organelle genetics of *Chlamydomonas* and yeast were carried out relatively independently, but inevitably the rationales soon paralleled one another. The isolation of antibiotic resistance mutations, such as streptomycin resistance of chloroplast ribosomes in *Chlamydomonas*, had their counterparts in mutations such as resistance to the respiratory inhibitor, oligomycin, and chloramphenicol resistance of mitochondrial ribosomes in yeast. These mutations were more localized than the generalized defects implied by acetate requirement in the alga or respiratory deficiency in yeast and much more useful than the rearrangements and deletions of DNA in petite yeast. Recombinational work in both organisms therefore accelerated. Yeast had a considerable advantage at this stage, owing to the small size of its mitochondrial DNA: 85,000 nucleotide pairs, less than half the size of *Chlamydomonas* chloroplast DNA.

In the 1970s, mutagens specific for the two organellar DNAs were discovered: fluorodeoxyuridine for *Chlamydomonas* chloroplast DNA and manganese ion for yeast mitochondrial DNA.[41] These compounds interfere with organellar DNA polymerases, leading to errors in replication. They greatly facilitated not only the genetic analysis then under way, but also the mutational analysis of a greater variety of organellar genes. Before that time, the search for organellar mutants was limited to specialized targets, such as the ribosomes. Other functions could be mutated, but a large background of nuclear mutants with similar phenotypes obscured them. These included segregational petites of yeast and nuclear mutants requiring acetate in *Chlamydomonas*. With organelle-specific mutagens and nuclear mutants affecting the formation and function of organelles, analysis of photosynthesis, respiration, organellar protein synthesis, and organelle biogenesis entered a new phase. In all of these areas, the theme of nuclear–organellar collaboration was common. Many discoveries have followed, especially in the area of targeting and localization of proteins, whether made in the organelle or the cytoplasm, into their proper places in organelles.

Research with both organisms wrestled with the problem of the approach to genetic homogeneity of the organellar DNA populations with mitotic growth. This proceeded much more rapidly than expected according to random segregation of the large number of copies of DNA in each cell. Bernard Dujon, with others, speculated that mechanisms for the clustering of DNAs derived from a common progenitor or that gene conversion among DNAs might account for rapid phenotypic segregation from the heterogeneous population, but without physical evidence.[42] The problem may be resolved by the recognition that mitochondrial DNA is not a group of independent circular molecules as assumed until recently, but linear, tandem arrays with substantial recombination among them, as discussed in the next paragraphs. Nevertheless, mapping studies yielded a more and more coherent, continuous map of yeast mitochondrial DNA up to 1975.[43] Mapping the organelle genomes by genetic techniques ceased in both organisms upon the advent of physical mapping techniques, restriction enzyme analysis of DNA, and DNA sequencing techniques.

Early attempts to visualize the DNA of yeast mitochondria with the electron microscope followed closely upon the discovery of circular DNA molecules in animal mitochondria. At that point, the case for the descent of mitochondria from bacterial symbionts became an easy assumption, and yeast workers expected, naturally enough, that their organism would also have circular mitochondrial DNA.[44] However, despite increasingly refined methods of isolation, few circular molecules were found. Because this form was not entirely absent, most workers assumed that the abundant linear DNA molecules represented circles broken during extraction. The substantial differences in length of these molecules (and some were quite long) required further assumptions: the long molecules were contaminating nuclear DNA; short molecules could be fragments originating either from nuclei or from mitochondrial circles. In the 1970s, the use of restriction enzymes (see appendix 3) for physical mapping of DNA molecules became possible, and to almost everyone's satisfaction, the techniques yielded a circular map of yeast mitochondrial DNA. So why were circular molecules so rare?

Molecular biologists generally forgot about this problem until a new technique of isolating and determining the contours of very large DNA molecules appeared. (The technique is an electrophoretic technique known as pulsed-field gel electrophoresis, or PFGE, combined with specific detection of mitochondrial DNA by hybridization with radioactive DNA probes.) The results again showed that few circular molecules appeared, and again that the linear molecules were quite heterogeneous in length.[45] The simplest explanation for these findings can be found in the "rolling circle" model of replication of plasmids (see fig. 7.3), in which a circular molecule can generate a long "tail," with multiple genomes in tandem, by replication of one strand peeling from the circle. As noted previously, this would keep descendant genomes connected and reduce the number of segregating units. The few circles seen in the electron microscope, in fact, often have a tail. Since the time of these recent discoveries, mitochondrial DNAs of many other lower organisms, including *C. reinhardtii* and *N. crassa*, have been found to have the same general organization.[46] In addition, some speculate that promiscuous recombination among the linear portions of these molecules may explain the branched DNAs often seen with the electron microscope. Physical (restriction) maps derived from circular DNAs and those derived from long, tandem arrays of DNA genomes will both yield a circular representation. This seems to have escaped many workers in the face of the virtually unassailable assumption that mitochondrial DNA, derived from a prokaryotic endosymbiont, must be circular and replicate by a "theta" mode, one circle at a time.[47]

Recently, a technique for DNA-mediated transformation of chloroplast DNA has been devised. It depends on coating microprojectiles with DNA (usually in plasmid form) and firing them at high velocity at a cell population. Sufficient DNA enters the chloroplast in this "biolistic" technique to effect transformation. Using the expression of a gene on the injected DNA as a selective marker, successfully transformed cells can be isolated. This has endowed *Chlamydomonas* with a bright future in the study of the functions of chloroplast (and mitochondrial) genes.

The mitochondrial RNAs of yeast (and *Neurospora*) revealed unusual phenomena with respect to their maturation into coding sequences. The DNA is transcribed into very large RNA molecules (primary transcripts) that are then processed into coding sequences for one or more mitochondrial proteins. Not only are these large transcripts cut into fragments, but they contain introns that are removed, the remaining parts of the transcripts being spliced together to construct coding regions. Of special interest is the finding that the introns of the RNAs themselves encode "maturase" proteins that initially help in the cut-and-splice mechanisms responsible for the introns' removal. A great deal of the early work on this problem relied on strictly genetic and biochemical techniques, well before the mitochondrial nucleic acids could be studied in detail.[48] This work is summarized in modern terms in Gillham's book, which points out the importance and widespread nature of this and related phenomena in organellar biology.[49]

The accounts above of yeast and algal organelle genetics illustrate a point remarked on before: as mutants accumulate and a genetic system is domesticated, mutations originally sought as markers become the entrée to intense functional studies of a variety of biological processes and structures in the model organism. My main purpose has been to indicate how and for what purpose these models were chosen and the major research efforts that illuminated mechanisms common to many organisms. These efforts expanded in their significance with the detection of DNA in mitochondria and chloroplasts. From that time, *Chlamydomonas* and yeast workers legitimized organellar genetics to those hostile to the idea of cytoplasmic inheritance, uncomprehending of genetic analysis, or both. The work on organelle genetics and the advance of bacterial genetics prepared the ground for the current acceptance of the symbiotic theory of organellar origins from bacteria.[50] Work on organellar genetics has contributed to our knowledge of how mitochondria and chloroplasts are assembled in a cooperative enterprise of nuclear and organellar genomes and how these organelles still function as partially autonomous entities, descended from prokaryotic invaders eons ago.

15

Yeast Becomes a Supermodel

Nevertheless, every major conclusion will have eventually to be tested directly in higher organisms. The point of any model system is to make these final tests possible, not to make them unnecessary.
—D. Botstein and G. Fink

Yeast Catches On

I define a *supermodel* as an organism that reveals and integrates many and diverse biological findings applying to most living things or to most members of a kingdom. Thus, *Drosophila* yielded many principles of heredity, development, and population genetics in eukaryotes. *E. coli* revealed many of the universal aspects of the structure, replication, mutation, and recombination of DNA and the regulation of its expression. Both organisms may reasonably be designated supermodels. In contrast, the bacteriophages domesticated after 1945 were quite diverse, and only collectively became models of viruses in general in their lifestyles and their relations to their hosts. *Neurospora* and *Aspergillus*, although they initiated the modern understanding of gene action and formal genetics and gave scientists the methodology of dissecting complex pathways by mutation, were preempted first by *E. coli* and eventually by another organism, yeast, that became a supermodel for eukaryotic cell biology and molecular genetics.

Yeast crept into this role slowly. As discussed earlier, Lindegren made complex contributions, both advertising the advantages of yeast and providing data that seemed to nullify them. Through the efforts of Ephrussi's group in France and of Roman, Hawthorne, and Mortimer in the United States, the domestication of yeast as a genetically tractable organism proceeded, but it was not a

199

prominent contributor until the 1970s. Seymour Fogel (1919–93), a corn geneticist who began work on yeast in 1958, later remarked dryly, "Most general geneticists took the position that when yeast researchers presented a consistent, understandable body of data, their reports would then receive appropriate attention and approbation."[1] That time arrived with the early work of Roman and Hawthorne, the isolation and genetic mapping of many mutants by Mortimer and his associates, and their distribution of mutants in a standard genetic background to both American and European geneticists. In Europe, the work of Ephrussi and Slonimski (France), Winge (Denmark), Giovanni Magni (Italy), Jean-Marie Wiame (1914–2000, Belgium), and Edward Bevan (b. 1926) and Donald Williamson (both in England) attracted associates to each laboratory. Members of these groups would meet in France with the American workers at Gif-sur-Yvette in 1963 at the second international yeast genetics meeting (the first was the 1961 meeting at Carbondale, Illinois). These meetings have continued on a biennial basis and were supplemented in alternate years by meetings at Cold Spring Harbor starting in 1975 under the title "Yeast Molecular Genetics."

The community of yeast geneticists took on a distinct flavor. As work on *E. coli* focused increasingly on the biochemical and molecular details of DNA, gene expression, and regulation, *E. coli* genetic techniques were used more and more as laboratory tools by bacterial molecular biologists innocent about classical, eukaryotic genetics.[2] In contrast, the yeast community was founded by individuals steeped in the traditions of corn and *Drosophila* genetics. Accordingly, their initial interests lay in classical problems such as the substructure of the gene, recombination, mutation, and cytoplasmic inheritance. The founders therefore imparted strong genetic sensibilities to the future of the field by training newcomers first in genetic approaches to research programs. The problems taken up included metabolic sequences, mitochondrial physiology, enzyme regulation, the steps of the cell cycle, intracellular signaling pathways, and cell morphogenesis and specificity. The early focus on genetic phenomena led naturally to early advances in genetic techniques. These included efficient mutagenesis, many clever selective techniques for obtaining mutants of all kinds in large numbers, facile dissection of asci for tetrad analysis, and the use of multiple mutant strains in complex genetic and physiological experiments.

Many biologists recognized the advantages of yeast as the attraction of *E. coli* and its phages began to wane. Like yeast geneticists, other biologists felt that yeast might be a model eukaryote, a counterpart of *E. coli*, and certainly more serviceable than *Neurospora*. Interest in the biology of eukaryotic cells grew strongly after 1970 at a time when the nature of the bacterial cell had become so well known. The problems of interest to molecular biologists also changed. In addition to looking at issues that were relatively easy to formulate intellectually, such as DNA replication, coding, or the behavior of the *lac* operon, molecular geneticists began to seek an understanding of even more complex features of the eukaryotic cell and how they were integrated functionally and structurally. These features included a distinct nucleus; linear, multiple chromosomes; a complex cytoskeleton; mitochondria; intracellular membrane systems; and the mitotic and meiotic divisions. None of these features had close counterparts in bacteria, although the many remaining prokaryotic workers also developed a taste for complexity and cell integration at the same time. The early problems for which *E. coli* was suited as an experimental organism were no longer so mysterious or exciting. However, the remaining fundamental questions could not fully occupy the interests of the army of bacterial molecular biologists or of the institutions that supported them financially. These molecular biologists therefore turned their focus to eukaryotic cells. They knew that funds would be available for those who found an *E. coli*-like model for the human condition at the cellular level.

Yeast finally stepped into that role in the mid-1960s, aided by an important institutional mechanism in 1970, the annual Cold Spring Harbor yeast genetics course. James Watson, who became director of the Cold Spring Harbor Laboratory in 1968, believed that yeast would be a suitable model eukaryote.[3] He invited Gerald Fink (b. 1940) and Fred Sherman (b. 1932, a student of Roman's) to organize a yeast genetics course in the summer of 1970. Fink, who had taken up yeast as a graduate student and had then worked on *Salmonella* as a postdoctoral fellow, became a proselytizer for yeast in the late 1960s. The course he and Sherman and their successors taught was modeled closely on Delbrück's Cold Spring Harbor phage course. Basic yeast techniques and recent improvements are demonstrated each year in exercises designed for the purpose. Students are driven to technical innovations and the use of the available techniques in original experiments. Those taking the course are often accomplished scientists seeking a simple organism for their own work. Therefore, the sophistication and heterogeneity of scientific discourse during the three-week series, involving students, instructors, and visiting lecturers, yields ideas and collaborative opportunities that benefit the attendees. The course provides strains and manuals that investigators take back to their home institutions and renders them and their students independent members of the yeast community if they wish to continue. The course became a training ground for a significant number of the principal investigators using yeast since 1970. In addition, a sustaining thread of yeast research has been the two series of international meetings mentioned previously. The older series became too large for Cold Spring Harbor and now, sponsored by the Genetics Society of America, attracts more than 1000 attendees.[4]

The Molecular Approach

A series of mutants affecting macromolecular synthesis had been isolated by Leland Hartwell (b. 1939) and Calvin McLaughlin (b. 1936) in the late 1960s, and Hartwell had then used a subset of these mutants to begin investigating the cell division cycle as a gene action sequence in the early 1970s.[5] This extraordinary work, which I describe later, served as an early demonstration of what yeast could do to resolve the components of a fundamental, complex process and to order them in time of action. The work was well in advance of comparable knowledge of the cell cycle of *E. coli* and inspired a number of people to enter the field of yeast molecular biology.

The transformation of yeast with plasmid DNA was accomplished in 1978. This and other molecular techniques deserve separate mention because of their importance to all studies thereafter. Yeast transformation is the fulcrum of the period between 1960 and the present, virtually distinguishing two generations. The manipulation of DNA by the use of restriction enzymes, DNA sequencing techniques, and the cloning of foreign DNA in *E. coli* (appendix 3) had advanced considerably since about 1974 when these techniques were introduced. The advances encouraged yeast investigators to apply these techniques to their organism. The steps leading to transformation are well described in a personal account by David Botstein (b. 1942), a phage geneticist at the Massachusetts Institute of Technology who, before his tenure decision, was told that the days of prokaroytic genetics were numbered.[6] In the article cited, he disputes this, especially given the later impact of new molecular techniques on bacterial research. However, a future in eukaryotic cell biology had its practical attraction. He and his postdoctoral student, Ira Herskowitz (b. 1946), took the yeast genetics course at Cold Spring Harbor in its second year (1971). Botstein said later:

> We, the prokaryote geneticists, brought with us an already sophisticated molecular thought process, following the intellectual paths illuminated by Jacob and Monod on the one side and Luria, Delbrück, and Hershey on the other. We brought with us an obsession with DNA and the central dogma, especially with the analysis of regulation at the level of transcription and, as we often hoped but rarely found, translation. We worried about messengers, ribosomes, and tRNA. Our favorite genetic tools were the selective cross and the *cis-trans* test.[7]

The combination of molecular and genetic sensibilities among yeast workers would enormously expand the scope of yeast studies once DNA manipulations became routine. A small cadre of investigators including Fink, Botstein, Herskowitz, and several others began another step in the domestication of yeast. A general goal of the new yeast molecular geneticists was discovering which molecular attributes of yeast were shared with prokaryotes and which were not.

The molecular technology of yeast still lagged that of *E. coli*, and a concerted effort was made over the period 1974 to 1979 to develop strains and tools that could make yeast "the *E. coli* of eukaryotes." A joint sabbatical of Botstein, Fink, and John Roth (b. 1939) at Cold Spring Harbor in 1974–75, together with Ray Gesteland (b. 1938), already on the staff of the laboratory, facilitated a concerted effort in this direction at the very time that restriction enzymes and recombinant DNA techniques appeared. Discussions among the group led to further standardization of wild type strains and to improvements of molecular techniques. The groundwork was thereby laid for making a DNA library of yeast, using small *E. coli* plasmids for cloning.

Libraries of yeast DNA are made by physically breaking or enzymatically cutting DNA extracted from yeast and inserting the DNA fragments into plasmids as parts of these circular molecules (appendix 3). The plasmids, introduced by transformation into *E. coli*, multiply as plasmid clones within the cell, and the cells multiply as colonies (clones of cells) on agar medium. From any one of these colonies, one may isolate many copies of a particular plasmid in pure form. The library is simply the large number of bacterial colonies that collectively contain all or most genes of the yeast genome. DNA libraries of yeast were prepared by several investigators, including John Carbon (b. 1933) shortly after the introduction of DNA cloning techniques and paved the way for tests of yeast transformation. Out of the 1974–75 sabbatical of the group involved in improving molecular techniques came plans for the first molecular biology of yeast meeting at Cold Spring Harbor in the summer of 1975, which drew 167 participants. Clearly, a modern yeast community had already formed, many of its members united by their common experience in the yeast genetics course started in 1970.

The development of transformation techniques for yeast was accelerated by Carbon's discovery of a plasmid in his yeast library carrying the yeast *LEU2* gene.[8] The library had been made in a strain of *E. coli* carrying the *leuB⁻* mutation as a marker to distinguish it from contaminant bacteria. To Carbon's initial surprise, one of the plasmids bearing yeast DNA overcame the biochemical deficiency caused by the *leuB⁻* mutation of the bacterium. A yeast gene, *LEU2* as it turned out, appeared to function in *E. coli* to carry out a step in leucine biosynthesis common to the two organisms.[9] Remarkable as that was (yet soon to be matched by other such cases), the significance of the discovery here is how this plasmid was used. The *LEU2* gene of yeast had now been identified and could be prepared in pure form (as part of the plasmid) in a vast number of copies. Fink's group used this preparation to test whether the plasmid could enter a yeast cell and overcome the biochemical deficiency of the *leu2⁻* mutant of yeast. This effort succeeded, thanks to the extremely tight selection for leucine-independence that detected about one transformant in 100,000 to 1 million cells tested. The results showed not only that the plasmid could enter cells but that it integrated into the yeast chromosome with high efficiency at the site of the *leu2* mutation by simple, homologous recombination.[10] Since that time, this remarkable property of exclusively homologous recombination into yeast chromosomes has been shown to be the rule for all yeast DNAs.

The plasmid used in the initial studies cannot replicate by itself in the yeast cell, but only in *E. coli*. If the plasmid DNA is to replicate in yeast, it must integrate into the chromosome, where it replicates as part of the chromosome. The integrating plasmids are not easy to retrieve but are useful in targeting genes to their proper location in the chromosomes. Soon, however, a variety of plasmids was artificially constructed for different experimental purposes, some of them yielding much higher frequencies of transformation. The two most significant of these newer types each replicate as plasmids in the yeast cell as well as in *E. coli*. One of them maintains itself in yeast in high numbers per cell; the other has a centromere and is maintained faithfully at about one copy per cell as a circular "mini-chromosome."

Libraries of the yeast genome made with plasmids (vehicles for DNA fragments) that can replicate in yeast permitted investigators to isolate individual genes easily. An entire library of plasmids collectively carrying all the DNA of a wild-type genome may be added to a population of a mutant yeast strain. The mixture is then plated on a medium selective for growth of transformants with the wild-type phenotype. Such a phenotype will be imparted to any mutant cell that acquires a copy of the wild-type allele of the resident mutant gene as part of a plasmid. Because these new plasmids replicate in the cell, a transformant selected in this way will yield the intact plasmid after extraction and simple fractionation of cellular DNA. The plasmid may then be cloned by introducing it into *E. coli*. This technique, called "cloning by complementation," is a primary method of isolating genes of yeast (and now, in other eukaryotic organisms). The integrating and the replicating plasmids, the former having to integrate into the yeast genome at a location homologous to the yeast DNA borne by the plasmid, give extraordinary experimental range to yeast studies. The techniques permit investigators to (i) find the chromosomal location of any cloned DNA, even if it has not been identified with a function; (ii) retrieve DNA sequences from the yeast genome; (iii) maintain and express any DNA sequence, yeast or foreign, at high or low copy number; (iv) integrate foreign DNA into the yeast genome if it is flanked by homologous yeast DNA; (v) disrupt or delete any gene for which one has a cloned copy; and (vi) replace one gene with an alternative form of that gene.

Additional techniques of isolating specific genes include isolation of a gene on the basis of its homology with a gene in another organism or by detecting with an antibody its protein product, produced by a clone in specially engineered bacterial or phage systems. The complementation technique mentioned earlier and the property of homologous targeting of transforming DNA provide opportunities to perform "reverse genetics." This rationale contrasts with the older methods of obtaining mutants by random mutagenesis and clever selection schemes. Reverse genetics starts with a protein or DNA rather than with mutations of the organism. The gene, once in hand, may be altered in specific ways or systematically mutagenized in the test tube before introducing it into the yeast genome and determining the effects of the changes. The techniques are potentially applicable to any transformable species. The elegance of yeast plasmid and transformation techniques therefore makes yeast the organism of choice to study any process it shares with eukaryotic organisms. From the discovery of transformation onward, yeast has become the most sophisticated biological system of any eukaryote for genetic and molecular work.

The Scope of Yeast Studies: Extensions of Genetic Work

Having described the major technical advances with yeast—the ease of isolating mutants, the genetic procedures that exploit the high rates of recombination, and transformation and gene-targeting methods—I summarize in this section some of the major studies in yeast biology. After cytological research showed that yeast had most of the attributes of eukaryotic cells, yeast genetics and molecular biology exploded in a variety of directions, and no single account can fully convey the resulting depth of knowledge.[11] Two large multivolume reviews of yeast molecular biology have been published by Cold Spring Harbor Laboratory Press, one of them (1981–82) obsolete before it appeared, the second (1991–97) already due for major updating.[12] In the interval, the yeast community kept reminding all biologists, without modesty, that its time had come.[13]

Gene Conversion, Meiosis, and Other DNA Transactions

Fogel switched his research in the 1960s from corn to yeast for the same reasons that his fellow graduate student, Roman, did. Fogel could not grow corn in New York City, and he believed yeast could answer interesting genetic questions. He developed a more efficient micromanipulator and began a program involving tetrad analysis that allowed him to detect gene conversion easily. Fogel's and Mortimer's groups studied the phenomenon in great detail. This led, in turn, to molecular models, proposed by others, for the mechanism of crossing over. Together with continuing work on *Neurospora* and other filamentous fungi, their more extensive data led to important refinements of knowledge of gene conversion and crossing over. The popular, DNA-based Meselson-Radding model of molecular recombination that prevailed from 1975 succumbed in the mid-1980s to the double-strand break-repair model that now holds sway.[14] This work is among the most celebrated contributions of yeast to genetics and applies potentially to all eukaryotes.

The study of meiosis began with the isolation of mutations that disturb the process in specific ways. Integrated with the studies of molecular recombination noted above, many steps of chromosome behavior are now better understood. In yeast, extremely high meiotic recombination rates prevail. Populations of diploid cells can easily be synchronized in the steps of meiosis for biochemical studies. They have revealed details of the pairing of homologous chromosomes in prophase I, the establishment and behavior of the synaptonemal complex (an apparatus required in the later steps of recombination), and the proper disjunction of chromosomes in the first meiotic division. Many of the genes involved encode specific proteins that can be studied in the test tube. Again, these are studies of universal significance in one of the most distinctive processes of sexually reproducing organisms, a process that simply cannot be pursued in such depth with any multicellular organism. The studies are impossible to pursue in filamentous fungi because homogeneous populations of meiotic cells cannot be prepared by any means, and in any case, recombination and gene-conversion rates are at least 10-fold lower. Even tetrad analysis is slower now in these organisms compared to vastly improved methods in yeast.

Study of DNA repair after application of DNA-damaging agents has also provided basic, widely applicable knowledge of this process. Mutations imparting sensitivity to such agents have been identified, and the roles of the proteins encoded by the corresponding genes have been elucidated. Many of these genes have roles in meiosis as well, much as repair and recombination functions overlap in *E. coli*. Here, some of the genes of yeast are clearly homologous to and functionally interchangeable with mammalian genes.

The themes of homology and similar function of yeast and mammalian genes have now entered many areas of research, solidifying yeast as a model eukaryote. Equally promising because of the facile genetic technology one can apply to it, yeast cells provide a biological milieu in which to study foreign proteins in familiar territory. The opportunity arose with the discovery of transformation of yeast, by which foreign genes, carried by plasmids, are introduced into the chromosomal material of yeast and are expressed by the yeast cell.

The Translation Process

After the genetic code had been fully determined (ca. 1963), Sherman began in the mid-1960s to study the major steps of gene–protein relations and translation in yeast. Sherman and his colleagues closely analyzed the initiation of translation of a gene, *CYC1*, encoding iso-1-cytochrome *c*, a mitochondrial protein encoded in the nucleus.[15] He showed that, as in *E. coli*, AUG, the nucleotide triplet (for methionine) was the only codon used for initiation of the synthesis of the polypeptide. The use of mutations near the initiation codon revealed that only the first AUG in an mRNA would serve as the initiator codon. These findings, preceding comparable studies in other eukaryotes, exploited both the mutational and the recombinational techniques possible with yeast. The studies were all the more remarkable in view of the lack of gene isolation and DNA sequencing techniques at the time. Instead, the mutational alterations of the N-terminal part of the protein were determined directly by sequence analysis of the isolated, mutant proteins. The changes were then used to infer the corresponding coding sequences. Later work in mammals and yeast showed that the biochemical steps required to initiate translation of yeast mRNAs (as opposed to the initiating codon) differ from those of prokaryotes.

A study of the coordinated control of amino acid metabolism in the 1970s initially promised insight into the mechanics of this global gene control system. The system, referred to as "general amino acid control," was initially surprising: starvation for any amino acid would lead to the derepression of enzymes in all or most amino-acid synthetic pathways. It was first described in 1965 in *Neurospora*, but investigations in the next decade were largely descriptive and physiological.[16] However, in yeast, two sorts of mutants affecting regulation of the comparable system were isolated in 1975. One was unable to derepress a suite of genes of amino acid biosynthesis (null, or *GCN*), the other was unable to turn them off (derepressed, or *GCD*). These mutants lived up to their promise in defining the general control pathway. They showed that a single protein, the product of the *GCN4* gene, activated a large array of genes encoding biosynthetic enzymes.

In the course of the study, however, insight into two other more fundamental issues arose. The first concerned the control of the synthesis of the *GCN4* protein. This protein is a straightforward transcriptional regulatory protein that binds to DNA sequences upstream (near the promoter) of many genes encoding enzymes of amino acid biosynthesis. As it does so, it exerts positive control, leading to derepression (or, better, activation). Strains mutant for this gene are, naturally, of the *GCN* class. However, the concentration of the *GCN4* protein in the wild type is determined not by the rate of transcription of the *GCN4* gene into its mRNA, but of the rate of translation of *GCN4* mRNA. Here, nucleotide sequences in the unusually long noncoding "leader" (the RNA sequence prior to the polypeptide-coding region) of the mRNA modulate the frequency with which ribosomes, scanning from the beginning of the mRNA, gain access to the *GCN4* coding region. As amino acid starvation proceeds, the process of translating the *GCN4* mRNA becomes more and more efficient. As more *GCN4* protein is made, amino acid biosynthetic genes become more and more active.

The ribosome movements along the *GCN4* mRNA are affected by the levels and interactions of proteins of the protein synthetic machinery, the second issue opened up by studies of general amino acid control. Mutations of these proteins (among which were numerous *GCN* and *GCD* gene products) led to their identification as fundamental participants in general protein synthesis.[17] Yeast now offers opportunities for the most detailed studies of this process in eukaryotes.

Metabolism and Enzyme Regulation

Many laboratories pursued gene–enzyme relationships by mutationally dissecting metabolic pathways, both biosynthetic and catabolic. As these studies progressed in the 1960s with the isolation of more mutants with increasingly subtle disturbances of function, interest in the regulation of the enzymes of each pathway became the focus of research. The reigning model of enzyme regulation was that of Jacob and Monod, especially because so many yeast biochemists had just left the area of prokaryotic molecular genetics. Thus the early gene–enzyme work in a number of biosynthetic and catabolic pathways provided a prebuilt foundation for derivative regulatory studies. The study of histidine mutants, in fact, would be the point of embarkation for Fink's major contributions to yeast molecular biology. Fink uncovered the general amino acid control system of yeast in 1975 using *his* mutants, demonstrated a three-enzyme protein encoded by the *HIS4* gene in the late 1970s, and achieved DNA-mediated transformation of yeast in 1978 with the *LEU2* plasmid. The last of these projects could not have been accomplished as easily without prior work in this period on the yeast leucine metabolic pathway and the mutant strains available at the time.

The advent of DNA sequencing and of transformation in yeast permitted detailed study of the structure of genes and their controlling elements. Both the *cis*-acting regulatory sequences (the upstream activation sequences, enhancers, and RNA-polymerase binding site, the promoter region) and the binding proteins themselves were subjected to manipulations of sequence. Many of the current models of transcription, transcriptional controlling elements, and the kinetics of gene regulation have emerged from these studies and have been applied to other organisms.[18]

Mitochondria

The study of mitochondria (chapter 14) began in the 1940s and continues still at a most sophisticated level. The work of the 1960s and 1970s dealt mainly with genetics of the organellar DNA and the steps of organellar biogenesis. The latter study became the origin of interests in targeting of proteins, once made, to their proper locations in the cell, in this case, the mitochondrion. Yeast and *Neurospora* share predominance in this area, for which they are both well suited in somewhat different ways.

The Scope of Yeast Studies: Complex Processes

Progress in disentangling the complexity of certain features of the eukaryotic cell enabled yeast to supplant the filamentous fungi definitively as a model eukaryote.

The Cell Cycle

Hartwell demonstrated spectacularly the potential of yeast for the analysis of complex cellular processes in the early 1970s in his study of the yeast cell cycle.[19] Hartwell, who had worked on bacteriophage, *E. coli*, and animal viruses before taking up yeast, obtained a series of temperature-sensitive mutants of yeast that could not progress through the cell cycle. They were different from mutants affected in protein, RNA, or DNA synthesis in that they ceased progress through the cell cycle at specific and characteristic points, revealed by microscopic monitoring of nuclear division and budding. By 1974, 35 different genes had been identified in this way, distinguished from one another by complementation and linkage analysis and by their apparent times of action in the cell cycle. More detailed measurements of the initiation and progress of DNA synthesis and of the distribution of DNA to daughter cells could be correlated with the point at which the cell cycle was arrested.

The coordination of nuclear events (DNA replication and nuclear division) and budding (cellular division) was indicated by their coupled response to certain mutations, among which were those of the *CDC28* gene. The function of the *CDC28* protein (abbreviated now as Cdc28p) is embodied in the term "start," the earliest step of cell reproduction as normally pictured, from which the steps of the nuclear cycle and budding proceed more independently in two obligatory sequences. More detailed cytological observations, many with the electron microscope, specified more exactly the points at which various mutations had their effects. For sets of phenotypically indistiguishable mutations lying in different genes, clever methods for determining their relative order of function and even their physical interactions in the cycle were devised. The insights about the cell cycle derived at each stage fed hypotheses that generated new types of mutants, biochemical information on the action of the corresponding genes, and a picture of how they fit into the web of gene action.

Two features of this work stand out. One is the success in ignoring distraction by suggestive mutants not strictly affecting the cell cycle. The other is the coherent formulation of a cell cycle model over 4–10 years without the use of recombinant DNA techniques. The success of Hartwell's laboratory and the universal importance of cell division greatly enhanced the reputation of yeast as a model for eukaryotic organisms after the early 1970s. The work inspired many studies of complex cellular processes in other organisms. Independent work on the fission yeast *Schizosaccharomyces pombe*, begun in the mid-1970s by Paul Nurse (b. 1949), generalized the picture to another and quite different eukaryotic species.[20] The work on both yeasts has influenced cell biologists to use the names of many yeast proteins for homologous proteins of the mammalian cell cycle. Hartwell shared the 2001 Nobel Prize in Physiology or Medicine with Nurse and Timothy Hunt (b. 1943) for their contributions to understanding the cell cycle.

Mating Type

Studies of the mating type loci of yeast opened a window into the area of cell type determination and the regulation of a set of genes involved in the transition of haploidy to diploidy. Yeast cells of different mating types (**a** and α) can recognize one another through pheromonal signaling, condition themselves for fusion, and turn off both of their mating capabilities as they fuse to form a diploid cell. The control of these steps lies in proteins encoded in the regulatory genes of the mating-type regions. These are the *MATa1* gene of the *MATa* locus and the *MATα1* and *MATα2* genes of the *MATα* locus. These genes regulate the transcription of many other genes, present in both mating types, some active in **a** cells, some active in α cells, some active in both kinds of haploid cells, and some active only in the **a**/α diploid. The *MATα1* gene, for instance, activates the α-specific genes required for the α phenotype; the *MATα2* gene is required to turn off the **a**-specific genes in the same cells. The *a1* gene is not needed in the **a** cells because the **a**-specific genes are active by default, and the α-specific genes are not transcribed in a cells because there is no α*1* gene to activate their expression. When the diploid forms, the *a1* protein and the α2 protein form a complex that blocks the synthesis of the α*1* protein (and hence expression of α-specific genes) and all genes needed only in the haploid cells. This extraordinary, coordinated picture of gene action provides a model for the complex genetic maneuvers seen in differentiation of cell type in higher organisms.[21] In particular, it demonstrates how few proteins are needed to govern a dependable switch between three specific cell phenotypes with large differences in their constellations of gene expression.

Protein Targeting

In another important program, Randy Schekman (b. 1948), modeling his studies on Hartwell's rationales, embarked upon a study of protein secretion.[22] Virtually all cells, both pro- and eukaryotic, display this process. Yeast does so as it secretes the proteins that become part of the cell wall and the digestive enzymes that diffuse to the medium where they break down complex carbohydrates, proteins, and phosphate compounds before assimilation of the products by the cell. More generally, however, the process is one of many examples of targeting proteins to specific destinations within the cell, not only to the vesicles that carry some of them outside the cell. The work, which began with the isolation of mutants for many genes necessary for enzyme secretion, has now provided a picture of the large number of steps by which proteins find their way to the proper places. The steps include translation from mRNAs located near the endoplasmic reticulum, entry of the translated polypeptides into the endoplasmic reticulum, the addition of carbohydrate side-chains (glycosylation) to them, and sorting the polypeptides in the Golgi apparatus to vesicles destined for secretion or to various organelles such as the vacuole.[23] Although these studies came after considerable biochemical work in mammalian systems on protein targeting, they greatly extended the understanding of the process in eukaryotes. Once again, the genetic approach and recombinant DNA techniques have shown that many of the proteins required in the process are not only homologous to their mammalian counterparts, but, in some cases, interchangeable with them.

A theme has played out in these studies. In the case of the general amino acid control work, the analysis of mating-type determination, and the placing of the *CDC* genes into a sequence in the cell cycle, ordered regulatory systems took form. The formal logic differs little from the rationales developed by Beadle and Tatum for determining the roles of genes in enzyme sequences. This theme has pervaded most of yeast molecular genetics as it is applied to chains of reactions required for responses of cells to internal or external signals. For instance, the protein encoded by the *CDC42* gene (Cdc42p) is a member of the family of rho-type GTPases, which can respond to internal or external chemical stimuli and activate a cascade of three protein kinases. In these cascades, one kinase phosphorylates the second kinase, which enables it to phosphorylate the third kinase.[24] The last kinase phosphorylates a final protein, enabling it to enter the nucleus and act as a transcriptional activator for the suite of genes appropriate to the original stimulus. Many cascade modules of this type, some participating crucially in the cell cycle and many of them interconnected in their action, behave quite similarly and are found in all eukaryotes. They are used by cells in different combinations to control diverse, multigenic functions and responses. Although the networks embodied by these proteins are extraordinarily complex, their roles can be judged by the effects of individual mutations affecting one or more specific cell phenotypes. Yeast has been and remains a model for the analysis of development in multicellular organisms, where many similar signaling cascades had been characterized previously with strictly biochemical techniques.

Morphogenesis might have seemed a particularly unlikely area in which yeast could contribute. However, if one thinks of cell morphogenesis, one immediately realizes that the determination of cell shape, the positioning of the bud, the organization of the cytoskeleton, the particulars of the budding process, and the building of the cell wall all have counterparts in higher eukaryotes. Many such cellular attributes are vital components of the morphogenetic processes in the development of multicellular eukaryotes, the establishment of cell polarity being a prime example. A large literature on these matters has accumulated in the last decade.[25]

The Yeast Genome

The year 1996 saw the publication of the entire DNA sequence of the 16 yeast chromosomes, the first genomic sequence of a eukaryote.[26] This accomplishment emerged from a collaboration of many yeast molecular biologists in Europe, the United States, and Japan. The rapid completion of the sequence reflected in part the small size of the genome, 12.8 megabases (Mb), about three times the size of the entire *E. coli* chromosome (4.6 Mb). It also reflected the willingness on the part of participating laboratories to set other goals aside temporarily to get the job done. This landmark has given biologists in general something quite special. For a start, it confirmed that less than half of the genes, recognized by computer as potential protein-coding regions (*open reading frames*, or ORFs), had been identified by mutation in previous studies. Most of the functions of these genes remain unknown, although homology of some genes with those known in other organisms has provided clues to their roles. The known genes could be allocated to various cell functions such as macromolecular synthesis, the cell cycle, meiosis and sporulation, small-molecule metabolism, and the like. The sequencing effort demonstrated straightforwardly that a free-living, eukaryotic cell needed no more than 6000 genes to survive in a competitive world. (This figure, low in comparison with some previous guesses, is now matched by an even more surprising figure: human cells appear to have only 30,000 to 70,000 genes, although alternative ways of expressing these genes and splicing their transcripts greatly increases the diversity of encoded proteins.)[27]

The value of the yeast genome sequence for further work has only begun to be appreciated. One technical advantage of having the genomic sequence lies in the opportunity to analyze the expression of all genes at one time. Methods were developed to determine which mRNAs are present at any one time under given conditions and to determine the sequence of change of the population of mRNA molecules in cells after transferring them from one environment to another. The technology to do such experiments depends on robotically made microarrays of the different genes. These arrays may have thousands of spots of DNA on a microscope slide, all in known arrangements. These spots act as hybridization traps for homologous mRNA samples extracted from the cell cultures of interest. With fluorescent labeling of the DNA copies of extracted cellular RNAs, their relative amounts trapped by the spots on the array can be estimated by optical devices that scan the arrays and feed the data into a computer. Such information indicates whether the amount of each mRNA at one time or in one condition is greater, less, or unchanged than in another time or condition. Computer analysis and ordering of the data allows one to visualize global changes in mRNA synthesis or stability.[28] Chapter 18 gives a fuller account of these techniques with a depiction of a microarray.

As one example, the genes active in cells grown in rich medium can be compared to those active in minimal medium. Or heat shock will evoke a response from a special set of genes as the cell initiates adaptation to high temperature. Finally, the program of meiosis and associated events may be inferred by identifying sets of genes active at different times in cells undergoing meiosis in synchronous fashion.[29] Such data reveal the activity of many genes previously unidentified by mutation. Those that seem particularly significant in a given condition can be isolated and studied further by reverse-genetic techniques to see what role they play in the process under study.

Another way of identifying the roles of unknown genes depends on gene disruption. Because yeast DNAs introduced by transformation target the homologous chromosomal gene and recombine with it, deficient gene copies can be used to target and replace the resident gene, a capability mentioned previously. Systematic, genome-wide gene disruption surveys now seek to determine which genes are indispensable and which are not.[30] The roles of dispensable genes can then be focused upon by less severe mutational studies. In these surveys, investigators have confirmed earlier indications that most (ca. 80%) yeast genes are dispensable for growth in a rich medium.[31] Gene dispensability is not well understood, but is plainly a property even of the stripped-down yeast cell, not to mention genomes of more complex organisms that have been sequenced since 1996. The observation suggests that the robustness of a highly evolved system may inevitably include extensive redundancy of function (see chapter 18).[32] The idea that a built-in reserve of alternative ways of achieving normal overall function extends the engineering analogies for living things. The observations seriously belie impressions created by the one-gene, one-enzyme phase of biochemical genetics.

The Fitness of Yeast

Yeast has many advantages as an experimental organism. Foremost, of course, are its considerable technical advantages. These were not obvious for some time, and considerable effort was devoted to overcoming biological impediments. These included the tough cell walls and that of the tiny ascus, the homothallism of most natural strains, and the genetic diversity of stocks. Perhaps as important in the rise of yeast biology were two factors I have also mentioned before. These were, first, the entrance of the modern yeast community from the field of prokaryotic molecular biology into a field already dominated by geneticists. The second factor was the rapid increase in the number of investigators in the field. The latter factor was a self-propagating, accelerating process arising from opportunities to study fundamental problems, the growing technical resources of yeast, and the success of the studies after 1970. With more investigators, more areas were investigated. As investigations proceeded, they overlapped with others on related problems until many large areas achieved confluence. One indication of this confluence is the increasing tendency of investigators on different subjects to isolate mutations of the same gene independently and only later recognize the identity. This has increased the appreciation of the interconnection of many genes' activities in more than one complex process. The connections that have developed among several levels of biological organization therefore provide a rich context in which any new observation or mutation can be related to previous work and quickly interpreted. Investigators are gratified by this increasingly familiar territory and by the approbation of colleagues. Generalizations have developed rapidly in an atmosphere of cooperation and competition.

After 1970, yeast workers not only initiated major programs in basic cellular and molecular biology but appropriated research problems initiated in other organisms. The fitness of yeast for study of these problems, hinted at long before by Lindegren, was recognized not only by those already working on the organism, but also by the wider biological community. As a result, yeast soon gained hegemony over the field of eukaryotic molecular and cellular biology, substantially eclipsing studies of similar problems in the filamentous fungi. In short order, yeast became a supermodel, challenging even *E. coli* in the visibility of its contributions.

To illustrate this point, I describe some biological problems originating in *Neurospora* and *Aspergillus* research that were appropriated by the yeast community. In doing so, I will pose the question of why so few of the workers on filamentous fungi changed to work on yeast.

Because modern studies of gene transmission and function began in *Neurospora* and *Aspergillus*, many important observations were first made in these organisms. Among the genetic phenomena is gene conversion. The discovery of gene conversion in 1955 by Mary Mitchell (chapter 3), and, it must be said, in 1956 by Roman in yeast, sparked a long series of analyses in *Neurospora*, *Sordaria*, *Ascobolus*, and *Podospora*, all filamentous fungi amenable to tetrad analysis. The advantage of these organisms was that ascospore color could be used as a marker, thereby obviating the need to dissect many tetrads to gather data. For a decade, observations prompted a number of molecular models and refinements, setting the stage for the yeast work in Fogel's laboratory. Here, technical advantages of yeast overwhelmed studies in the other organisms and led to the major advances I have discussed. Relevant work continues in the filamentous fungi, in which significant variations in gene conversion and recombination mechanisms prevail, though without threatening the basic picture drawn from the yeast work.[33]

General amino acid control was first defined as a global regulatory mechanism in *Neurospora* in the mid-1960s. The observations showed that starvation of an auxotroph for a single amino acid led, surprisingly, to the derepression of many enzymes of all amino-acid biosynthetic pathways. However, the work in *Neurospora* stopped after the few workers on the phenomenon had defined it with physiological observations. No early studies succeeded in deranging the process by mutation. By 1979, when the first such mutations of *Neurospora* appeared, mutational studies were well under way in yeast. Indeed, the yeast work became paradigmatic for further studies in *Neurospora*. The yeast work soon outstripped that of *Neurospora* due to the facility with which mutants could be isolated, put in proper order of action, and characterized at the molecular level. (The imagination and focus of the major investigator, Alan Hinnebusch, should not be overlooked.) Once transformation techniques became available in yeast, the program became a major contribution to the knowledge of protein synthesis. The comparatively small *Neurospora* community found more promising problems to pursue; the promise of these problems resided in many cases in the fact that no one working with yeast was pursuing them.

Study of the ATPase found in the plasma membrane of many lower eukaryotes followed a similar pattern. The energetics of the plasma membrane and its novel, proton-pumping ATPase was inferred with microelectrode studies of *Neurospora* in the laboratory of Clifford L. Slayman (b. 1937).[34] This ATPase was isolated and characterized by the laboratories of Carolyn W. Slayman (b. 1936) and Eugene Scarborough (b. 1940).[35] Yet C. W. Slayman switched her studies to yeast because of the ease with which mutational and molecular studies could be pursued after 1980. She was one of the few to make the switch.

The physiology of the yeast vacuole had been explored before 1975, particularly as a site of protein degradation and the accumulation of certain small molecules. The role of the organelle as an amino-acid and polyphosphate store was defined in yeast and *Neurospora* almost simultaneously in 1972 and 1973.[36] *Neurospora* vacuoles were purified to homogeneity, and their contents analyzed in the early 1980s. The purity of the preparations allowed B. J. Bowman (b. 1946) and E. J. Bowman (b. 1939) to isolate and define a new membrane ATPase, an enzyme found in the equivalent organelle of animals, the lysosome.[37] And yet this enzyme, together with other functions associated with the vacuole and its membrane, are better known in yeast. Why? Because more than one yeast group, some quite large, worked on the organelle from several standpoints: as a straightforward physiological entity, as a site of degradative proteases, as an organelle to which proteins were targeted, and as an organelle displaying energy-requiring membrane transport of many ions and small organic molecules.[38] In the last three of these programs, detailed mutational studies, initiated in the 1970s, were the foundation for vital further progress at the molecular level when transformation techniques became available in 1978.

Other problems originating in *Neurospora* and taken over by yeast can be cited. Among these are amino acid transport through the cell membrane, mutagenesis and DNA repair, mitochondrial biogenesis, many areas of metabolic regulation, and several systems of stress response. By a process of natural selection, as it were, yeast has attracted many post-1970 investigators with strong training in all three of the vital disciplines: genetics, biochemistry, and molecular biology. Their ability to integrate these disciplines in their research and in their training of newcomers to yeast has led to the connectivity of the many levels of organization that defines a model organism. This ability has also led to the explosive growth of the yeast community, which has been so self-sufficient in defining life's fundamental attributes that its members are no longer obliged to read the literature on other fungi.[39] A serious asymmetry prevails in this matter: *Neurospora* and *Aspergillus* investigators ignore the yeast literature at their peril.

Neurospora and *Aspergillus* workers found themselves in 1975 still in a relatively classical environment—the more purely genetic traditions of Beadle and Tatum. The community of filamentous fungal biologists remained small. A large fraction of this community was less sophisticated in biochemistry and molecular biology. The more sophisticated workers carried on in tightly focused, significant research programs, though these programs were much less connected to one another than those of yeast. The remaining groups working with filamentous fungi pursued a number of problems up to a point with genetic and microbiological techniques alone. Thus this community survived, even if it did not thrive. Another factor operated strongly as well. Over many years, a social cohesion had developed among the *Neurospora* (and, separately, *Aspergillus*) workers, abetted by the instruments of the community such as shared mutants, the *Neurospora Newsletter*, and the Fungal Genetics Stock Center. This cohesion in turn reinforced an affection for the filamentous fungi as biological entities and the reluctance of members of the community to jump a growing chasm to another organism. The next chapter describes the fate of this community.

16

The Filamentous Fungi

Eclipse and Renewal

Veterans such as myself may well look back with nostalgia to the time when it was possible to draw novel and important conclusions from simple growth tests. . . . But for those who have the new molecular technology at their fingertips the excitement is as great as it has ever been.
—J. R. S. Fincham

The Early 1950s

In chapter 6, I presented briefly the scope of biological research on model fungi before 1955. These studies, which broadened greatly thereafter, became vital to further progress in plant pathology, medicine, and the commercial sector. We therefore should be curious that modern researches with fungi, the majority of them filamentous forms, did not develop as a more coherent field. During the years that *E. coli* and yeast became prime model organisms, *Neurospora* and *Aspergillus* researchers seemed almost unaware of the large world that their organisms represented. Instead, these workers continued, at first happily, in the wake of *E. coli* as participants in the molecular revolution. I want to describe here some of the continuing areas of investigation with fungal systems and to evoke the process by which, increasingly separate from studies of *E. coli* and yeast, the area of molecular genetics and biology of filamentous fungi took its modern form. In this period, *Neurospora* and *A. nidulans* came to serve as models for studies of this more restricted form of life, although they retained high relevance to the wider world in some areas.[1]

By 1955, workers on *Neurospora* and *A. nidulans* had become committed to these organisms, intoxicated by the ease with which they could be used to solve biochemical questions with genetic techniques. At that time, some of the simpler chemical techniques required—how to handle low molecular-weight intermediates, radioactive tracers, and enzyme assays—could be learned on the job and did not require extensive training in chemistry and biochemistry. The *Neurospora* community was already tightly knit because they worked in a few, prominent institutions such as Caltech, Yale, and Stanford. A Nobel prize to Beadle and Tatum (with Lederberg) in 1958 dignified the fungi as important models for microbial genetics. David Bonner remained with *Neurospora* even as Yanofsky left it for bacteria. Bonner organized the first of the biennial *Neurospora* conferences in 1961, continuing even now, at which the *Neurospora Newsletter* was also conceived. The *A. nidulans* community in Glasgow, given life in 1953 by the comprehensive Pontecorvo-Roper paper, propagated several additional foci, one being that of John Pateman (b. 1926) and David Cove (b. 1937) in Cambridge, England in the early 1960s.[2] With the excitement aroused by investigations of bacteria and phage and the novelty of the structure of DNA, the fungi appeared to have great promise in answering new questions, even in 1960.

To fungal geneticists, bacterial and phage genetics was initially an expansion, not a diversion, of their field of interest. They therefore felt part of the erupting field of molecular biology. But these fungal workers remained geneticists at heart. This was a crucial difference from the growing community of prokaryotic molecular biologists. Many of the latter had entered bacterial genetics from physics, physical chemistry, biochemistry, and microbiology. Few had backgrounds in the Mendelian genetics of higher organisms, and if they became bacterial geneticists, it was well after they finished their formal academic training. Their interest in genetics was in many cases simply in isolating mutants and constructing strains required in specific biochemical or molecular projects. The prokaryotic molecular biologists working on specific genetic attributes of DNA, however, developed and maintained a genetic orientation that would lead them to redefine the yeast community in due course (chapter 15). The specific subject matter of interest to both sorts of prokaryotic molecular biologists diverged from the themes of those working with filamentous fungi, and this slowly isolated the latter from mainstream molecular biology. The prokaryotic biologists had a much greater taste for reducing problems to their simplest form in organisms they could not see individually. In doing so, they thereby lost, notwithstanding Jacob, the feeling for the organism that fungal geneticists would retain in greater measure. This feeling was a strong factor deterring many of the latter from defecting to the prokaryotic world or, later, to yeast. Yeast molecular biology became big science after the even bigger science surrounding *E. coli*. All the while, the filamentous fungal researchers remained few, independent in their experimental programs, and scattered geographically.

Continuing Investigations, 1955–75

What else, then, were fungal geneticists doing during this time, and how did they survive as a scientific group? The answer provides a striking contrast to the development of the yeast community. Briefly, as the yeast community focused heavily on molecular studies of *S. cerevisiae* and the fission yeast, *Schizosaccharomyces pombe*, investigators of filamentous fungi continued with a more biological attitude, culminating in the mid-1980s in the formation of the inclusive field of fungal genetics and biology. This field now embraces work on all filamentous fungi, much of it having its origins in the early 1900s, and from which it had progressed for some time out of the view of the *Neurospora* and *A. nidulans* communities. Let us look at how the present emerged from the past.

Genetics

In the 1960s, strict geneticists focused on the details of meiosis. In work on both yeast and the filamentous fungi, tetrad analysis remained fashionable. The discovery of gene conversion in Neurospora in 1955 by Mitchell and in yeast in 1956 by Roman (chapters 3 and 5) made clear the relevance of fungal investigations to genetics in general. The discovery showed that gene conversion was correlated with crossing over in the same chromosomal region and suggested that the phenomenon might lead to a model for recombination. Indeed, Mitchell's work, done at Caltech, was quickly seized upon by Delbrück's phage group, also at Caltech, as a possible case of "copy choice," a DNA-based model of recombination then being considered in the case of phages T2 and T4. While investigations of phage λ first revealed the molecular nature of recombination, work on gene conversion in fungi became the most significant point of entry into the mechanism of recombination in eukaryotes.[3]

A hint of the diversity that would become a hallmark of fungal molecular biology can be found here. The popularity of tetrad analysis and the rarity of gene conversion in *Neurospora* drew or drove a number of investigators to fungi other than *Neurospora* in hopes that they could find a more efficient experimental system. The other genera, *Ascobolus*, *Sordaria*, and *Podospora*, all Ascomycetes like *Neurospora*, offered ascospore color markers that revealed gene conversion visually and obviated the need to dissect many normal, uninformative tetrads. This simple attribute brought these new species into a community of gene-conversion specialists in the 1960s, although they would be overwhelmed by progress in yeast in the 1970s. While it lasted, gene conversion and the hope of understanding crossing over sustained many laboratories, some of which remain active in the specifics of the meiotic process.

The *Neurospora* and *A. nidulans* communities also deepened the knowledge of classical genetics. *Neurospora* genetics became and remains to this day the most fully developed among the filamentous fungi. Tetrad analysis confirmed most of the classical theory of chromosome behavior in meiosis. An extraordinary knowledge of chromosome aberrations, their meiotic behavior, and their uses in a variety of investigations has also emerged.[4] In parallel, work with *A. nidulans* yielded information and mechanisms about the parasexual cycle and mitotic recombination. In both organisms, early examples of cytoplasmic inheritance were explored as genetic phenomena, although a clear molecular picture to explain them did not emerge for some time.[5]

Metabolism

The fungi, including yeast, had an early prime in the study of metabolism. The field of biochemistry became more sophisticated as the Beadle–Tatum program continued. By the late 1950s, the fungi were ripe with information and mutants, many of which could be investigated with new biochemical techniques such as paper chromatography, radioactive tracers, ion exchange separation, cell fractionation, and new protein purification methods. As this sophistication developed, three trends took hold of the area, both with *Neurospora* and *A. nidulans*. The first was a study of the pathways of uptake, catabolism, and assimilation of fundamental sources of carbon, nitrogen, sulfur, and phosphate.[6] The investigations uncovered or clarified a host of new biochemical activities for the first time, many shared widely with other free-living organisms.

The second trend emerged from this work with the finding that many mutants unable to use various sources of nitrogen, phosphate, and sulfur were not simply deficient in key catabolic enzymes. Some revealed themselves as regulatory mutants, and positive regulatory mutations at that. Their discovery contrasted with information arising from the study of E. *coli* operons and suggested that the fungi might yield insights comparable in detail to those offered by Jacob and Monod. However, good as they were, the genetic techniques—especially mutant selection and genotype manipulation—could not match those in *E. coli*. The filamentous fungi would have to wait for recombinant DNA techniques before rapid progress was possible. In the meantime, workers on the filamentous fungi retained an intense (if unreciprocated) awareness of yeast work as yeast investigators began to develop their own *E. coli*-like rationales.

The third trend concerned metabolic organization. The compartmental character of eukaryotic cells contrasted sharply with the relative homogeneity of prokaryotic cytoplasm. The latter feature sustained a "bag of enzymes" view that prokaryotes controlled their metabolism mainly with regulatory mechanisms and the kinetic parameters of enzymes. Workers on eukaryotes, however, could not ignore the physical organization and sequestration of metabolic sequences within cells. Organellar compartments and aggregations of proteins, the former already exemplified by the mitochondrion, became common subjects of study.[7] Indeed, the structure, biogenesis, and function of mitochondria and vacuoles were explored extensively in *Neurospora*. *Neurospora* and yeast became early, prominent models for the study of mitochondrial structure and physiology, supplanting mammalian liver. Thus in bioenergetic research the fungi diverged strongly from prokaryotes, each applying specifically to its own kingdom. Studies of fungi retained their coherence and importance in this area throughout the latter half of the 20th century.

In contrast to the clustering of genes in operons in bacteria, metabolically related genes in fungi usually lie at scattered locations in the genome. However, a striking additional difference between the two groups of organisms arose in studies of the rarer cases of related genes that were, from genetic evidence, clustered in fungi. These looked like the remnants of multigene operons held over from prokaryotes as they gave rise to eukaryotes, and they were initially treated as such. To the surprise of many biochemists, these clusters were revealed in the late 1960s as cases of single genes encoding polypeptides with several enzymatic activities.[8] The "cluster-genes," as they are called, reflect the fusion of several ancestral, monofunctional genes into multifunctional genes. This organization represents an alternative to the operon, in which only the mRNA is a multidomain entity, the proteins being separately translated. Both gene organizations facilitate coordinated expression of different enzymes.

Multidomain polypeptides fed continuing speculations about "metabolic channeling," the concept that biochemical intermediates are bound to a multifunctional protein as they are conducted through a series of reactions.[9] Even without multidomain proteins, channeling might take place on aggregates of separately synthesized polypeptides. With these speculations, such work in fungi began to reach its limit, or at least a plateau. The chemistry of multidomain proteins could not be pursued easily due to the difficulty of obtaining sufficient undegraded material from extracts from the fungi and the tendency of weakly interacting proteins to disaggregate upon extraction.[10]

As knowledge of gene–protein relationships deepened, another phenomenon of great interest arose: the ability of two mutant alleles of the same gene to complement one another. This phenomenon, which violates the rule that alleles do not complement, was defined best in *Neurospora*. In that organism, a normal phenotype is occasionally restored when nuclei bearing different missense mutations of the same gene are put into heterokaryotic association—that is, in the *trans* relationship. The observations showed that two inactive, mutant polypeptides, products of allelic genes but having amino acid substitutions at different positions, could cooperate in the cytoplasm to produce enzymatic activity, albeit usually abnormal and weak activity. The phenomenon, when pursued biochemically, is now known to reflect the ability of the polypeptides to aggregate and thereby mutually to alter their conformations toward the normal, thus regaining enzyme activity.[11] The discovery of allelic complementation contributed significantly to later studies of protein structure.

Cell Biology and Development

From the 1960s on, *Neurospora* and *A. nidulans* geneticists found themselves largely in zoology, botany, or biology departments, teaching genetics. This environment endowed them with a more biological orientation than they might otherwise have had. An obvious attraction of the fungi over bacteria, in the context of general biology, was the possibility of analyzing simple developmental sequences in a tractable, easily visible organism. From the earliest genetic research on both organisms, color mutants and morphological mutants affecting hyphal growth, colony morphology, conidial formation, and sexual development had arisen repeatedly in searches for mutants. These mutants were conspicuous to the naked eye, and for that reason many such variants became standard genetic markers. *A. nidulans* genetics depended heavily on color markers as indicators of hybrid cleistothecia and as aids in linkage studies (chapter 4). Certain colonial mutants of *Neurospora* were used in work requiring platings in the absence of the colony-inducing sugar, sorbose. An aconidial strain of *Neurospora* (fluffy) became the standard mating type tester because it did not produce contaminating asexual spores in Petri dish cultures. In the heady, early days, some geneticists attempted to link these morphological alterations of *Neurospora* to underlying biochemical deficiencies. Curiously, this provided a great deal of information about intermediary metabolism, even though it failed to achieve its initial goals.[12]

In *A. nidulans*, a more formal approach yielded models of conidial and cleistothecial development that would inform later molecular studies.[13] Much of this work concentrated on close observation of the morphological steps in development, isolation of mutations interrupting the sequence, and the chemical and physical factors required to induce conidia and cleistothecia.[14] No attempt was made to link fundamental steps of metabolism to morphological development, as was done for *Neurospora*. Work on *Aspergillus* conidiation gained widespread attention in the 1970s because of the ease of defining these steps by mutation—a counterpart to *Drosophila* studies in how development might be analyzed. Around the late 1970s, the laboratory of N. Ronald Morris (b. 1933) began a study of conidial germination, septation, nuclear division, nuclear distribution along hyphae, and their relation to growth in *A. nidulans*. This groundwork supported extensive molecular studies when molecular techniques became available.[15] Until then, work on both *Aspergillus* and *Neurospora* development and growth paused. When DNA-mediated transformation and molecular approaches became available, the entire field of fungal growth and development resumed. It is now a substantial and growing field in which, significantly, molecular understanding of new data in a variety of other species invokes prior, if incomplete, work on yeast, *Neurospora*, and *Aspergillus*.[16]

The Natural History of Fungi

As yeast invaded the intellectual territory of *Neurospora* and *Aspergillus*, the community working on the latter models began to see better research opportunities in higher levels of organization, some peculiar to filamentous fungi. The genetic analysis of cell recognition, both in heterokaryon formation and in mating, had engaged biologists working with other fungi for some time. The interest in heterokaryosis in *Neurospora* and *Aspergillus* initiated studies in those species and others on the biological and evolutionary significance of the phenomenon. In addition, interest in fungal sexuality had emerged in the 1930s and1940s from the tetrapolar mating systems of Basidiomycetes and how these systems might have evolved from the bipolar systems (*A* and *a* mating types in *Neurospora*, for example). An awareness of other species with respect to these biological issues furthered a taste for diversity and comparative work among some *Neurospora* and *Aspergillus* investigators.

Most evolutionary biologists of the 1940s and 1950s had a blind spot: an inattention to the population genetics of haploid organisms. David Perkins made systematic, worldwide collections of sexual *Neurospora spp.* from the wild during the 1970s and 1980s that initiated work on the population genetics of the organism. Those who worked with these collections, including Perkins's laboratory, described the distribution of genetic diversity in natural populations, the degree of genetic exchange locally and globally, the extent of heterokaryon-incompatibility genes, and the existence of mitochondrial plasmids. The work also explored the species barriers of the genus, revealed a new *Neurospora* species, and began to define the ecological niches for the five sexual species now known. The intriguing questions remain of how the several known asexual *Neurospora* species evolve and whether they entirely lack a means of genetic recombination. Allelic diversity within the species, appreciated fully in higher organisms only in the mid-1960s, has provided the raw material for charting the course of divergence and evolution of the separate *Neurospora* species and of individual genes. It has also revealed interesting cases of lateral transfer (i.e., via transient cell fusions) of mitochondrial DNA, plasmids, and infectious elements. The natural population work is almost unique among the fungi (yeasts included) and may become a model for such studies in haploid eukaryotes hereafter. It is notable that similar global studies of *E. coli* populations, also motivated by the discovery of molecular diversity in natural populations of higher organisms, were started at about the same time.

The *Neurospora* collections of exotic strains have become an important resource for molecular biologists. Besides studies of molecular evolution and the discovery of mitochondrial plasmids noted above, the genetic polymorphism of *N. crassa* provides molecular markers for the genome. These markers are known in the trade as restriction-fragment-length polymorphisms (RFLPs), and such markers were first harnessed to map the human genome and those of other higher organisms. The markers are variations of short nucleotide sequences of DNA that happen to be targets of bacterial restriction endonucleases (appendix 3) that can recognize and cut the DNA at those points. (Eukaryotes do not have restriction-modification systems like those of bacteria.) Variations in these targets arising by mutation will lead to variation of the lengths of DNA between them; these can be visualized by molecular techniques and may therefore be used as genetic markers. The markers are closely spaced enough not only to make a fine-scale map of chromosomes, but also to facilitate isolation of particular genetic segments for cloning and sequencing. The molecular characterization in the form of RFLP maps has greatly facilitated the recent sequencing of the entire *N. crassa* genome.

The Impact of Yeast Research and Molecular Techniques

Biochemical and genetic work with filamentous fungi reached a plateau in the late 1970s as yeast came into its own. The promise of the model filamentous fungi in contributing to molecular biology paled significantly. If important contributions were made by these organisms, they were not associated with being a model organism in the world of molecular biology. Instead, like most other experimental organisms, they served as useful systems for work in specific fields that might involve other diverse organisms. One would find *Neu-*

rospora and *A. nidulans* researchers at conferences devoted to mitochondria, phosphate metabolism, circadian rhythms, arginine synthesis, genetics, and the like. The presence of the filamentous fungi in biology was maintained in this way even as the subjects of *Neurospora* and *Aspergillus* meetings became more diverse and less coherent.

Research with filamentous fungi revived in its modern form with the advent of new molecular techniques in the late 1970s. The development of transformation techniques in yeast[17] led *Neurospora* and *Aspergillus* workers to duplicate their success. Similar techniques were introduced soon thereafter for many other filamentous fungi.[18] Initially, those using *Neurospora* and *Aspergillus* looked jealously at yeast biologists, well ahead of them, and over their own shoulders as they competed with one another. The techniques had importance to both academia and industry because they opened other fungi to real molecular biology. Priority goals included successful, easy transformation techniques and the discovery and development of autonomously replicating plasmids that would greatly facilitate gene cloning. Such plasmids were already available in *E. coli* and yeast. The scramble to meet these technical goals led to an unseemly series of premature publications, a common feature of exciting times in any field. The domestication of filamentous fungi for serious molecular work made many species besides *Neurospora* and *A. nidulans* prominent research organisms. Thus, while *Neurospora* and *Aspergillus* made great contributions, they were accompanied by many other fungi now open to study.

In the late 1970s, a crucial event took place. Charles Yanofsky, who had abandoned *Neurospora* years before and had devoted himself since then to bacterial molecular biology, allowed one of his graduate students, Eric Selker (b. 1953), to work on *Neurospora* for his dissertation. The success of Selker's project encouraged Yanofsky to take new postdoctoral fellows into his laboratory for continued research with *Neurospora*.[19] Yanofsky attracted a burst of recruits in the 1980s and early 1990s that opened up many new investigations with *Neurospora*, such as studies of mating type genes and those controlling conidial differentiation.[20] Equally important, the laboratory undertook systematic improvement of molecular techniques and the creation of DNA libraries crucial to further progress with the organism. These advances took place across the hall from the neighboring Perkins laboratory, which provided essential genetic resources and expertise. The activity of these laboratories accelerated progress in the next decade and was critical in the eventual sequencing of the *Neurospora* genome.[21] Few events could illustrate better the effect of having a critical mass reassembled in one place to energize a common effort, even as the members of the laboratory pursued separate projects. It echoes the effect of the continuing work with the yeast group in Seattle, or the year that Fink, Botstein, Roth, and Gesteland spent at Cold Spring Harbor in 1974–75 (chapter 15).

Transformation and molecular techniques for the model filamentous fungi had another important effect on the mycological community. The new techniques provided the *lingua franca*—molecular biology—of the wider field. Before 1980, mycologists of all sorts specialized in particular groups of fungi and in biological processes in ways that insulated them from one another. This reflected in part the diversity of fungi and the arcane specialization of form, habitat, and behavior of the fungi. Other programs took place in industry, a separate culture from academia. In addition, many entrenched institutions and societies of taxonomy, plant pathology, and medical mycology, had become so organism oriented or problem oriented that they could ignore studies of model fungi entirely. Understandably, they resisted overdue changes up to the late 1970s. With molecular techniques, these changes came about.

By 1985, the *A. nidulans* community had remained coherent due to the inclusion of standard stocks at the Fungal Genetics Stock Center and a running bibliography by A. John Clutterbuck (b. 1937), who still works in Glasgow. Indeed, the *Neurospora* conferences and the *Neurospora Newsletter* had earlier and informally welcomed the *A. nidulans* workers. In 1985, the biennial *Neurospora* Information Conference opened its meetings to all those who worked with filamentous fungi. The new meetings were named the Fungal Genetics Conferences. The decision was taken with the recognition that the *Neurospora* community had diminished, defined then by only a handful of major laboratories. The elders neared retirement with apprehension that the next generation was too small and too dispersed, even with the Yanofsky laboratory's contribution, to sustain a critical mass. In addition, the agreeable social ties of this long-standing community had blinded it somewhat to the need for new scientific energy. So, in the reconstituted 1986 meeting, instead of the two model fungi, 14 species appeared in the abstracts; 142 people attended. In 2001, 104 species appeared in the much thicker program, and 670 people attended, with 100 on the waiting list. Between those two meetings, the edgy, competitive spirit among the various groups dissipated.

In parallel, in the mid-1980s, other mycologically oriented meetings were established. These included a meeting initially called Fungal Metabolism as part of the Gordon Research Conferences in the United States and regional meetings of fungal geneticists and biologists in continental Europe, Great Britain, and Latin America. Each has its particular flavor and organismic diversity, but all have converged on new and standard methods, with increasing overlap among the problems of interest.

A striking willingness emerged on the part of previously insular scientists (among them the *Neurospora* and *A. nidulans* communities) to hear about the wider world of filamentous fungi. The fungal biotechnology industry, which developed rapidly after 1985, entered the community and supported meetings and much of the research, both in academia and in house. Many former academics now led new commercial enterprises based on fungal molecular biology. The journal *Experimental Mycology* was renamed *Fungal Genetics and Biology* in 1996 and now competes for papers with other more general or more specialized, discipline-oriented journals. The Fungal Genetics Stock Center ties the community together with the *Fungal Genetics Newsletter*, a web site, and many sources of information and stocks.

As noted previously, the communities working with *Neurospora* and *A. nidulans* had been almost unaware of the possibilities dormant in the garden of filamentous fungi. The more or less formal establishment of the new field revealed that *Neurospora* and *A. nidulans* had been models all along but now could contribute aggressively and with clearer sense of relevance in their new environment. The conflicts among the community, initially based on priority, visibility, and contests about the significance of findings, began to focus on the issue of the academic purity versus the commercialization of their enterprise.

New Directions

The problems of interest to workers on yeast and filamentous fungi in 1980—quite complex, quite biological, and bearing less and less resemblance to the reductionist programs using bacteria and phage—could now proceed at the molecular level.[22] The traditions and the differing histories of the fields of yeast and filamentous fungi over the preceding 15 years led to yeast appropriating many programs of cell biology, regulation, and genetics, exploring them in exquisite detail. Investigators of filamentous fungi saw new opportunities in exploring the diversity of the taxon Mycota. Thus, as yeast workers sought to integrate and universalize their findings to other organisms, many working with filamentous fungi adopted comparative approaches to fungi. (I elaborate on this in chapter 18 in the context of new directions in general molecular biology.)

The appreciation of the comparative approach among filamentous fungal researchers should not obscure the importance of focused research programs and discoveries. *Neurospora*, in particular, because of its wealth of information and mutants, is being used in these programs as a model system for specific programs even as it remains a model for other fungi. The state of knowledge about the organism—high connectivity of many levels of organization—suits it especially well to sophisticated study in these modern programs. This is the destiny of model organisms once they have played out their broadest, universal themes.

Gene Silencing

The advent of transformation techniques and gene cloning in *Neurospora* revealed a quite messy biological process. Unlike yeast, which integrates transforming yeast DNA at homologous locations in the genome, *Neurospora* integrates particular transforming DNAs almost anywhere on the seven chromosomes, including the homologous location. Integration at nonhomologous locations is called *ectopic* integration. Moreover, no plasmids that can replicate autonomously (that is, without integrating into the chromosomes) could be found. The molecular analysis of *Neurospora* transformants was therefore quite difficult, but an important phenomenon emerged from further studies. They showed that both the ectopic and the resident copy of a gene in the nucleus of a transformant acquire many mutations before fusion with the other parental nucleus in matings. (The name of the process, RIP, derives from the term given to the process, "repeat induced point mutation.") The two copies of DNA that emerge are not only mutated, but often highly methylated as well. These phenomena, almost unique to *Neurospora*, have revealed a great deal about DNA methylation, a process of great importance in the development of higher organisms. Unrelated mechanisms of gene silencing, initially confused with the RIP process, were discovered in *Neurospora* thereafter and were shown to be similar to processes called *trans*-gene silencing in plants, mammals, and the nematode *Caenorhabditis elegans*.[23] In both cases, the opportunity to isolate mutants defective in these processes has rendered *Neurospora* a model system and thus a prime contributor to the gene silencing field.

Regulation of Metabolism

Aspergillus and *Neurospora* investigators continue to explore in greater depth systems already well developed in the past by genetic techniques. Their work can stand as models either because many of the systems have not been explored much elsewhere or because these organisms still hold the lead over yeast. In the first category are the analysis of the *aro* gene cluster and the nitrogen catabolite repression systems of *Neurospora* and *Aspergillus* and the striking polyprotein gene encoding two enzymes of arginine synthesis in *Neurospora*.[24] Where the work proceeds in parallel with investigations in yeast, substantial comparative information enhances the understanding of the findings and the evidence for their universality. The comparative tastes of filamentous fungal workers have been rewarded by some important findings that resemble phenomena in higher organisms more than do similar findings in yeast. A growing sense that yeast is atypical of other eukaryotes in certain ways presents the *Neurospora* and *A. nidulans* communities opportunities to fill in the blanks.

Development

Analysis of growth and differentiation of *N. crassa* and *A. nidulans*, discussed above, could not realize its potential through dedicated mutational and biochemical analysis. This area benefited in the 1980s not only from molecular techniques but from a modern knowledge of regulatory phenomena, signal transduction cascades, and the details of hyphal growth, nuclear movement, septation, and cell wall biochemistry. Indeed, *A. nidulans* has yielded understanding of nuclear migration applicable to human disease.[25]

Knowledge of hyphal growth and cell wall biochemistry had developed by the 1990s in a number of fungi from the Phycomycetes to the Basidiomycetes. The diversity of organisms used in acquiring this knowledge explains the slow start of systematic molecular studies, finally taken up in *A. nidulans* and *N. crassa*. These organisms, with yeast, have now yielded a broad understanding of the steps and genes required in the formation of conidia, but the molecular details are still obscure.[26] Similar studies of hyphal growth and the role of the cytoskeleton and the secretory apparatus have followed, facilitated by spectacular improvements in fluorescent staining that allows observations of subcellular structures in living cells.[27] As I have shown in the

preceding chapter, yeast has been the most prominent contributor to knowledge of molecular details of cell determination, regulatory cascades, and polarity in fungi, preceding or superseding *Neurospora* and *A. nidulans* as a
model.[28] Although related studies of *N. crassa* and *Aspergillus* progress, they
are products of only a few laboratories, and most use information from yeast
as a starting point.[29] Yeast has been the leader in understanding cell determination and polarized growth because, in contrast to heterogeneous efforts on
many filamentous fungi, so many researchers work with yeast on a host of
related problems.

Biological Rhythms

Neurospora has taken a prime place in the study of circadian (daily) rhythms
and photobiology. Like most of the efforts described in this chapter, the analysis of the circadian rhythm in *Neurospora* began with physiological observations, soon complemented by genetic approaches. Mutant strains displaying
longer or shorter "day-lengths" (rhythms of conidial formation during growth
in constant darkness) were painstakingly isolated by J. Feldman (b. 1942) in
the early 1980s.[30] The function of the genes identified in this way could be
inferred only in a general way by further genetic tests of dominance, multiple
mutants, and physiological manipulations. Once molecular techniques became
available, the understanding of the regulatory proteins involved in rhythmic
phenomena advanced rapidly. The work displaced for a time even the well-
developed programs on biological rhythms in the other prime model organisms, *Drosophila* and the mouse. *Neurospora* is an organism of choice in the
study of this one last universal attribute of living things, to which it still contributes new insights.[31]

Cell Recognition

In fungal biology up to 1950, the theme of sexual mechanisms and the related
phenomenon of heterokaryosis had unusual prominence. The cell-recognition
phase of these phenomena was beyond understanding before molecular techniques became available. Yeast was the first species to offer a paradigmatic,
detailed picture of fungal mating-type determination and mating, and it continues to lead with studies of the regulatory phenomena that follow cell-to-cell
recognition. Molecular techniques have now yielded a detailed picture of the
gene products used in these processes in filamentous fungi, including well
developed studies in the Basidiomycetes *Coprinus* and *Schizophyllum.*[32] The
products include regulatory and other proteins needed to accomplish specific
cell-to-cell recognition, incompatibility reactions, compatible fusions, and post-
fusion events. Not surprisingly, many proteins involved in these steps do not
resemble gene products of yeast, and a substantial diversity prevails among
the filamentous forms. This realization emerged with the study of many pathogenic fungi, in which recognition of the host and change of cell type during
infection has uncovered unique details. In this area, no one organism stands

out, partly because these matters have been so ripe for so long in so many organisms that molecular approaches began simultaneously on all of them. The
many recognition mechanisms have impelled the field, now well defined, to
embrace diversity rather than to ignore it by intense concentration on a single
model organism.

The Genome

The sequencing of the *N. crassa* genome, the first among filamentous fungi,
was accomplished early in 2001. Its public posting on the Internet in March
2001 drew, within the first three weeks, more than 15,000 hits from all over
the world.[33] *Neurospora* thus continues to define itself as a role model in the
new century. (As an indication of changing times, the *A. nidulans* genome,
roughly sequenced as early as 1999 by at least one commercial firm, is available only with conditions for its use. This may compromise the model-organism
status of the organism in the future.) The *Neurospora* sequence will probably
stimulate a phase of research comparable to that of the 1950s and 1960s, when
hunts for mutants ignited investigations of many new problems. Even before
the genome was sequenced, expressed sequence tags (ESTs; cDNAs derived
from mRNA populations of cells) and other sequence databases provided evidence for many more orphan genes in *N. crassa* than in *S. cerevisiae*—those
that lacked homologues in other species.[34] Because many of these genes may
be specific to filamentous fungi, they offer great promise for understanding
the fungal lifestyle. In addition, homologues of the orphan genes may soon be
found in higher organisms and confirm the closer relation of filamentous fungi
than of yeast to the higher forms.

A Concluding Remark

This chapter demonstrates that work on model filamentous fungi never died,
and indeed many careers will be devoted to further research in these organisms. My description is a partial antidote to thinking that model status is particularly significant to most of the biologists who do not work on a model
organism. But for those that do, and particularly for those who have seen their
organism lose the attention of the wider world, it is (or was) a considerable
motivation in their work. In the halls of traditional universities, one may find
older *Drosophila*, *E. coli*, *Neurospora*, *Aspergillus*, and *Paramecium* researchers commiserating with one another about their fall from grace and looking for ways to capture some of the old glory. The least they can do is redefine
the boundaries of their worlds, as *Neurospora* and *Aspergillus* workers did after
discovering the rest of the filamentous fungi.

17

The Role of Biochemistry

God invented time so everything wouldn't happen at once.
—Anonymous

Biochemistry: A World of Its Own

I noted in chapter 7 a tendency of molecular biologists and the early historians of molecular biology to ignore the role of biochemistry in the development of the field. Molecular biologists were divided with respect to the ownership of their field. Those emerging from X-ray crystallography, later dubbed the "structural" school, contested with the "informational" school, whose origins lay in genetics, for title to molecular biology well into the 1960s.[1] This issue of disciplinary authority has intrigued historians recently as they explore the proceedings of professional societies and representations of the various groups to national funding agencies. Of more interest to us here is a larger question of the role of biochemists in the origin and development of what we now call molecular biology.

I have deferred explicit discussion of this matter to this point so that we can look back on the contributions of biochemistry to the research covered in the preceding chapters. Until 1950, biochemists considered themselves the custodians of biology at the molecular level. The discipline had a distinguished origin in chemistry, medicine (as physiological chemistry), agriculture, and industry. Its membership was grounded in organic and physical chemistry and had developed quite orthodox empirical methods devoted to a reductionist, antivitalist analysis of cell and organismal chemistry. However, the discipline had foundered on a major issue in the early 1900s: the structure of proteins and other macromolecules.[2] Despite spotty evidence to the contrary, biochemists accepted for a time the notion that macromolecules were aggregates of small

229

molecules held together by weak (unspecified) forces similar to the colloidal state of iron compounds, sulfur and the like. This idea delayed the proper characterization of proteins by biochemists until the mid-1920s. By then, newer techniques and advances in X-ray crystallography gave clear evidence that proteins could have molecular weights greater than 40,000 daltons.[3]

Nevertheless, even without proof of the identity of the catalysts, biochemists made great progress in the study of enzyme reactions, cell energetics, and metabolic sequences. Characterization of enzyme reactions relied on chemical standards of proof. Such investigations proceeded through prescribed steps, starting with the development of a specific and rapid assay of the reaction, dialysis to remove small molecules from a cell extract, and determination of the dependence of reaction progress upon time and enzyme concentration. This was followed by optimization of reaction mixture ingredients, chemical evidence for the identity of the products, and consistent accounting for the use of substrate(s) as the product(s) accumulated. A description of the kinetics and energetics of reactions in physical-chemical terms was a consistent goal of such experiments. Purification of the enzyme assured that only one reaction underlay the transformation of substrate to product. The biochemist would then pursue further details of the enzyme reaction regarding substrate affinities, reaction equilibrium, pH and temperature optima, substrate analogues, inhibitors, and reactive chemical groups of the enzyme.

Most of the proteins that had been crystallized up to 1925 were abundant proteins without standard enzymatic activity. These included hemoglobins, plant storage proteins, egg-white albumin, and serum albumin. Techniques for purification of enzymes developed along the same lines and accelerated thereafter, especially following the purification of the providentially abundant jack-bean urease to the crystalline state by James Sumner (1887–1955) in 1926. Sumner's claim that the urease reaction was catalyzed by this protein, rather than by a contaminant, was not generally accepted until the 1930s. The reason lay in a disturbing fact. During purification of enzymes (detected only by their catalytic activity), extraneous proteins are successively removed by various fractionation techniques. In many cases, the catalytic activity remained in highly purified preparations, which by that time seemed to contain no protein at all. The problem was reconciled with the advent of more sensitive chemical means of detecting proteins, which typically revealed one of them as a consistent accompaniment of catalytic activity in physical separations. Only when biochemists prepared other crystalline enzymes, the first being the crystallization of pepsin by John Northrop (1891–1987) in 1930, was protein accepted as the substance of enzymes.[4] By 1945, purification had become a major aim in defining particular enzyme reactions as unitary processes and in characterizing the catalysts as macromolecules. The standards of purity included the use of a number of chemically distinct methods of purification, together with criteria based on behavior in the ultracentrifuge and electrophoretic apparatus, newly developed in the 1930s. Researchers struggled until recently to eliminate the possibility that a protein that copurifies with a catalytic activity is not simply a major impurity in the preparation.

During the period between 1945 and 1960, independent programs of genetics, physics, macromolecular structure, and microbiology (especially virology), converged under the umbrella heading of molecular biology. Only in
microbiology and biochemical genetics had orthodox biochemistry been a major
component, and genetic work with microorganisms actually developed in some
medical school microbiology departments as a marginalized field. As the momentum of molecular biology increased, spokespersons began to think of the
field as "biology at the molecular level." This was the very phrase by which
biochemistry had proclaimed itself a discipline at the turn of the century, when
it differentiated itself from chemistry and physiology. Leaders of both biochemistry and molecular biology have described their views of the apparent bound-
ary between the two fields in rueful or triumphant autobiographical accounts,
and historians and historiographers have amplified it further. But the main
events had to do with a confused contest of authority.[5] Biochemists stressed
the field's continuity with molecular biology in its attempt to absorb the new
discipline, but at the same time biochemists were suspicious of the rationales
of molecular biology. Molecular biologists, in contrast, stressed their field's
differences from biochemistry, its appreciation of protein sequence, protein
three-dimensional structure, and genetic information, at the same time a bit loose
with biochemical standards. The contest has been analyzed from several points
of view, but all agree that the traditions and interests of the two fields were
distinct.[6] For our purposes, the characterization of Gilbert is apt: biochemistry
concerned itself largely with the flux of metabolism in the steady-state system, while molecular biology focused on macromolecules, information, and
replication of living things.[7] Even this characterization ignores the contribution of the Carnegie group, whose study of *E. coli* metabolism (chapter 11)
demonstrated that God made occasional exceptions (see epigraph).

Biochemistry was healthy, if a bit baroque, during the heady days between
1950 and 1975 and an essential, if initially reluctant contributor to progress in
molecular biology. The reluctance lay in a frustrating dependence, by 1968, of
biochemistry on the insights of molecular biology, and an unwillingness of
molecular biologists to acknowledge their methodological debts to biochemistry. By 1965 molecular biologists and biochemists became bedmates, each
trying to incorporate the other into its professional rubric, but with different
approaches to the same material. Biochemistry's loosened grip over biology
at the molecular level allowed unlicensed scientists into the field. I take up
three important areas in which continuing progress in molecular biology relied heavily on frankly biochemical orientations.

Biochemical Genetics

Obstacles to understanding the structure of proteins prevailed until about the
1950s and were overcome by crystallographic work, Pauling's proposal for the
structure of the α-helix, and Sanger's determination of the amino acid sequence
of insulin in 1952. Well before that, an understanding of enzyme catalysis and

intracellular metabolism were developing so rapidly that these attributes had become fundamental properties of life in every biologist's mind. This endowed Beadle and Tatum's finding that genes governed metabolic reactions with enormous significance. Tatum's biochemical expertise and his understanding of microbial nutrition were vital to that enterprise. The choice of both *Neurospora* and *E. coli* as model organisms reflected the need for simplicity in the experimental material, in which only a cell envelope lay between the investigator and the machinery of growth.

Thus, not only did Beadle and Tatum depend on the considerable biochemical knowledge available in 1940, but they introduced methods that quickly extended this knowledge. Some biochemists felt initially that the rationales of biochemical genetics—easy and effective in identifying the steps and intermediates of a metabolic pathway—were tantamount to cheating. But on the other side, the goals of *Neurospora* and *Aspergillus* workers obliged these geneticists to become competent biochemists. Several interesting features of the scientific groups characterize the following years. First, the people recruited by Beadle and Tatum included solid, chemically oriented biochemists such as Herschel Mitchell. But more of them were drawn from biological backgrounds in genetics, embryology, and other fields. If they were to have influence in the wider field of biology, they had, as best they could, to adopt the standards of biochemical work (fairly primitive at the outset) when they published. In return, the biochemical profession readily accepted even some of the earliest papers from the Beadle–Tatum laboratory in premier journals such as the *Journal of Biological Chemistry*.[8]

The second interesting feature of the period is the disdain of some geneticists such as Pontecorvo (chapter 4) for biochemistry. The distaste was expressed not only by "pure geneticists," but was even shared by some investigators doing solid biochemical work. This curious attitude lay in the ease with which biochemical information could be derived from studies of mutants and the irrelevance of meeting all the fussy chemical standards of the classical biochemist. Those educated mainly in genetics sacrificed precision and certainty to speed and efficiency in getting the overall picture of a metabolic system. In the same spirit, they ignored Delbrück's criticism that they could not eliminate the possibility of "multivalent" mutations (of genes affecting more than one enzyme) with their methods. They used genetic reasoning in demonstrating that a single enzyme catalyzed a given reaction: if one mutation blocked it, they might say, the reaction was catalyzed by one enzyme. The identification of the position of a metabolic block could often be established by a simple comparison of crude extracts of mutant and wild-type strains. Gene interaction studies, carried out by comparing nutritional behavior of multiply mutant strains, gave rise to towering pictures of biochemical relationships within the cell.[9] Speculations, however, often grew into unjustified certainties in the mind of some biochemical geneticists. They might forget that a mutation blocking an enzyme reaction demonstrated that this enzyme was simply necessary, not that it was sufficient for the reaction. They might forget that a mutation could cause the accumulation of metabolites that simply inhibited an intact enzyme or that a spurious enzymic or nonenzymic reaction in cell-free extracts might mask the absence of the biologically relevant one. Recognition of such problems quickly drove most early biochemical geneticists to become biochemists if they could, even as they pursued their goals with unorthodox rationales. The incomparable benefits of mutational analysis and the simplicity of the organisms used allowed them to retain their advantage over classical biochemists as they did so.

The biological material and the mutational approach equipped more sophisticated biochemical geneticists to study subtle effects of partial impairments of enzymes caused by certain mutant alleles. They thereby imparted a deepening appreciation of the structural features of enzymes devoted to catalysis, stability, aggregation, feedback inhibition, and cellular location. These criteria interested geneticists and molecular biologists who foresaw the eventual determination of the amino acid composition and sequence of the polypeptides with which mutant attributes could be correlated. This is a clear point of union between the biochemical and the molecular disciplines, and both were pretty happy about it at the time. By 1970, professional biochemistry had adopted many of the rationales of biochemical genetics. A natural corollary was the adoption by biochemists of the model microbes used by biochemical geneticists. As the universality of metabolism became clearer, *E. coli* became a laboratory tool for biochemists, and *Neurospora* and yeast began to dominate the area of mitochondrial physiology and biogenesis and other eukaryotic specialties.

Biochemistry thereby remained a part of the new field of molecular biology, still identified as biochemical genetics. Despite the major contributions of biochemical geneticists to bacterial metabolism in the 1950s, however, their field became marginalized thereafter as work with DNA, RNA, polypeptide synthesis, and gene regulation evolved. The exceptional biochemists emerging from the Beadle–Tatum tradition and moving to prokaryotic systems, such as Yanofsky and Ames, became identified as molecular biologists because of their research on the informational or regulatory aspects of genes and proteins. Their biochemical training, expertise, and insight were vital to their contributions.

Nucleic Acids

Johann Friedrich Miescher (1844–95)—for all intents and purposes a biochemist of his day—first described nucleoprotein (then called "nuclein") in 1876. Soon thereafter, nuclein was resolved into protein (largely histones and protamine) and nucleic acid, localized to the nucleus of eukaryotic cells, and renamed chromatin.[10] Phoebus Levene (1869–1940) characterized DNA in biochemical terms, committing it for some years to a genetically uninteresting status: he proposed that DNA had a low molecular weight (ca. 1600 daltons) and a monotonous, information-poor, tetranucleotide structure in which the four bases were equal in their proportions. Oswald Avery, professionally a biochemist and microbiologist focused on pneumonia bacteria, adopted high biochemical standards in purifying and characterizing the DNA that could transform a rough *Pneumococcus* strain to smooth. The composition of viruses became clear through purely biochemical approaches during the 1930s (with tobacco mosaic and other plant viruses), and the nucleoprotein composition of the T bacteriophages was known to Ellis and Delbrück even in their initial paper of 1939.[11]

Seymour Cohen (b. 1917), who had worked on the composition of tobacco mosaic virus, demonstrated DNA as the only nucleic acid of the T phages. He characterized unusual modified forms of the nitrogenous bases of some phage DNAs (e.g., 5-hydroxymethyl-cytosine) and showed that many enzymes required for phage growth were induced in the host after infection.[12] This work set the stage for the Hershey–Chase experiment of 1952, which strongly suggested that DNA was the genetic substance of the phage. The importance of the prior biochemical work lies partly in the demonstration that only protein and nucleic acids need be considered in this experiment—a key virtue of phages as model organisms. This naturally suggested an experimental protocol based on uniquely labeling these components with specific isotopes (^{32}P and ^{35}S). At this time, Delbrück was still touting the virtues of a black-box approach to phage replication, which constrained his own studies toward the formal and mathematical. Had Delbrück (and Luria) welcomed biochemistry rather than eschewing it, many of the inner phage circle (and their historians) might remember biochemistry as a normal part of its history. Indeed, the contrast of these phage workers with the Pasteur group is striking, inasmuch as the Pastorians (particularly Lwoff and Monod) readily applied their background of biochemistry, microbial nutrition, and cell physiology to their genetic rationales.

Classical enzymologists contributed more than most geneticists to the understanding of the synthesis of nucleic acids and the breaking of the genetic code.[13] Severo Ochoa (1905–93), Arthur Kornberg, H. Gobind Khorana (b. 1922), and Marshall Nirenberg (b. 1927) received their training in biochemistry and had contributed to enzymology, cell energetics, and the chemistry of biological molecules. Ochoa and M. Grunberg-Manago (b. 1921) discovered an enzyme in 1955 that catalyzed polyribonucleotide (that is, RNA) synthesis, although the activity was later shown to be the biologically less relevant, reverse reaction of an RNA degradative enzyme. However, the enzyme was later used in crucial studies, many by Ochoa's group, in providing artificial RNA templates for in vitro polypeptide synthesis.

Kornberg described the first enzyme able to polymerize deoxyribonucleotides using a DNA template in 1958. Kornberg, a confirmed biochemist by origin, noted later that a template-directed enzymic reaction was "unique and unanticipated by any evidence in enzymology, and was for some biochemists, even after a number of years, very hard to believe."[14] Here we see another important connection between biochemistry and molecular biology, with an illustration of a cognitive barrier at the outset. Kornberg contributed greatly in these studies to knowledge of DNA structure—in particular, to the antiparallel relation of the nucleotide chains of the double helix. His studies of enzymes of DNA replication continued with the model bacterium, *E. coli*, from which he could make suitable extracts for his studies. Among his rationales were tests of purified components from the wild-type strain to see whether they would complement (i.e., make up for) particular deficiencies of extracts of mutants impaired in DNA synthesis. Khorana was among the first to make oligonucleotides of known sequence in vitro. These materials were indispensable for the final phase of determining the nucleotide triplets corresponding to particular amino acids in protein synthesis, studies carried out by Nirenberg, who was also a biochemist.[15]

Protein Synthesis

The nature of proteins became clear through the studies of Pauling on the α-helix, of John Kendrew (1917–97) and Max Perutz (1914–2002) on the crystallographic determination of the structure of myoglobin and hemoglobin, and of Sanger on the determination of the amino acid sequence of insulin. Before the last of these advances, however, Jean Brachet (1909–88) and Torbjörn Caspersson (b. 1910), an embryologist and a cytologist, respectively, showed that RNA was highly concentrated in the nucleolus and in parts of the cytoplasm of animal cells. Moreover, the RNA content of cells appeared to be roughly correlated with their rates of protein synthesis.[16] Purely biochemical studies of protein synthesis by Paul Zamecnik (b. 1912), Henry Borsook (1897–1984), and others followed in the late 1940s. They used mammalian cell extracts to demonstrate the incorporation of isotopically labeled amino acids into proteins in vitro. Philip Siekevitz (b. 1918) and Zamecnik showed that the reactions required an energy source, various soluble enzymes, and a sedimentable cell fraction known as "microsomes." Microsomes were found later, with the help of electron microscopic studies of Kieth Porter and George Palade (b. 1912), to consist of ribosomes attached to fragments of the endoplasmic reticulum. The composition and the two-subunit structure of ribosomes became clear in the years 1957–59, again through biochemical studies innocent of genetic approaches.[17]

Fritz Lipmann (1899–1986), a biochemist of enormous insight into cell energetics, had clarified the role of ATP as a general energy currency of cells in the early 1940s, including the suggestion that amino acids might have to be activated by phosphorylation (using ATP) before they were used in protein synthesis. This prediction was realized in enzymological studies of mammalian extracts in the mid-1950s. At this time many molecular-biological ideas began to penetrate the biochemical studies of gene expression. Using biochemical information available at the time, in 1958 Crick described a model of protein synthesis involving RNA as an informational intermediate between DNA and protein.[18] While he had to assume that ribosomal RNA acted as the template for protein synthesis, he was able to propose an "adaptor hypothesis," calling for amino acids to be attached to RNA trinucleotides (the anticodons) before they could read the RNA template (appendix 2). At the time Crick could not reconcile the difference in size between his proposed trinucleotide adaptors and the actual transfer RNAs (tRNAs) to which the amino acids became attached. The tRNAs had been roughly characterized in 1956, and they seemed simply too large (ca. 75–100 nucleotides) to serve the purpose. However, the fusion of biochemistry and genetic theory in Crick's article, in which the term "Central Dogma" first appears, represents as well as any the interactive contributions of both fields at the time. The article served as a template of discovery in the next few years.

In 1956, Elliot Volkin (b. 1919) and Lazarus Astrachan (b. 1925) found a species of RNA, formed and degraded rapidly in *E. coli* cells after infection with phage, that had a nucleotide composition resembling the phage DNA more closely than bacterial DNA. This early finding was recognized in retrospect in 1961 as supporting a model of gene expression involving a messenger from DNA to the site of protein synthesis. In 1960, Brenner, Jacob, and Meselson successfully tested the model, chemically defining mRNA in studies with *E. coli*.[19]

Purely biochemical work on protein synthesis continued, both in mammalian cells and in *E. coli*.[20] Nirenberg and J. Heinrich Matthaei (b. 1929) perfected *E. coli* as a source of cell-free extracts for their initial studies of polypeptide synthesis, using artificial RNA polymers as templates. It is of interest that bacteria were less suited initially to these studies than were animal cells. Rat liver extracts displayed slow and relatively few metabolic activities, while extracts of bacteria rapidly catalyzed highly diverse reactions. This compromised the use of radioactive tracers in bacteria because extraneous activities made isotope incorporation specifically into proteins hard to discern. Nirenberg and Matthaei overcame these obstacles mainly by separating cell fractions by ultracentrifugation and combining certain fractions in particular proportions for use in their experiments.[21]

By this time, such biochemical studies had been absorbed under the heading of molecular biology, and we may take the late 1960s as the time by which no clear distinction could be made between the rationales of biochemists and nongenetically oriented molecular biologists. Some classical biochemists, with their contempt for molecular biological hotshots, would persist in many institutions, but the insularity of biochemistry had effectively come to an end.

In referring to nongenetically oriented molecular biologists, I call attention again to the entry of chemists, calling themselves molecular biologists, who entered the field without experience in genetics. Their techniques were strictly biochemical, and if they used mutant material, it was supplied by others. Molecular studies proceeded in such laboratories without special regard to cell phenotype. In the 1990s, the model prokaryotic organisms, including their plasmids and their phages, became the workhorses of genetic engineering. The attitude has progressed to the point that plasmids and phage have become vehicles for cloned DNA, and *E. coli* has become the medium for replicating them.

Conclusion

The often understated contributions of biochemistry to molecular biology were actually substantial and in many cases indispensable to rapid progress. The mutants of Beadle and Tatum would have yielded much less information were it not for parallel, independent progress in enzymology. Many historians point only to the discovery of DNA as they extend their accounts from a brief mention of biochemical genetics. The nature and replication of phage could not have been appreciated without early biochemical characterizations. And the Central Dogma would have remained a poorly examined dogma indeed had it not gained the flesh that orthodox biochemists put on it between 1955 and 1965.

A second relevant point is the convergence of biochemistry and molecular biology on *E. coli* as a model organism. *E. coli* gave experimental groups of both disciplines a common venue in which to compete, to cooperate, and finally to merge. Because of the intense study of this bacterium by so many investigators from so many standpoints, it achieved iconic, almost divine status. The connectivity of information at all levels of organization—even in population genetics—made it a supermodel. It remains a model of model organisms, from its use as an intellectual reference for the world of prokaryotic research to its role as a lowly reagent for most of genetic engineering.

18

Genomics

There's a lot of things in this world that aren't worth knowing.
—H. M. Raup

The Beginnings of Genomics

Lily Kay has documented the conceptual change of orientation, from structure and specificity of proteins among crystallographers and related biochemists, to the idea of molecular geneticists that proteins and DNA carry information. In her postmodern words, "Specificity was bounded in matter, while information was mobile, transporting the memory of form beyond material bounds."[1] Her comment refers to her concern that biologists have lost sight of organisms as real, wet, complex entities and have reduced life to a cybernetic analogy. The focus on sequence information has led, through a series of methodological advances, to the rise of genomics. This in turn has reversed the strongly reductionist character of the field as it developed in the last century, albeit in a way Kay may not have welcomed.[2]

The advent of genomic sequencing may be the most dramatic institutional step underlying the theme of this book, a transition of biology from an academic to a postacademic science, as Ziman calls it.[3] The Second World War evoked a commitment of government to scientific research in universities, greatly supplementing support by private foundations. Government support expanded the new instrumentation available to biologists and encouraged transdisciplinary contributions of physicists and chemists. The success of science led to collectivization of efforts in certain research institutions, in which interdisciplinary research programs flourished well before they had names. Model organisms became a natural focus for these interests, and the new scientific fields using them did not fit into existing departments of universities even as government grant-review panels codified the fields. The mushrooming success and expense of the enterprise made science ever more dependent on government support. The loudly proclaimed benefits of biological discoveries to medicine, commerce, and agriculture led the public and the government to think more seriously about societal applications of academic research. And government support, problematic even then, began to be supplemented by commercial support with increased political and ethical complications for biological research. Whatever freedom individual investigators once had (a good deal less than they might have thought) is now more restricted. University biologists perform contract work for industry, write grant proposals to the National Institutes of Health on disease processes, set up companies to exploit their latest diagnostic kit, and engineer a tastier tomato. Genome sequencing has ushered in a consortium-based approach to functional biology, which now requires the continued, collective effort of many scientists. Where a geographically dispersed group of scientists can't do the job, a commercial venture located in a single, well-equipped building will.

In the 1980s, tools to isolate and to determine the sequence of specific segments of DNA and to study the arrangement of genes in the genome of any organism became widely available (see appendix 3). The ambition to determine the actual nucleotide sequence of entire genomes soon followed. The implementation of this ambition is a story of political persuasion and will, not only within the field of molecular biology, but in the arena of U.S. national politics. In the late 1980s, a proposal was put forth to determine the nucleotide sequence of the entire human genome. The project would be carried out by the U.S. Department of Energy (in need of new activities as the demand for atomic bombs and nuclear power waned) and the National Institutes of Health. It was launched as the Human Genome Project (HGP) and soon became coordinated with genome-sequencing efforts in a number of other countries under the umbrella of the Human Genome Organization (HUGO). A few molecular biologists justified the HGP with its promise of fully revealing the genetic endowment of the species, leading to an understanding of human biology and to successful intervention in genetic disease. The proposal was rash for its time because many knowledgeable biologists felt that having the genomic sequence would simply be a body of information, not an advance in our understanding. Efforts to isolate and sequence only selected parts of the genome, they believed, would lead more directly to the promised scientific and medical benefits. But the proposal was even bolder on the political level because the effort would cost an estimated $3 billion. In the minds of many, the effort might drain funds from other biomedical research efforts or national investments, and this invited opposition from many scientists and politicians.[4] Some frustrated biologists grumbled that to find out what wasn't worth knowing, we had to know everything.

The skepticism of many biologists was well founded, given the cost and the time needed for DNA sequencing at the time. As in the case of the moon landing in the 1960s, the U.S. government was persuaded by an influential constituency, led this time by James Watson, among others. It promised that sequencing and bioinformatic technology would improve greatly and quickly enough to get the job done by 2003. This bold promise was realized in 2001, in fact, with enormously improved robotic, high-throughput sequencing systems and massive computer programs able to handle billions of bytes of sequence information as fast as they were produced.[5] Of interest here is the strategy the HGP adopted toward the goal of human genome sequencing. Funds would be spent first on the genomes of simpler organisms to test various strategies and practical aspects of sequencing. Immediately—especially as the need for full sequencing of model organisms' genomes was felt more strongly—biologists prominent in the study of candidate model organisms abandoned their opposition to the idea of genomic sequencing and embraced it. Politicians, in turn, responded to the medical promise of the project and the more unified lobbying of biomedical scientists.

An important additional factor speeded the completion of the human genome sequence: the entry of a competing, ultimately commercial sequencing group. J. Craig Venter (b. 1946), under the aegis of The Institute for Genomic Research (TIGR) took a different approach to sequencing efforts than the methodical one adopted by the HGP. The HGP had decided to map the genome physically with molecular markers (RFLPs among others), clone segments of the genome, and then sequence these individually. Venter surprised almost everyone by promising "whole-genome shotgun" sequencing, in which the entire genome would be cut up into small bits and each fragment sequenced. To reduce the dangers of missing any segments of the genome, at least 10 genome-equivalents were processed. Computer searches for overlaps in the sequences of the fragments would yield an alignment that represented the whole genome. Venter demonstrated the effectiveness of this strategy by rapidly sequencing the entire genome of the bacterium *Haemophilus influenzae* and publishing it in 1995. With a massive infusion of venture capital to sequence the human genome, he set up a company, Celera Genomics, and did it in three years. The race for the "first draft" of the human genome (ca. 90% complete) ended in a rather mean-spirited tie in 2001, with the public HGP having speeded its efforts in order to present a publicly available genome sequence simultaneously with the private one.

Along the way, the genomes of many microbes and multicellular organisms have been sequenced. They have changed the complexion of much of functional biology from a hypothesis-driven science to a discovery phase, in which handling and integration of massive data sets drawn from many levels of organization has become a major feature. I discuss in the final sections of this chapter, dealing with comparative and functional genomics, two components of this area as they illuminate the fate of model organisms. I show how protein and DNA sequence information from a variety of organisms has informed the study of the origin of major domains of life and the process of more local evolutionary change. Second, I explore the use of genomic information in studying the functional complexity of cells and the patterns of gene expression during environmental or developmental change. These efforts have drawn biology back (or forward) to a holistic appreciation of diversity and complexity that preoccupied biologists at end of the 19th century. My thesis, developed further in the last chapter, is that the return of biology to these attributes of living things could not have been achieved without an intervening age of model organisms.

The Roots of Molecular Evolutionary Studies

In the 1960s, molecular biologists developed methods, cumbersome by present standards, for determining the ribonucleotide sequence of small RNA molecules. These methods gave them the ability to discern changes over evolutionary time in the sequence of the abundant 16S rRNA, an RNA found in the small subunit of the ribosomes of all organisms. Ribosomes arose early in the history of life and perform the essential chemical steps in protein synthesis. Evolution does not tinker idly with such a fundamental feature of cells, and consequently the ribonucleotide sequence of the small 16S rRNA is one of the most conserved pieces of information in the biological world. In 1977, Carl Woese (b. 1928), in comparing the sequences of 16S rRNAs of organisms from prokaryotic to eukaryotic, demonstrated the homology of this molecule (based on the similarity of sequence) over the whole evolutionary tree of living things. In the course of his study, he found characteristic differences in the sequences of 16S rRNA among major groups of organisms. Strikingly, the prokaryotes (informally named for their lack of a nuclear membrane) appeared to fall into two major categories based on these sequences. One was the more familiar *Eubacteria*, which includes *E. coli* and most other known bacteria; the other he named the *Archaea*.[6] The last group includes "extremophiles," those with unusual metabolism (sulfur-oxidizing; methane-producing) and habitats (deepsea vents, hot springs, and highly saline waters). The Archaea had distinctive enough 16S rRNAs that they could be considered a third domain of life, with the Eubacteria and the *Eukarya* (eukaryotes). This opened a new phase of studies of the origin of life and of the evolution of organisms.

The term "molecular evolution" arose in the 1960s as the tools of sequencing proteins and RNA and, in the 1970s, DNA, became widely available. The earliest approaches to the direct study of changes of amino acid order in polypeptides and of nucleotides in RNA and DNA began with the demonstration of the basis of sickle-cell anemia. This revealed for the first time that a single mutation could cause a single amino acid substitution. Independently, population geneticists discovered that a huge amount of allelic diversity (i.e., heterozygosity for many individual genes) prevailed in natural populations. This was documented by the use of simple electrophoretic apparatuses that demonstrated differences in the mobility of protein molecules in an electric field. Such differences implied diversity in the amino acid sequences of the proteins analyzed and of the genes that encoded them. This discovery came at a time when such diversity, called "genetic polymorphism," could not be reconciled with the long-held conviction that natural populations could not tolerate such a large "mutational load," a term that imputes a selective disadvantage to most mutations.

Methods to determine the sequence of amino acids in long polypeptides improved rapidly in the 1960s. The new tools demonstrated amino acid substitutions in individual proteins among members of a species and between representatives of different species. The work was led by the laboratories of Émile Zuckerkandl (b. 1922), studying animal hemoglobins and of Emanuel Margoliash (b. 1920), studying cytochrome *c*.[7] The work was confined to these abundant proteins, present continuously in evolutionary lines of descent. From these data, workers in the growing field developed the idea of evolutionary distances based on the number of amino acid substitutions in the homologs of these proteins in various species. The field of molecular phylogeny has developed on the basis of these findings, since the pattern of divergences of lines of descent can now be rendered as evolutionary trees.[8] The studies also led to the controversial hypothesis of an evolutionary clock, a clock that seemed to yield a steady rate of substitutions per year, rather than per generation, for many proteins.[9]

The two areas, the amino acid polymorphism in natural populations and the evolution of amino acid sequence, deal with two magnitudes of difference. In the case of natural populations of a single species, variants are found in a small, significant fraction of individuals, but one form predominates. The amino acid substitutions are found at only a few positions in various parts of a protein. In evolutionary lineages, on the other hand, one sees consistent differences among the majority forms of a protein in members of a major taxon (fig. 18.1). Investigators in both areas, however, now draw a similar conclusion about mutations in nature. Most mutations are neutral in effect. They may become lost, they may maintain a polymorphic presence for some time after they appear, or they may eventually become fixed (i.e., uniformly present in all descendent members of the species) due to the random drift of allele frequency over time in sexually and asexually reproducing populations.

The Impact of the Human Genome Project

By 1985, cloning and sequencing of DNA had become practical and was widely practiced. Indeed, the nucleotide sequences of several bacterial and mammalian viruses had been determined. No longer were amino acid sequences of proteins determined directly. Instead, molecular biologists cloned the genes that encoded them, determined their nucleotide sequence, and deduced the amino acid order according to the genetic code. This carried the additional benefit of unambiguous knowledge of the codons underlying each amino acid in a sequence. Such knowledge could not be garnered from amino acid sequences because of the multiple codons for the same amino acid in the genetic code (fig. A2.6, appendix 2). Public databases of nucleotide and amino acid sequences (e.g., GenBank, maintained at the National Institutes of Health) grew rapidly, together with capabilities of performing rapid sequence searches and comparisons. The sequences deposited in these databases were accompanied by annotations of the role of the genes and proteins in cells. Therefore, a sequence newly determined in a laboratory could be compared to sequences in the database and recognized as a homolog (a gene with a similar evolutionary

```
                    MSSFTKDFDC--HILDEGFTAKDIIDQKINEVSSSDDKDAFYVADLG               46  Mm
MVMPTVVSDRMGTIDFIDYTNNHVFSKCQTDSLNTVNNGSLKHDDYLHGLANGKLVAKQMIGDALRQRVESIDSEFCEPGDEDTFFVADLG     91  Nc
MSSTQVGNALSSSTTTLVDLSNSTVTQKKQYYKDGETLHNILLELKNNQDLELLPHEQ--AHPKIFQALKARIGRINNETCDPGEENSFFICDLG   93  Sc
                                                              *    *  ***


DILKKHLRWLKALPRVTPFYAVKCNDSRAIVSTLAAIGTGFDCASKTEIQLVQGLGVPAERVIYANPCKQVSQIKYAASNGVQMTFDSEIELMK    141  Mm
EVIRQHLRWKLNLPRVKPFYAVKCHPDERLLQLLAALGTGFDCASKRAEIEQVLRMGVDPSRIIYAQPCKTNSYLRYVAQQGVRQMTFDNADELRK   186  Nc
EVKRLFNNWKELPRIKPFYAVKCNPDTKVLSLLAELGVNFDCASKVEIDRVLSMNISPDRIVTYANPCKVASFIRYAASKNVMKSTFDNVELHK     188  Sc
      *  ***     **      **  *   *    * * ** ***  **    * *  **   *  ** *


VARAHPKAKLVLRIATDDSKAVCRLSVKFGATLKTSRLLLERAKELNIDVIGVSFHVGSGCTDPDTFVQAVSDARCVFDMAT-EVGF-SMHLLDI   234  Mm
IARLYPDAELFLRILTDDSSSLGRFSMKFGASLDSTDGLLGLARQLGLNVVGVSFHVGSGASDPTAFLKAVQDAHVFQQAA-AYGY-SLKTLDV    279  Nc
IKKFHPESQLLLRIATDDSTAQCRLSTKYGCEMENVDVLLKAIKELGLNLAGVSFHVGSGASDFTSLYKAVRDARTVFDKAANEYGLPPLKILDV   283  Sc
      *    ** *** * *        *       **  *    *    *********      **  **   ** *     *         **


GGGFPGSEDTKLKFEEITSVINPALDKYFPSDSGVRIIAEPGRYYVASAFTLAVNIIAKKT------VWKEQPGSDDEDESNEQTF-MYYVND    320  Mm
GGGF-CSDDS---FEQMANVLRAALDRYFPAHTGVNLIAEPGRYASSAFTLACNIIARTIQDGSAVSVSDSSSMSDDGSVNNGDARYMVVVND     370  Nc
GGGF-------QFESFKESTAVLRLALEEFFPVGCGVDIIAEPGRYFVATAFTLASHVIAKR------KLSENEA------------MIYTND    351  Sc
****          *      **    **      **  *** ***   *  ****        *                      * * ** **


GVIGSFNCILYDHAH-VKALLQKRPKPDEKY-----------------------YSSSIWGPTCDGLDRIV-ERCNLPE-MHVGDWMLFENMGAYTVAA   393  mm
GLYGNFSSIMFDHQHPVAKILRAGGRTMY----NSVAAHESSAEDAIEY----SIWGPTCDGIDRIT-ESIRFREILDVGDWLYFEDMGAYTKCS     456  Nc
GVIGNMNCILFDHQEPHPRTLYHNLEFHYDDFESTTAVLDSINKTRSEYFYKVSIWGPTCDGLDCIAKEYYMKHDVI-VGDWFYFPALGAYTSSA    445  Sc
  *  *  **              *                       *         ********* *  *    *            ****  ****


ASTFNGF-QRPNIIYVMSRPMWQLMKQIQSHGFPPEVEEQDDGTLPMSCAQESGMDRHPAACASARINV   461  Mm
ATTFNGFSNEHDVIYVCSEFGAMALLGL   484  Nc
ATQFNGFEQTADIVYIDSELD   466  Sc
  *  *   *   * ****
```

origin and probable function) of those already known. From such information, it has become increasingly easy to recognize DNA sequences that code for polypeptides (open reading frames, free of nonsense codons), the nature of the encoded protein, and many architectural and regulatory features of DNA that do not code for polypeptides. As the databases grew, so did ambitions to sequence entire genomes.

Members of TIGR were the first to sequence the genome of a free-living microbe, *H. influenzae*, as noted previously. With the demonstration that sequencing by the shotgun approach was practical, at least for bacterial genomes, interest turned to genomes of other species. The second sequenced genome was that of *Mycoplasma genitalium*, partly because it was the smallest known among bacteria. Not only was *M. genitalium* a medically important bacterium, but genomic analysis promised to yield an estimate of the minimum number of genes needed by an organism that can be cultured free of its host. The next genome sequenced, in 1996, was that of an Archaean, the deep-sea autotroph, *Methanococcus jannaschii*. Here, comparative genomics could provide a more detailed knowledge of the "essential" differences among the three domains of living things. This hope was sustained with the publication in 1996, by an international consortium of yeast workers, of the complete genomic sequence of the eukaryotic *S. cerevisiae*. The genomic sequence of *E. coli* was published in 1997 by a group that prided itself on accuracy over speed, thereby solidifying the role of this organism as a model microbe. Since then, sequencing of the eukaryotic genomes of *Neurospora crassa*, *Aspergillus nidulans*, and the model multicellular organisms *Drosophila melanogaster*, the nematode *Caenorhabditis elegans*, and the small weed *Arabidopsis thaliana* have been completed. A number of other multicellular organisms' genomes have been sequenced to satisfy more restricted biological, medical, agricultural, or commercial interests.

By the end of 2002, well over 100 microbial genomes, most of them prokaryotic, had been sequenced. This is a spectacular departure from the idea of model organisms as I have described them and their uses in previous chapters. In fact, it marks the end of an era when molecular biologists were confined by mating and recombinational systems to single species and the beginning of highly detailed, new studies of diversity of life on earth.

Comparative Genomics

It is impossible to describe or even list the activities of the new field of comparative genomics, but the appreciation of diversity is at its heart. Many who study evolution want to carry on with quantitative estimates of similarity and evolutionary distance among related forms and to develop phylogenies of the pro- and eukaryotes.[10] Others want to deal with the poorly known Archaea of the extremophile category. Other groups focus on the relative prevalence of clonal evolution (i.e., strictly asexual propagation and mutation) in bacteria versus processes involving conjugation and gene exchange via phage and plasmids. Still others want to study major genomic duplications and reorganizations during evolution.[11] The last processes can be recognized by hunting for similar, if divergent, sequences within a single genome and by comparing the arrangements of similar genes on the chromosomes of related bacterial species.

Two unusual phenomena deserve special mention. Evidence for substantial lateral transfer of genes from one species to quite a distant species has accumulated.[12] Indeed, some genes of Eubacteria appear to be derived directly from the Archaea, and vice versa. This process of gene swapping has made phylogenetic trees of bacteria more difficult to derive from simple comparisons.[13] Investigators must take care to base evolutionary distances on the genes most continuously retained in the main lines of descent.[14] But the phenomenon raises the question of the prevalence of this process, presumably accomplished by infective agents or natural transformation, over evolutionary time. The related process is the loss of genes. As new, foreign genes are acquired, others drop out, leaving a mixed genome. And to know whether a gene is absent, generally one must have a complete genomic sequence. These processes may expand our sense of how prokaryotes evolve well beyond what we might expect from the study of higher forms.

As more sequence information becomes available, a second unusual phenomenon comes into view: the existence of the more than 95% of bacterial species that cannot now be cultivated. Such bacteria are recognized, if at all, by detection of their DNA sequences in soil or water that clearly differ from the genes of known forms.[15] Norman Pace (b. 1942) argues that quantitative measures of diversity will be available only with considerably more genomic information and that the interactions of bacteria in nature—most of them as yet unknown—cannot be appreciated with the use of just a few species in the laboratory.[16]

The neglect of diversity during the reign of a mere few model microbes in molecular biology has given way to its allure—its origin, nature, distribution, and maintenance—for its own sake. Because of the avalanche of sequence information, many investigations are now data driven, a state of the field in which the imagination to formulate and pursue programs of interest is at a premium. The area of comparative genomics is reminiscent of the state of biology as a whole at the end of the 19th century, when collectors and classifiers were in the saddle, and when a desire to understand evolution motivated most functional biologists. Clearly, none of these problems (old or new) could be pursued now if research were to remain confined to a narrow group of model organisms.

Functional Genomics

Some Tools of the Trade

The availability of the entire genomic sequence of an organism enables a biologist to monitor gene expression on a genome-wide basis and to manipulate any gene of the organism to determine its effect on expression of all other genes. The term *transcriptome*, the entirety of RNAs produced by a cell or organism, has become common in the field, and the study of the transcriptome of cells is now possible through the use of microarrays, or "gene chips." Microarrays place as many as 400,000 tiny dots of specific DNA sequences, each one (or combinations thereof) representing a transcribed gene uniquely, in grids on glass no larger than a microscope slide (fig. 18.2).[17] The position of each sequence is known. These sequences can capture RNAs (or DNA copies of them) extracted from cells of various sorts or in various conditions, by complementary base pairing. The RNAs are labeled with a fluorescent tag, and images of the grids are digitized and manipulated by computer. The RNAs from two different cell populations may be labeled with different fluorescent labels and applied to the same microarray, their colors separately recorded and compared in intensity, thus comparing changes in gene expression in different cells or in the same cells in different conditions.

Tools to visualize the *proteome*, the ensemble of all proteins of a cell or organism, have also been developed. These tools depend on clever tagging and sequence-detection with the use of a mass spectrometer.[18] In addition, methods to study effective contact and binding of various proteins to other proteins on protein microarrays have been developed.

Finally, the manipulation of the genomes of bacteria and yeast has become quite sophisticated. One of the most important tools is the ability to disrupt any gene of these organisms at will.[19] These "knockout" strains, if viable, are invaluable in determining the roles and redundancies of genes by determining

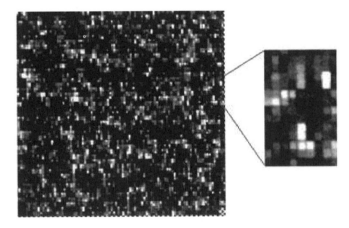

Fig. 18.2.   A small portion of a mouse DNA microarray (left), with a still smaller portion of it magnified to show individual spots (right). A grid of different, short nucleotide sequences has been formed as square spots on a glass slide, and the variation in brightness of spots reflects the variation in the amount of different mRNAs, extracted from a mouse tissue, that bind to them. The spots, which are actually groups of pixels, are read by a scanner and digitized for data processing. Detail from an original image kindly supplied by J. Denis Heck, University of California, Irvine.

the effect on the gross phenotype, growth rate, or genome-wide transcriptional activity.

With these tools, molecular biologists have acquired enormous control over cell behavior and efficient methods of collecting vast amounts of data. In the latter respect, the massive experimental data sets appearing now are as overwhelming as the descriptive data in the field of comparative genomics. Indeed, a name has been given to the emerging field: "systems biology."[20]

Emergent Properties and Emerging Principles

The ability to screen the transcriptional activity of yeast with genomic microarrays quickly led to a fuller appreciation of modularity in gene expression.[21] Among conditions surveyed with respect to transcriptional activity in yeast were various nutritional states, the induction and progress of mitosis, meiosis, and the activity of protein signaling pathways that adjust and integrate responses of genes to the external environment.[22] Computer analysis of the spots on microarrays includes the ability to group genes into modules by their common increases and decreases in expression during environmental changes, yielding patterns of their coordinated expression. The value of these observations is clear enough. First, increases and decreases of expression of many known yeast genes could be observed. Second, changes in expression of many of the 2000 unidentified yeast genes (out of 6300) were also seen.[23] The expression patterns of both classes of gene give investigators clues regarding the interactions inherent in genomic activity and identify the functional areas to which unknown genes may be assigned.

Experimental intervention in these modules by gene knockouts and other means suggest the manner in which the members of the modules interact. Recent information suggests that the modules are organized in a hierarchical fashion, with lower-level, discrete sets of functions integrated into higher-level modules. In such an organization, more connections prevail among the members of the lower-level modules than they have to higher levels of integration.[24] Extending the notion of modularity, a descriptive approach to the interactions of genes (connectivity) can be made at the level of the proteins.[25] Only on this genome-wide scale can the vast number of interactions of proteins be appreciated, and it is becoming harder and harder to represent them diagrammatically in the literature.[26] The goal of systems biology is an understanding of the connections at all levels of organization: gene, mRNA, protein, protein modification and contacts, metabolic pathways, and signaling networks. It can only be done by perturbations whose effects can be monitored comprehensively and then modeled in an iterative process (as more perturbations are studied) until the model predicts the behavior of the system over a wide range of conditions.[27]

Mutational studies that resolve the phenotype into specific elements (gene activities) have been a staple of genetics since the time of Morgan. However, only with genomic manipulation can one appreciate the degree of dispensability of genes. The methods for systematic deletion of genes in bacteria and yeast have developed to high precision.[28] The surprise in such studies is the large fraction of genes whose deletion is wholly without effect on the phenotype. A number of cases of this sort had been known previously and had been attributed to some redundancy of genes in the genome. The scale of this phenomenon and the number of conditions in which such deletions seem to have no effect give an entirely different impression than the studies of the mid-1900s. Now, the concepts of robustness and buffering, attributes displayed strongly even by simple genomes, have descended from a previous generation with some force upon molecular biologists. These are not new principles. The complex connection between genotype and phenotype had been stressed by embryologists and quantitative geneticists for more than 100 years. Their rather vague formulations, however, were obscured by the clear mutational dissection of the phenotype by the Morgan school and especially by the direct gene–enzyme relationships discovered and espoused so emphatically by Beadle and his followers.

The large amount of gene disruption that a genome can sustain without serious phenotypic effect indicates that the genome does not control the phenotype as directly as once imagined.[29] Instead, genomic activity must be integrated by hierarchical interactions among proteins at several levels that endow it with robustness.[30] Robustness depends not only on simple genetic redundancy, but on the evolved and much more complicated web of interactions by which the system has become highly sensitive to individual perturbations at relatively few critical nodes. And the localized sensitivities can evoke compensatory responses in alternate pathways if the one more commonly called upon is disabled. These interactions, in embryogenesis, canalize development toward a particular end point despite the inevitable mutational and enviromental variation seen in nature. The limits to this buffering can be appreciated in selection experiments. If one selects the few members of a population that display a defective phenotype in extreme conditions and use them as parents for the next, the defective phenotype may eventually emerge among the descendants even in normal conditions. This suggests that populations have diversity among genes that normally strengthen functional networks against perturbations. The understanding of such interactions relies heavily on computer analysis and simulation of global, hierarchical interactions impossible to visualize directly.

Conclusion

We may now return to a serious issue noted in previous chapters: the perception of the role of nuclear DNA. Before Morgan, few doubted that the cytoplasm carried heritable information required for the proper progress of development. Argument continued well into the 20th century about this matter, and Sonneborn was one of the most articulate defenders of the view that the cytoplasm could not be ignored in inheritance, much less in development.

Against this view, the Morgan school called attention to the very few cases of mutable, heritable information—even developmental information—that was not carried by nuclear genes. Of course this was not an open-and-shut case because redundancy of cytoplasmic genes might militate against ever seeing the phenotypic effect of a cytoplasmic mutation. But the case grew stronger with Beadle and Tatum's work, which strongly reinforced the reductionist, Morganist position. The argument for nuclear hegemony over all of life's functions appeared conclusive to many following the discovery of DNA structure and the implications of Jacob and Monod's model of gene regulation. The upshot was the popular view promulgated by molecular biologists that DNA was the master molecule, the program directing all the details of structure, function, and replication of the cell.

Could it be argued that the cytoplasm is the master, and the genome simply a source of data needed for a variety of tasks?[31] The confusion between program and mere data implied by these alternative views is acute and reveals the poverty of computer analogies. Living cells are the smallest autonomous entities that can replicate. No nucleus ever built the cytoplasm around itself from scratch; going further, we realize that the fertilized egg is a highly structured entity, choosing, as it were, what to do next by virtue of the distribution of molecules within it. In particular, even at that early stage of a plant's or an animal's origin, the cytoplasm is an environment that calls forth appropriate genetic activity from the DNA and even tells the genome when to replicate.

This is a different perception of the role of the genome than the one promoted by molecular biologists of the late 20th century. The cytoplasm-as-master conception would, of course, have serious limitations if it were to downplay the role of genetic information. The cell, in fact, builds itself, unlike any machine built by man or woman. It is as silly to ask who is the master and who is the slave as it is to ask which is more important, heart or brain? We have, in the genomic era, converged on the idea that causation goes both ways, upward from DNA and downward from cytoplasm and environment as we see cells function and organisms develop. Even the distinction between "genetic" and "epigenetic" fails to convey this interdependence.[32] Molecular biologists could so easily convince others that DNA is a master molecule because of the very simplicity of the cells they used—those of model prokaryotic and eukaryotic microorganisms. For many years, they and their rapt listeners could think of the cytoplasm as a passive medium and not an environment that, in the proximate sense, directs gene activity.

To conclude, the genomics era has returned biologists to the matter of diversity and complexity with the tools to explore them. These tools are much more than vague descriptions and metaphorical models, which were the weakness of many prior studies of development (exemplified in the previous paragraphs).[33] With this revisitation, an expanded mindset that no longer pits reductionism against holistic thinking now prevails. The goal of reducing biology to physics and chemistry has been abandoned as biologists turn toward evolution, which began as a slowly unfolding chemical accident, and which has given us a vast bouquet of surprises, one species after another.

19

The Age of Model Organisms

The dustbin of history will have to be emptied more frequently now.

Model Organisms, Model Systems

Writing about science often forces one to define terms with greater specificity than they deserve and to offer slippery metaphors in attempts to communicate.[1] Nevertheless, the concept of model organisms, developed in the preceding chapters, should now be clear. Briefly, model organisms have these key attributes:

- They begin as model systems for a genetic attack on a particular biological problem. Researchers simultaneously isolate mutants, map them on chromosomes, and pursue the problem of interest.
- Mutants isolated in the initial research program include many whose phenotypes suggest new research programs.
- Stocks, media, and methods become standardized.
- A community of researchers becomes bound to the organism not only by the confinement of intraspecific crosses, but to one another through newsletters, meetings, and stock centers.
- The investigations ultimately cover many aspects of the biology of the organism. The resulting integration of knowledge at many levels of organization provides a rich context for the interpretation of any new finding or mutant.
- The organism becomes a reference point (a model) for investigations of other organisms with which it shares features of interest.

The use of model organisms as I have defined them is inescapably a 20th-century phenomenon in biology—inescapable because modern genetics was born in 1900. Indeed, the model status of the simple model microorganisms originated in the exploration of new genetic systems—in the dreams of Dodge, Sonneborn, Lederberg, Pontecorvo, Sager, Lwoff, Delbrück, Hayes, Wollman, and Jacob. We now have models ranging from the most complex eukaryotes to microorganisms, with molecular biologists following phylogeny back to the simplest living things (fig. 19.1).

Let us consider again some of the choices of model organisms I have described. Dodge discovered the sexual cycle of *Neurospora* and promoted it as an ideal system for the study of genetics. The early work of Beadle and Tatum on *Neurospora* heavily favored isolation of nutritional mutants for their exploration of the gene–enzyme relationship. Gene interaction became a basic tool in understanding the sequence of enzymes in pathways, the regulation of gene expression, and the subcellular organization of metabolism. New observations based on tetrad analysis grew into studies of gene conversion and recombination on the one hand and mitochondrial inheritance, biogenesis, and

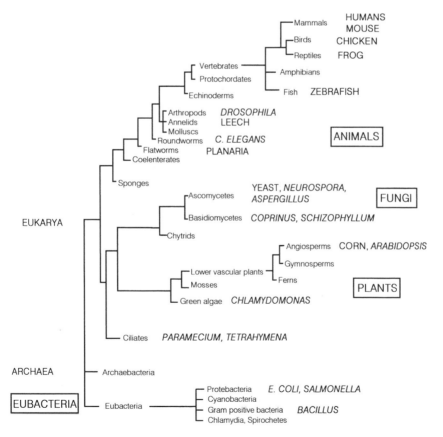

Fig. 19.1.   Rough phylogenetic relationships of life forms, with names of major groups in boxes. Model organisms (including some not discussed in this book) are shown in capital letters.

function on the other. Heterokaryosis, a tool of biochemical genetics, became a subject of interest in its own right. The study of ultraviolet and chemical mutagenesis gave us an early picture of the mutational process. Morphological mutants isolated early in the 1940s are exploited even now in molecular studies of hyphal growth, development, and the biological clock. An interest in the three main *Neurospora* species has opened the door to the population genetics of the genus, the discovery of transposons, and spore-killer genes. The global collections of *Neurospora* spp. yielded resources for RFLP mapping of the genome and many evolutionary studies of ensembles of genes, both mitochondrial and nuclear. All of this was done by a relatively small group of investigators that nevertheless leaves *Neurospora* virtually unchallenged as a model for the large world of filamentous fungi.

E. *coli*, initially used to explore the mutability of its metabolism, yielded auxotrophs like those found in *Neurospora*. These in turn enabled Lederberg to test for bacterial gene exchange. The happenstance discovery of the F factor and of Hfr strains led to rapid development of our knowledge of bacterial conjugation and the structure of the bacterial chromosome. The discovery of transduction in *S. typhimurium* led to a fundamental tool of bacterial genetics. More important, transduction enabled fine-structure studies of bacterial genes and gave us many examples of operons, identified as gene clusters by mutational studies. The operon theory and the model of the regulation of the *lac* gene complex emerged quickly from this. And, of course, the accidental discovery of λ phage gave insight into the temperate phage lifestyle, a striking contrast with virulent phages such as T4. By 1958, *E. coli* was the preferred organism for most nucleic-acid–based work on the genetic material, its replication, recombination, repair, and expression. Mutant banks and genetic techniques became working tools for all studies.

Yeast genetics, joined by studies of the petite colony phenotype, developed slowly. When it came of age, biochemical genetics of yeast led to isolation of temperature-sensitive mutations of macromolecular synthesis. Mutants specifically affected in the progress of the cell cycle were found among them. These became the roots of knowledge of cell division in eukaryotic organisms. At about that time, molecular biologists touched off a self-propagating explosion of knowledge about the organism by introducing transformation, plasmid manipulation, and finally genomic sequencing. The exploration of the details of the cell cycle, signaling cascades, protein targeting, mitochondrial biogenesis, meiosis, and a host of other phenomena were laid open for study like a frog on a dissection board. Yeast research has proved that fundamental information about all eukaryotes can be derived from an organism as tractable technically as *E. coli*.

It is hardly necessary to enumerate comparable steps in the use of *Aspergillus*, *Paramecium*, phages T4 and λ, and *Chlamydomonas*. The differences among the model organisms lie mainly in the scope of their model status, especially as this status changed with time. *Neurospora* was a model for both biochemical genetics and for formal genetics of all eukaryotes for 20 years or more; it now is the reference organism for filamentous fungi. *Aspergillus* geneticists could initially represent their organism as a model for fine-structure analysis of eukaryotic genes and for the possibility of parasexual genetics in higher organisms. It is now a sister to *Neurospora*, different enough to be of great comparative value, but similar enough that it shares much of its genetic, regulatory, and developmental phenomenology. *Paramecium* and *Tetrahymena* have a similar relationship, but their model status in the nonciliate world was lost with the realization of the uniqueness of their macronuclei, the demise of the metagon theory, and the discovery that kappa was a symbiont masquerading as a cytogene.

The relative contributions of the model organisms can be seen in the number of research papers published in certain journals in which their names appear in the title over the years 1940 to 2000. The three index journals I chose are top-tier publications in which papers on biological subjects appear without bias regarding organisms: *Science*; *Proceedings of the National Academy of Science, USA*; and *Genetics*. Figure 19.2 shows that *E. coli* and yeast virtually eclipse all other model organisms, a predominance that justifies their supermodel designation. This point is even clearer when we take into account the growth in the page numbers of these journals. Over the 60-year period the sum of pages per year in the three journals has grown 14-fold, whereas in half that time, the last 30 years, yeast titles have increased 16-fold. During the same time, *E. coli* titles increased by 9-fold.

The figure also shows that the two filamentous fungi and the phages λ and T4 maintained themselves on an absolute basis. This suggests that the number of laboratories using these organisms held rather steady over the last 30 years despite the expansion of biological research. This is probably true even if we recognize that more specialized journals founded in the meantime competed successfully for publications about these organisms.

In the decade 1990–2000, *E. coli* titles have plainly declined for the first time, making it likely, as we have inferred, that many *E. coli* workers and their descendants have migrated to eukaryotic organisms, certainly benefiting yeast in part. Whether the slight decline in the rate of growth of yeast literature in the last five years will be sustained in the next decade, we can only speculate. I would predict, for reasons explained below, that the decline will continue as biologists, equipped with microbial paradigms and genomic techniques, rush to explore a wider spectrum of the living world, concentrating on multicellular organisms.

The success of studies with the microbial models of molecular biology has emboldened biologists to develop the molecular biology of *Drosophila melanogaster*, *Caenorhabditis elegans*, *Aribidopsis thaliana*, the zebrafish, and other multicellular forms as molecular models of their taxa. Work has progressed so well that the genomes of many of these organisms have been or will shortly be sequenced. In addition to these new models, the mouse continues its model status, initiated with a commitment of early 20th-century geneticists to this organism. Owing to the history of the use of mice, the mouse genome has also been sequenced. The widespread use of mice as model mammals throughout the 20th century supports the principle I have espoused—that such a status depends on genetic domestication and development.

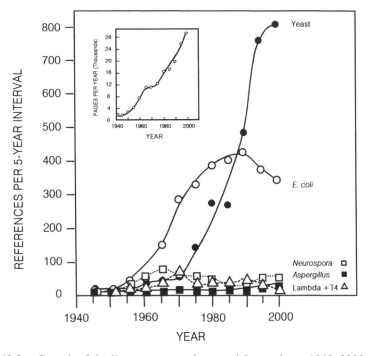

Fig. 19.2.   Growth of the literature on various model organisms, 1940–2000. The number of titles containing the species names that appeared in *Genetics*, *Science*, and *The Proceedings of the National Academy of Sciences of the USA* (*PNAS*) are summed for each five-year interval preceding the date. ("Yeast" includes titles with "*S. cerevisiae*," "*Saccharomyces*," "*Schizosaccharomyces pombe*," and "fission yeast") (*Inset*) Increase of journal pages (sums of the three journals) over the same interval, in thousands per year. The short plateau in the 1970s may reflect a change in 1971 to a larger page format by *PNAS*, which accounts for 40–50% of the total pages.

The rise of model organisms has had an ironic effect. The reductionist approach, in favoring simple systems, brought scientists to model microbes, and genetic rationales trapped its professionals in their use. The result was a concentration on a few organisms explicitly used to discover generalizable principles. Indeed, geneticists, in standardizing stocks and media, provided artificial representatives of various species living in an artificial environment. As we have seen, the appreciation of diversity and complexity was lost for some time among molecular biologists. A curious construction of biology arose in which many biologists spoke of "the cell" or "the organism" when in fact they were studying *N. crassa* or *E. coli* or *S. cerevisiae*. A Platonic view of life emerged in which model organisms were the reality and the rest of the living world was a chaos of variants and exceptions. The irony is that even as the neo-Darwinian views of evolution freed us from such thinking, molecular biology acquired an even more rigorous typological stance than that of Linnaean taxonomists of the early 19th century.

By way of contrast with the model organisms, a great variety of organisms that qualify only as model systems for the pursuit of individual research problems can be named. Many of them are ideal for the problem at hand, but knowledge about them as organisms acquires little of the depth that arises from intense genetic domestication. The sea slug, *Aplysia*, continues to be used for the study of the potentiation of nerve circuits; the squid axon is used for study of nerve transmission; the sea urchin is used for study of fertilization; and songbirds are used for research on nerve regeneration in the brain. Monkeys and cats are favorite organisms for the study of perception and brain function, and rats have been used for some time in studies of aging. Many animals, subjected to infectious or chemical agents, serve as models of human disease. The *Avena* coleoptile has been used for generations for the study of plant growth hormones. Whatever understanding of these organisms lies beyond the needs of particular research programs depends on prior and specific study of other features, together with the principles arising from related model organisms. None of the model-system species has been subjected to the phenotypic dissection that only genetics can perform. And before genome sequencing, spontaneous or induced mutations best provided the phenotypic surprises that lured biologists into undreamt-of paths of discovery. But we should not forget that biologists continue to sample a host of organisms as systems for study of specific problems simply because they offer great experimental advantages that can be exploited without genetic rationales.

Molecular Biology and the Rest of Biology

Molecular biology, starting with genetics, macromolecular structure, and biochemistry, and brought to a head by Beadle and Tatum in 1941 and by Watson and Crick in 1953, created a discontinuity in 20th-century biology. Genetics in particular was utterly transformed by the microbial approaches I have described, and fresh interpretations spread retroactively, as it were, into most of what had been learned of life cycles, sexuality, the chromosome theory, developmental biology and evolution. Similarly, genetic and molecular techniques became dominant in prokaryotic microbiology.

How did the rest of biology respond to microbial genetics and molecular biology? The discontinuity here was not as severe as many molecular biologists would have us think. Indeed, a sizable minority of biologists even now finds the language of macromolecules, information, and genetic analysis foreign.

As I noted in chapter 17, biochemistry did not start with Beadle and Tatum; biochemists had done much to help Beadle and Tatum pose their major question. Biochemists also provided the chemical intuitions that gave life to the black-box abstractions of Delbrück's phage school. Biochemistry continued happily after 1940 in the exploration of energy metabolism, glycolysis, the chemiosmotic theory of membrane energetics, and photosynthesis well before molecular genetic techniques were applied to them. At the next level of organization, much of our present knowledge of cell biology had come into view with the advent of phase-contrast and electron microscopy. These advances, soon accompanied by biochemical analysis, had given biologists a good feel for the cytoskeleton, the fluid mosaic model of membranes, mitosis, intracellular membrane systems, targeting of proteins to membranes, exo- and endocytosis, cell communication, and the biosynthesis of cell walls of plants and bacteria. All of these problems, too, were well formulated before molecular biology got hold of them. The crucial fact is that model organisms rarely began as model organisms; they began to serve that role by appropriating research programs established earlier in other species. For these problems, genetic and molecular techniques and the use of model organisms simply quickened the pace of research.

The study of whole-plant and whole-animal function (including human; among the oldest subjects of interest in biology) had started as a science in the Renaissance and continued well into the 20th century without overwhelming contributions from genetics. This reflects the fact that physiology, development, neurobiology, disease, and other preoccupations of organismal biologists comprise integrative themes shunned by the reductionist approach. Until recently, these areas were not ready to abandon holistic thinking as a primary tool. Therefore the study of the structures and organ systems of plants and animals, their nutrition, sensory systems, hormones, tropisms, homeostatic mechanisms, reproductive systems, and growth had preceded molecular biology and have continued until recently in many cases without the aid of molecular techniques. Indeed, the central concept of homeostasis, promoted in the 1930s by Walter Cannon (1871–1945), predated by some years its molecular counterparts: allosterism and negative control of gene expression. And study of feedback systems based on hormonal control, as worked out by Cannon and many others, could be pursued into the 1980s without molecular principles as working tools. Only when biochemical and molecular probes became available through commercial development of antibodies, nucleic hybridization probes, and many other detection systems (often available in diagnostic kits) did the more integrative fields convert seriously to the language and techniques of molecular biology. These new resources have had a major impact on the knowledge of complex organisms and, in particular, disease processes.

In the study of evolution, geneticists had formulated the neo-Darwinian theory as a genetic problem, starting with the ways in which Mendelian variants were distributed, inherited, and maintained in populations. With the recognition in the mid-1960s of vastly more variation in natural populations than imagined even by the variation-conscious Dobzhansky,[2] the findings then compelled the adoption of biochemical, molecular, and finally genomic approaches, usually as a supplement to older methods that continue to be used.

Finally, ecology has given us an appreciation, deepening every year, of the interdependence of organisms with one another and their environment. (And, more darkly, of the impact of humans on the condition of the earth and its biota.) Molecular biology per se has had least impact on this field of all others, but it proceeds with a more modern flavor thanks to comprehensive use of physical, chemical, and mathematical techniques and a fuller understanding of plant, animal, and prokaryotic life.

In sum, molecular biology and its model microbes are simply parts of a large science to which they have contributed understanding and unification. Much new knowledge has originated almost explosively in molecular biology, but within a few short years of the ascendancy of prokaryotic genetics, the rest of biology has contributed most of the questions. In their married state, the new and the old biology have developed a partnership that has swept us back into the study of diversity and complexity, now approachable with new tools.

Is Sex Necessary?

As it happens in many marriages, sex will become less important with the growth of interests shared in the new biological partnership. Model organisms, achieving their status through intense genetic development, entrapped their users by the requirement of intraspecific crosses. This sent investigators down narrow paths to a limited view of biology through the study of single species. However, the representative quality of a model organism can be tested only by comparisons with related forms. The success of *D. melanogaster* as a model organism is partly a function of parallel studies of the evolution and adaptation of the many other *Drosophila* species to different habitats and of the exploration of the chromosome mechanics, gene expression, and development of other dipterans. Similarly, *Neurospora* and *Aspergillus* complemented one another and benefited greatly from what was already known of the genetics, sexuality, and biology of other filamentous fungi. *E. coli*, even at the beginning of its genetic career, was already perceived as representative of bacteria. Such comparative knowledge gave credence to generalizations, much as broad knowledge of biological structure gave credence to the cell theory. But in many cases, the sexual systems, if not the genetics, of the companion species had to be at least crudely in hand before valid comparisons could be made.

One can now test for the existence of particular homologous genes in any species, document their variation where found, and even replace one species' genes with homologs from another, often quite unrelated, species. One may isolate genes from an organism, mutate them in any way one wants at will, and restore the altered forms to the organism for a test of the corresponding phenotype. Comparative and functional genomic approaches, in short, have freed genetics from the constraints of sexual reproduction. Biological complexity and variety can be explored with the same rigor with which 20th-century genetics endowed the reductionist approach. This is not to say that the use of model organisms is at an end, since investigators will use "model genomes" as points of departure for investigation of biological variety and evolution.

Conclusion

I have charted progress in a circumscribed facet of biology in this description of the uses of model microbes. Those who played a part in this can be forgiven if they exaggerate the extent of their contributions to biological understanding. Onlookers can be forgiven for similar perceptions because molecular biologists have not been shy in propagating their view of their importance.[3] The popular press and many early accounts pictured a path to the double helix that has all the qualities of the Yellow Brick Road.[4] Despite exaggeration, however, and recognizing that the importance of discoveries is a value judgment, Western science welcomes most those discoveries that unify our knowledge and give us a more widely shared language. Unquestionably, molecular biology has done both for biology.

I observed in the previous chapter that the "Age of Model Organisms"—in effect, the 20th century—links the 19th-century awe at the diversity and complexity of organisms with the beginnings of the 21st century understanding of these attributes. The 20th century is distinctive in the intense application of reductionist approaches. These approaches, involving the use of the simplest biological systems, the avoidance of concern with emergent properties, and the universalization of findings, were by no means new. The successful application of such approaches to whole organisms, however, was introduced by genetic rationales and the discovery of DNA structure. This in turn created the discontinuity I have described. Reductionist approaches will not be abandoned in the face of complexity and diversity. Much is yet to be learned about the detailed workings of organisms, and resolving the logical elements of this complex and diverse world requires reductionist thinking. But to contend that the style of reductionist analysis of the recent past will be adequate for analyzing integrated biological systems in the future is no longer defensible. Molecular biology now has an appreciation of robustness, redundancy, connectivity, and modularity, even if it still lacks the predictive capabilities that accurate modeling will provide. To seek formal reducibility of biological systems to physics and chemistry is already out of the question, given the multiple, unpredictable routes to equivalent fitness by which various organisms have evolved.[5] An intuitive understanding of complex organismic properties may never be possible; our new knowledge already consists of what we see on a computer screen, not on the earth beneath our feet. Research will be driven by arrays and algorithms managed by software, indispensable supplements to the human mind. The truths that we extract with the approaches of systems biology will force upon us the fact that we're not in Kansas anymore.

Some commentators see in the rise of molecular biology many unfavorable consequences ranging from historical myth-making, egregious simplification, a neglect of multicellular forms, the increasing confusion in the definition of the gene, an essentialist approach to human diversity; the commercialization of universities, and gene therapy to the cloning of mammals and human embryos.[6] Such critics argue that the era has deprived us of a vital appreciation of living things—a "feeling for the organism"—as it has plunged us into dangers and ethical dilemmas encountered only in fiction 50 years ago.[7] That said, we must ask rhetorically whether any understanding of complexity and diversity could emerge in the 21st century without a prior understanding of the fundamental, shared elements of living things. We no longer hide our ignorance with words like "*élan vital*," "colloids," "*Entelechy*" or "polyphasic systems." An appreciation of just how complicated nature is has been restored to molecular biologists. They have properly muted the slogans of the 1960s (e.g., "What is true of *E. coli* is true of the elephant"). With undiminished aggression, they attack the complexity of the human genome with the same vigor with which they once attacked phage T4. Because the appreciation of complexity was achieved with the new genomic tools now being used to pursue it, we may be confident that we are not at the end of the line. We will continue to hear words like "autopoeisis" that impart little real understanding, but they will be no more than the names of new research programs.

So how should we answer the postacademic, postmodern regret that molecular biology has fundamentally changed (for the worse) our view of ourselves and our relation to the rest of the living world? The subjective reaction to the possible public uses of our knowledge is in keeping with those that met gunpowder, Darwinian evolution, the automobile, eugenics, atomic energy, recombinant DNA, and in vitro fertilization. We have learned from these "advances" that danger accompanies every such opportunity, and misgivings are as appropriate as they are inevitable. I leave to others the task of predicting the social problems occasioned by genetic and genomic research.[8]

But the point at hand, the perception of ourselves and our biological companions as mechanical and unmysterious, coded and wired, the sum of our genes, is another matter. We must see the current pictures of life as we do modern paintings, placed in museums or forgotten entirely within 25 or 50 years. Persuasive scientists and their supporters continually touch up the unfinished portraits and occasionally abandon them for a new start. Even now, molecular biologists have turned from two-dimensional canvases to digitized forms of representation most of us cannot fathom. While the picture of life in 2050, if we could recognize it at all, might appear to deprive us of the last shred of our own human mystery and importance, most philosophical regrets we now have about progress are simply wasted; they are our regrets, not those of our descendants. For the present, we are free to construct any view of what our current knowledge might mean, if anything, and to be dour, detached, or delighted about the future. But our great-grandchildren will share little of what we predict and even less of what we feel about what, in the end, are their own daily lives.

Appendix 1. Life Cycles
and Genetic Principles

The Alternation of Generations

Two lifestyles have emerged among nonbacterial organisms. In one of these, the vegetative cells (those conspicuous in nature) are *haploid*. The nuclei of haploids have only one chromosome of each type, and therefore only one copy of every gene. These organisms include many lower organisms such as one-celled algae, filamentous fungi, yeast, and mosses. *Diploid* organisms such as mice, higher plants, *Drosophila*, and humans, in contrast, have two of every type of chromosome and thus two copies of every gene. However, the sexual life cycles of haploid organisms have a phase in which certain cells are diploid, and diploid organisms have a brief haploid phase. Both types of organisms, in fact, progress through the same events in the same order. These events are, first, *fertilization*, in which haploid cells called *gametes* fuse to form a diploid cell called a *zygote*. The second important event is *meiosis*, a series of two cell divisions of a diploid cell that produce four haploid *meiotic products*. The two life cycles differ only in the phase (haploid or diploid) in which cells undergo simple division, or *mitosis*, to form many separate cells or the multicellular organism we see in nature. The alternation of haploid and diploid phases is often called the alternation of generations. All sexually reproducing eukaryotic organisms display this sequence of events.

Genetics in Diploid Organisms: Mendel's Crosses

In diploid organisms, the haploid, gametic stage is transitory, existing only as egg and sperm in animals or as ovule and pollen nuclei in plants. The gametes

261

are produced from diploid cells of the gonads (testes, ovaries, and anthers) that, unlike most of the cells of the body, are specialized to undergo meiosis. Genetic analysis begins with parents that differ in one or more obvious ways. Mendel's crosses started with parents that differed in pea color; one line had yellow peas, the other had green peas (see fig. 1.2, chapter 1). When crossed, the fertilized parent yielded pea pods derived from the parental ovary. The pea pods contained the progeny peas, called the F_1 generation. These peas were all yellow. When these peas were planted, grown up, and allowed to self-fertilize, the pods of this second (F_2) generation contained, on average, three-quarters yellow peas and one-quarter green peas.

The parents, the F_1 peas, and the F_2 peas are diploid, all having two copies of the determinant (gene) for pea color. We call the alternative forms of the gene in the two parents *alleles*: one parent contains two alleles for yellow (*AA* in Figure 1.2); the other parent has two alleles for green (*aa*). Because the diploid parents carried two of the same allele, they are called *homozygous* for the particular allele *A* or *a*. The gametes of each parent form in meiosis and are therefore haploid. One cannot tell which gamete has a particular allele, since the trait is expressed only in mature peas. These gametes are used in a cross to form the zygotes of the F_1 generation. The F_1 zygotes, each having one allele of each kind are *heterozygous*, with the *genotype* designated *Aa*. The diploid zygotes develop into peas, the earliest stage at which F_1 traits, or *phenotypes*, might be observed. Observation of other characteristics of the F_1, such as plant height, requires that these peas be grown into adult plants.

The yellow color of the F_1 *Aa* peas reflects the *dominance* of the yellow allele over the *recessive,* green allele: the presence of *A* masks the presence of *a*. If these peas are grown to adult plants, the gonadal diploid cells (ovary, anther) of this F_1 plant undergo meiosis. One-half the mature pollen nuclei will contain the *A* allele, and half will contain the *a* allele. The same will be true of the ovules. This important point in the series of crosses is the stage at which a 1:1 *segregation* of alleles of the same gene (in this case, a determinant of color) occurs.

All allelic differences segregate in this 1:1 fashion at meiosis. In Mendel's hands, the adult F_1 plants produce F_2 peas by self-fertilization. In the process, as we saw in chapter 1 (fig. 1.2) the random combinations of pollen and ovules and the dominance of yellow over green yield the ratio three-quarters yellow and one-quarter green. Because the peas are diploid and because the color trait involves dominant and recessive alleles, students have considerable difficulty understanding the process. Indeed, it is one of the several reasons that the rules of genetics waited so long to be discovered.

Mendel's fortune lay in part in the availability of highly inbred strains of peas. The pure lines differed in various traits such as pea color, pea texture, plant height, and leaf shape. Mendel found that the F_2 3:1 ratio of one-gene traits involving dominant and recessive alleles prevailed regardless of the sex that had contributed the dominant allele in the first cross. Mendel formulated the notion, widely misunderstood at the time, that each parent contributes equally to the offspring. Mendel and his contemporaries knew nothing about chromosomes or meiosis, so his formulation had a quite abstract character.

Moreover, he did not symbolize the traits as we do now, nor did he really distinguish the phenotypes (the traits yellow and green) and the genotypes (the genetic constitution, in terms of what we now call genes). The events in meiosis are explained below.

The main text gives one way in which we can formulate and predict *independent assortment* in the F_2 of the diploid *AaBb*, often called a dihybrid or double heterozygote. In more detail, we may consider the two phenotypic differences determined by the two respective genes *A/a* and *B/b*. If we know that each displays a dominant–recessive relationship, and if we expect each gene to segregate in a 3:1 fashion in the F_2, we simply multiply the two 3/4 : 1/4 distributions to predict the outcome of the self-cross *AaBb* × *AaBb*. In doing so, we assume that the two distributions are independent of one another. The assumption is confirmed if we obtain a 9/16 : 3/16 : 3/16 : 1/16 ratio of the four phenotypes in the F_2, which we symbolize with the allele designations: *A_B_*, *A_bb*, *aaB_*, and *aabb*, respectively. The blank after the dominant allele symbols indicates that the second allele in such genotypes may be either the dominant or the recessive. Independent assortment, often called Mendel's Second Law, is expected if the two genes involved lie on different types of chromosomes. A departure from the 9:3:3:1 ratio may reflect the association of the two genes on the same chromosome; we say that such genes are *linked*.

Mendel worked out, by way of proving his hypotheses about gene transmission, the strategy of test crosses. Such crosses use a doubly recessive tester parent (*aa*) by which one could determine whether a particular plant with the dominant character (*A_*) had a homozygous (*AA*) or heterozygous (*Aa*) genotype. The testcross would yield green progeny only in the case of the heterozygote. Mendel did not discover linkage and recombination of linked genes. This was left to Morgan, who thereby revealed how so many different combinations of traits could be found in an organism, like *Drosophila*, with so few chromosomes.

The Life Cycle, Meiosis, and Genetics of Haploid Organisms

Genetic analysis in haploid organisms, which include most of the fungi, is much simpler. The vegetative cells of the organism contain only one set of chromosomes and therefore one set of genes. For simple genetic differences, the vegetative cells reveal their genotype because the complication of dominance is absent: haploid cells cannot contain two copies of a gene and thus cannot be heterozygous.

Sexual reproduction requires the differentiation of special cells called *gametes*. Meiosis is not the immediate source of gametes in haploid organisms because vegetative cells are already haploid. Gametes, specialized for fusion with other gametes, may be restricted by morphological sex (e.g., ovum and sperm or spore) or by physiological mating type (e.g., *A* and *a* of *Neurospora*) or, as in *Neurospora*, by both attributes. The *zygote*, the product of gametic fusion, is a diploid cell. In haploid organisms, the diploid zygote will undergo only a limited number of cell (or nuclear) divisions, if any, before it undergoes meiosis.

Meiosis is a sequence of two divisions of a diploid cell that follows only one division of the chromosomes. Thus in the four cells that result, the haploid condition—one set of chromosomes per cell—is restored. The two divisions yield a *tetrad*, the group of four products of a single meiosis. Meiosis is a complex process and is at the heart of genetic analysis. If the chromosomes of the two gametes that fuse carry different alleles of a particular gene, such as mating type, these differences will segregate among the meiotic products of every tetrad in a 2:2 fashion.

Figure A1.1 follows a single pair of chromosomes, one derived from each parent, through the entire meiotic process. The discipline of this process depends on the fact that for each type of chromosome in the diploid there are two homologous chromosomes, or *homologs*. Homologs carry the same genes in the same order, except for the alternative forms (the alleles of particular genes), that have arisen by mutation from a common ancestor. Before meiosis, the chromosomes all divide, but the two copies (*chromatids*) remain held together by the centromere. The *centromere* is the point of attachment of spindle fibers, which emerge from the poles of the cell and pull chromosomes apart in meiosis and mitosis. Before the first meiotic division, the homologs pair with one another, point for point, each pair forming a four-stranded entity called a *bivalent*. At the first division, one half of each bivalent (one duplicated homolog) goes to each pole, thus leading to a *segregation* of allelic differences. At the second division, the two copies of each homolog separate from one another so that there is only one copy of each chromosome in each of the four meiotic products.

Each organism has a characteristic number of chromosomes. In the haploid phase of humans (gametes), it is 23; in *Neurospora* cells it is 7; and in haploid budding yeast cells it is 16. If allelic differences occur on two different types of chromosomes when gametes fuse to form a diploid cell, the meiosis that follows will assort the pairs randomly (fig. A1.2). Thus in a cross *Ab* × *aB*, where genes *A/a* and *B/b* lie on different types of chromosomes, meiosis will distribute the alleles (represented by upper- and lowercase) of each gene randomly to the meiotic products. Some tetrads (parental ditypes, or PDs) will have only parental genotypes (*Ab Ab aB aB*); an equal number (nonparental ditypes, or NPDs) will show two of each nonparental product (*AB AB ab ab*), and some tetrads (tetratypes, or T) will have one of each kind of product (*Ab, aB, AB, ab*). The T tetrads form because of an exchange of chromosome parts, as indicated in fig. A1.2, before the first meiotic division. PD and NPD tetrads will be equally frequent, and T tetrads will occur at different frequencies, depending on how far the gene (in this case *A/a*) lies from the centromeres of its bivalent. (This pattern, T, reduces the PD and NPD tetrad numbers equally.) The overall behavior of many tetrads, in which the frequency of PDs equals that of NPDs signifies *independent assortment* (fig. A1.2). Because the PD and NPD tetrads are equally frequent and because T tetrads have equal numbers of each genotype, there are equal numbers of *Ab*, *aB*, *AB*, and *ab* meiotic products overall.

The behavior of chromosomes in meiosis in diploid organisms when they form gametes is the same as in the haploid. Random combinations of the ga-

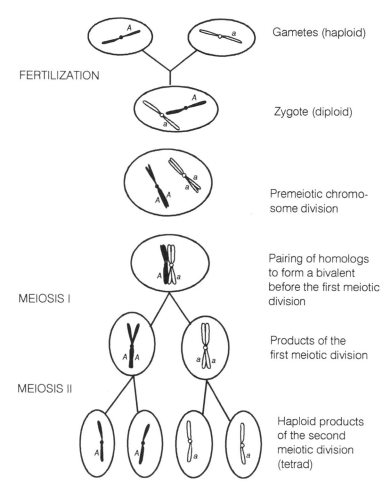

Gametes (haploid)

FERTILIZATION

Zygote (diploid)

Premeiotic chromo-
some division

Pairing of homologs
to form a bivalent
before the first meiotic
division

MEIOSIS I

Products of the
first meiotic division

MEIOSIS II

Haploid products
of the second
meiotic division
(tetrad)

Fig. A1.1.   The behavior of a single pair of homologous chromosomes during
fertilization and meiosis. All eukaryotes have more than one pair of chromosomes in
the diploid phase, and, as pairs, they behave independently of all others during
meiosis (see fig. A1.2).

metes formed by doubly heterozygous diploids underlie the phenotypic distribution of the F_2 dihybrid cross explained above. To prove this, one must take
into account four equally frequent types of gametes of one sex fusing with four
equally frequent types of gametes of the other; 1/16 of every combination is
found, but many of the combinations are identical in genotype.

Linkage of Genes on the Same Chromosome

Genes on the same type of chromosome are *linked*, but the linkage is not absolute. Meiosis provides a stage during the first prophase of meiosis called

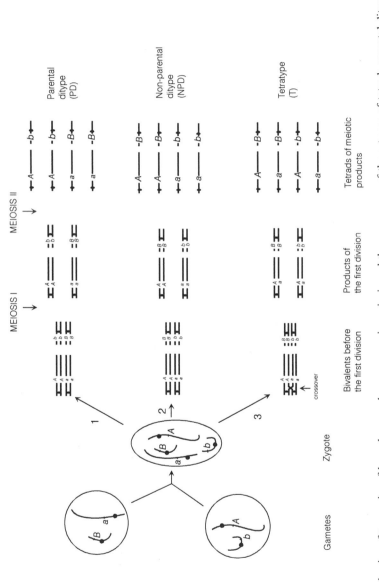

Fig. A1.2. The behavior of two pairs of homologous chromosomes in meiosis and the emergence of three types of tetrad: parental ditype, nonparental ditype, and tetratype. Each pair of homologs differs with respect to a single gene (*A/a* or *B/b*), and each chromosome, after duplication, pairs with its homolog to form a bivalent. The long chromosomes with *A/a* genes and the short chromosomes with *B/b* genes orient with each other differently (arrows 1 and 2) prior to the first meiotic division (meiosis I). In the lowermost figure, a crossover leads to an exchange of the alleles *A* and *a*. The outcome is determined by separation of the four chromatids of each bivalent in two nuclear divisions as they are distributed to the four meiotic products.

266

pachytene at which the copies of the homologs (chromatids of the bivalent) may exchange parts and then go on to segregate in the two divisions that follow. The exchange process, which was explained in connection with second-division segregation (fig. A1.2, tetratype tetrad), is called *crossing over*, which takes place infrequently along the bivalent during meiosis. Geneticists detect linkage by finding a preponderance of parental genotypes among the meiotic products. In the mating *Ab* × *aB*, we would see more *Ab* and *aB* (parental) meiotic products than *AB* and *ab* (recombinant) meiotic products when looking at a large number of tetrads. The percentage of recombinant meiotic products among the total is taken as the genetic *map distance* between the genes. This figure will be greater for genes far from one another on the chromosome than for genes close together. In long distances, however, multiple crossover events may occur. Some of these nullify the effect of others, restoring the parental genotypes and thus rendering the individual exchanges undetectable. Therefore, accurate *linkage maps* must be derived from crosses involving genes separated by small intervals, permitting at most 10–20% recombination and permitting very few multiple crossovers.

Morgan's group discovered linkage and recombination of genes that lay on the same type of chromosome in *Drosophila* (see below), but these phenomena are easier to understand in the context of a haploid organism such as *Neurospora*, *Aspergillus*, or yeast. Let us consider two genes on the same chromosome and cross two parents that carry allelic differences in these traits (*XY* × *xy*). The haploid progeny of the cross (meiotic products), if collected from many tetrads, might have the following percentages: 41.5% *XY*, 41.5% *xy*, 3.5% *Xy*, and 3.5% *xY*. The numbers are arbitrary: as noted above, the strength of linkage varies among different pairs of genes. However, we note two characteristics of the progeny population: (i) the parental combinations (*XY* and *xy*) outnumber the *recombinant* progeny (*Xy* and *xY*) and (ii) the two classes have equal numbers of complementary genotypes. Thus, each gene still displays a 1:1 segregation of alleles. The percentage of the recombinants, 7% in all, is the measure of the strength of linkage and is expressed as 7 map units. The map unit is often called the centi-Morgan, or cM.

Let us bring a third linked gene, *Z/z*, into the picture and see how far it is from the other two. A cross *YZ* × *yz* yields a map distance between the two genes of 5% (2.5% of each of the two recombinants *Yz* and *yZ*). A cross of *Xz* × *xZ* yields a map distance of 2% between *X/x* and *Z/z*. The striking feature of such interrelated crosses is that, in short intervals, these distances are *additive*. The longest distance, 7.0 map units, separates *X/x* and *Y/y*. This is the sum of the two distances separating *Z/z* from the other two markers, and therefore a map may be drawn as follows:

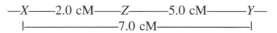

$$—X\text{———}2.0\text{ cM———}Z\text{————}5.0\text{ cM————}Y—$$
$$\vdash\text{————————}7.0\text{ cM————————}\dashv$$

The first genetic map was made by Sturtevant in Morgan's laboratory, and this method has been a staple of genetic analysis ever since. Geneticists with well-developed experimental organisms now always use "three-point" crosses to make linkage maps. For example, in the analysis above, one could mate strains *XYZ* × *xyz*. The cross would yield eight possible genotypes, and standard methods—given in any elementary genetics text—are used to determine the order and linkage distances between adjacent genes. I stress that linkage distances are actually probabilities of recombination, and these are in many cases poorly correlated with the actual amount of DNA that separates the two genes in question.

Linkage and Sex Linkage in *Drosophila*

Figures A1.3 and A1.4 illustrate sex linkage and linkage determinations in a diploid organism, *Drosophila*. Figure A1.3 illustrates the difference in the pattern of inheritance in the case of a nonsex chromosome (*autosome*) and in the case of sex chromosomes. Sex linkage has two properties of interest. First, the outcome of *reciprocal crosses* (as shown) are different. Second, a recessive gene may be expressed by males in the F_1 generation because they have only one copy of any sex-linked gene. For this reason, recombination of sex-linked genes is easy to detect and measure. Figure A1.4 diagrams how the two X chromosomes of a female, carrying two differences, might in rare cases have a crossover between them that recombines the two genes' alleles. The rare gametes (ova) that are recombinant will be easily seen in the sons of matings to any male because the Y chromosomes contributed by this male will be unable to mask the contribution of ova.

Genetic Fine Structure

Muller, Pontecorvo, and others found evidence that recombination might occur within a gene, between two sites of mutation. In their studies, independently isolated mutants of the same gene ($a^x × a^y$), when crossed, occasionally yielded a wild-type recombinant (a^+). This observation was surprising at the outset. It gave rise to the term "pseudoallelism," since mutations of the same gene (alleles) were not expected to recombine and thereby reconstruct a wild-type gene. The frequency of recombination in these *intragenic* crosses is extremely low and can be detected only by screening very large progenies—perhaps 10,000 to 100,000 for accurate measures of genetic distance. For that reason, microbial systems such as *Aspergillus* and *Neurospora*, in which such large progenies are obtainable, became the focus of fine-structure analysis of genes. Indeed, maps of genes in which a number of different mutations had been isolated were developed by the principles of linkage analysis above, even though the maps were a bit undependable for then-obscure reasons. The outcomes of intragenic crosses (those involving recombination tests within a single gene), when studied intensively in later years in the fungi, became the source of most of our information about the molecular basis of recombination and gene conversion in meiosis.

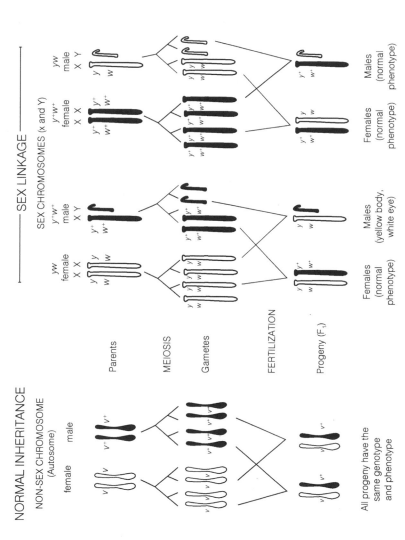

Fig. A1.3. Autosomal and sex-linked inheritance in *D. melanogaster*. At the left, the normal Mendelian pattern of inheritance in a single cross between homozygous parents is shown. At the right, reciprocal crosses of *Drosophila* parents, one of which carries two recessive genes (y, recessive to y^+, and w, recessive to w^+) on the X chromosome, are shown. The F_1 progeny differ because the Y chromosome of the F_1 males cannot mask the expression of the recessive genes contributed by the female parent in the first of the two crosses.

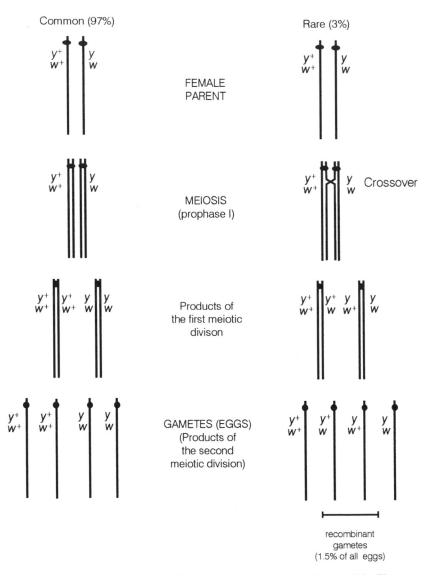

Fig. A1.4. Crossing over on the X chromosomes of a female *Drosophila*. The recombination of chromosome parts is rare (3% of all meioses in this particular case), and from such events two of the four meiotic products emerge as recombinants. Map distance is calculated as the percentage (1.5%) of recombinant gametes emerging from all eggs produced in the cross.

Appendix 2. Macromolecules and the Central Dogma

Macromolecules

Macromolecules include proteins (consisting of one or more *polypeptide* chains), polysaccharides, and nucleic acids. Each has characteristic repeating units: polypeptides are linear chains of amino acids, polysaccharides are linear or branched chains of sugars, and nucleic acids, either deoxyribonucleic acid (DNA) or ribonucleic acid (RNA), are linear chains of *nucleotides.*

Proteins

Twenty amino acids are used for protein synthesis, and the length of polypeptides may be very short (< 30 amino acids) or very long (> 2000). All amino acids have an acid (carboxyl) group and a basic (amino) group, and these are linked together in linear chains by attachment of the carboxyl group of one amino acid to the amino group of the next. When incorporated into a chain, an amino acid is called a *peptide*, with the bond between successive peptides known as a *peptide bond* (fig. A2.1). These chains are called polypeptide chains, or simply, polypeptides. The sequence of amino acids in a polypeptide chain, the *primary structure*, is determined by the *coding region* of the corresponding gene, a principle embodied in the one-gene, one-polypeptide relationship. The cell makes polypeptides from one end to the other, and as the chain forms, it folds into a specific, three-dimensional shape. The folding depends on interactions among the chemically distinct amino acids that make up the chain. These interactions, in the environment of the cell, are precise, and thus the amino acid sequence dictates the final shape, called the *tertiary structure*, of the polypeptide. Because of the chemical variety of amino acids and the variation in pri-

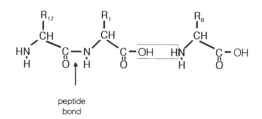

peptide
bond

Fig. A2.1. The peptide bond. This bond is formed between two amino acids through many reactions (see fig. A2.7). Ultimately, the bond represents the removal of water. The letter R signifies any of the 20 variable chemical groups of the protein amino acids.

mary structure of polypeptides, proteins vary enormously in their properties and biological roles. Polypeptides may become mature proteins simply by folding into their natural tertiary structure. However, most proteins consist of aggregates of several polypeptide chains. These are often identical to one another, but many proteins are aggregates of different types of polypeptides, using information from more than one gene. The number and type(s) of polypeptides in a protein is called the *quaternary structure*. The bonds holding polypeptides together are generally weaker individually than peptide bonds.

Polysaccharides

Polysaccharides are chains of one or more types of sugar and have considerable variety. However, many familiar polysaccharides such as starch and cellulose consist of only one type of sugar. Both linear and branched structures form, often in the same polysaccharide. They therefore have little sequence specificity. The shapes of complex carbohydrates, however, often vary enough to elicit specific antibody responses. Such specificity is found in the capsular polysaccharides of different strains of the *Pneumococcus* bacterium, used by Avery in his identification of DNA as genetic material.

Nucleic Acids

Four nucleotides are found in DNA: adenylic, guanylic, cytidylic, and thymidylic acids. Each nucleotide has three parts (fig. A2.2a): a phosphate group ($-PO_3$), a sugar, deoxyribose, and one of the four nitrogenous bases, adenine (A), guanine (G), cytosine (C), or thymine (T) (fig. A2.2). In RNA, similar nucleotides are found, but each has ribose in place of deoxyribose, and thymine is replaced by the chemically similar uridine (U) in corresponding positions. The nucleotides are polymerized into linear chains by linking the phosphate of one nucleotide to the sugar of the next (fig. A2.2b). The chain is therefore an alternating sugar-phosphate-sugar-phosphate backbone, with the nitrogenous bases attached elsewhere on the sugars. The polarity of nucleic acid chains

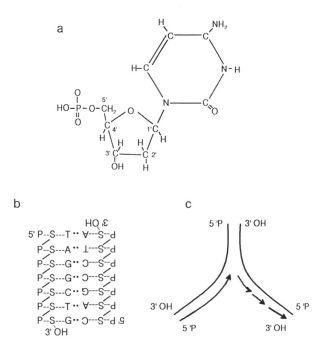

Fig. A2.2.   DNA structure and replication. (*a*) A deoxyribonucleotide (cytidylic acid), showing the phosphate (left), the deoxyribose (middle), and the nitrogenous base cytosine (right). Cytosine is linked to the deoxyribose at its 1' position; the phosphate is attached at the 5'. Linkages between nucleotides are between the 3' carbon of the sugar (via the oxygen of the OH group) to the 5' phosphate of the next nucleotide. The sugar, ribose, of ribonucleotides has an OH group rather than H on the 2' carbon. (*b*) The antiparallel structure of the double helix, with the sugar-phosphate backbone of each chain running in opposite directions. The chains are held together by hydrogen bonds between complementary nitrogenous bases of the two chains. (*c*) Replication of the double helix. The extension of each chain takes place only at the 3'OH end of the chains. This requires that while one new strand grows into the fork, the other is synthesized discontinuously, in segments, each segment extended away from the fork. The earlier segments are linked together after they are formed.

(5' to 3' or 3' to 5') is defined by the carbon atoms (3' or 5') of the sugar to which the phosphate groups are attached. Each free nucleotide has its phosphate on the 5' carbon; this phosphate will attach to the 3' carbon atom (via an oxygen atom) of the nucleotide at the end of a growing chain of nucleotides during nucleic acid synthesis. DNA consists of two chains of deoxyribonucleotides, running in parallel, with opposite polarities, and wound around one another in what we now call a double helix (figs. A2.2b and A2.3). In contrast, RNA normally consists of only one chain of nucleotides. The nucleotide sequence of a DNA or RNA chain may be of any length and of any sequence of the nucleotides A, G, C, and T (for DNA) or A, G, C, and U (for RNA).

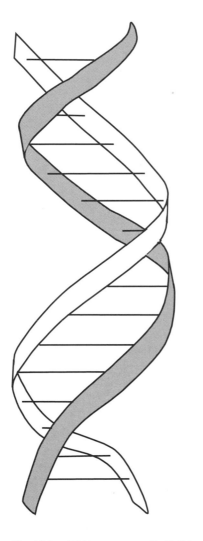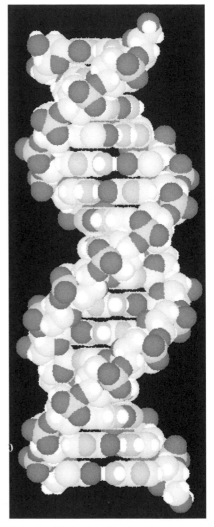

Fig. A2.3.   DNA structure. (*Left*) Diagram of the double helix, showing the sugar-phosphate backbone (gray and white ribbons) and the base pairs as steps in the spiral ladder. Each full turn of the helix comprises 10 base pairs. (*Right*) Space-filling model at the same scale as the diagram at left. Figure at right courtesy of Donald Senear, University of California, Irvine.

DNA and DNA Synthesis

DNA molecules have a special property that allows the cell to duplicate them faithfully. The two chains of the double helix each carry all the information of the entire molecule, but in complementary forms. The double helix is held together by weak bonding (hydrogen bonding) between specific pairs of bases: where one chain has A, the other has T; where one chain has G, the other has

C (fig. A2.2b). These pairs (A-T and G-C) are the only ones that can form these bonds. Moreover, A and G (the *purines*), are larger than T and C (the *pyrimidines*). The nucleotide pairs are therefore about the same size, which gives the double helix a uniform diameter and physical regularity (fig. A2.3). The base-pairs of the double helix lie in the center of the molecule, and the sugar-phosphate backbone lies on the outside. The two nucleotide chains separate during replication, with each one directing the sequence of a new partner based on the rules of complementary pairing (fig. A2.2c). This process takes place on the surface of the enzyme *DNA polymerase*, which provides for selection of the proper complementary nucleotide at each step and for the chemical bonding of these nucleotides to the end of the growing chain when they are in position.

RNA

RNA molecules are single stranded and normally much shorter than DNA molecules. They are made as copies of short segments of DNA corresponding to individual genes or operons. Most RNA molecules are copies of genes that encode the amino acid sequences of proteins. The role of these RNAs as the intermediary between the genetic information (DNA) and the process of protein synthesis led them to be called *messenger RNA* (mRNA). RNA molecules are less stable chemically than DNA, which makes them suited to the need for organisms to change the complexion of gene expression rapidly in response to changes in their environment. The synthesis of mRNA and two other types of RNA is detailed in the next section.

The Central Dogma

The fanciful term, "Central Dogma," was coined by Crick to designate the principle that the information embodied in the DNA nucleotide sequence is copied in the form of RNA, and the RNAs are then used as a source of information for polypeptide synthesis (fig. A2.4). A corollary is that the flow of information is irreversible; changes in proteins cannot be transmitted back to DNA, the ultimate source of sequence information. The Central Dogma simply comprises the two steps of gene expression: *transcription* and *translation* (fig. A2.4). In the first of these steps, DNA serves as a template for mRNA synthesis. In the second, mRNA serves as a template for polypeptide synthesis.

Transcription

The first process, transcription, is straightforward (fig. A2.5). One of the two strands of DNA at a point on the molecule becomes available, by unpairing from its complementary strand, as a template for the polymerization of ribonucleotides. RNA nucleotides are locally positioned such that an A ribonucle-

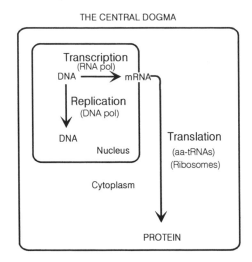

Fig. A2.4.   The Central Dogma. Three irreversible processes are shown: the use of DNA in replication, the transfer of information in DNA to mRNA via transcription, and the use of information in mRNA in the synthesis of protein. Since the promulgation of this picture in 1958, the use of certain RNAs as templates for DNA synthesis has been discovered. Nevertheless, proteins cannot be used as sources of information in the synthesis of nucleic acids.

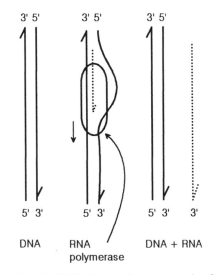

Fig. A2.5.   The synthesis of mRNA. One of the two strands of DNA is used as a template for the synthesis of a ribonucleotide chain. The enzyme that accomplishes this, RNA polymerase, reads the template in a 3' to 5' direction, adding ribonucleotides to the 3' end of the growing RNA chain. The choice of which chain of DNA to use and where to start and end mRNA synthesis is directed by the nucleotide sequence of the DNA. The mRNA is released from the template and can diffuse to the site of protein synthesis.

otide binds with the DNA nucleotide T; a U ribonucleotide binds with the DNA nucleotide A, and so forth. As these binding steps take place on or in the enzyme *RNA polymerase*, the enzyme attaches each ribonucleotide, one by one, to the end of the forming chain. The final mRNA molecule that is released is complementary in sequence to its template (and identical except for substitution of U for T to the other DNA strand). The nucleotide sequence of the DNA double helix contains signals for which strand is chosen by RNA polymerase as a template, where the starting point is, and where to stop. The DNA strands pair up again as the mRNA chain is being released from its template.

Two other types of RNA molecules of the cell are not translated. One of these is *ribosomal RNA* (rRNA: three or four types of molecules, found as part of ribosomal structure) and the other type is a set of approximately 44 *transfer RNAs* (tRNAs), adaptors for amino acids as the latter are used for polypeptide synthesis, described below.

Translation

The second step of gene expression, translation, is much more complex. Here, mRNA, made in the nucleus and carrying sequence information copied from DNA, specifies the sequence of amino acids in a polypeptide. mRNA, with its four types of nucleotides, must be translated into a different sort of "language," which uses the 20 amino acids as its basic elements. The mRNA is read not one nucleotide at a time, but in sets of three, starting from near one end of the molecule and proceeding to the end of the message. The sets of three nucleotides are often called *triplets* or *codons*, of which there are 64 types—more than enough to encode each amino acid uniquely. There are, in fact, several synonyms for most amino acids, as shown in the table of the universal genetic code (fig. A2.6)

The entire process of translation occurs on the ribosome, a compact, two-subunit aggregate of three to four kinds of rRNA and more than 50 specific proteins. All ribosomes of a cell are identical and provide special surfaces for mRNA and the amino acids that must be polymerized into a polypeptide chain. The manner in which each amino acid recognizes its proper codon in the mRNA template is not direct. The process requires that every amino acid be attached first to a particular tRNA, a short RNA molecule with a folded structure with several loops. At one end of the folded structure, an amino acid is attached. On one of the loops, a specific *anticodon* triplet is found, capable of complementary base-pairing with one or several synonymous codons in the mRNA. Each amino acid has one or more "cognate" tRNAs. The anticodon of each tRNA corresponds to the amino acid it will carry at the other end. The attachment of amino acids to their specific tRNAs is catalyzed by a highly specific set of enzymes, each recognizing one amino acid and one or more corresponding types of tRNA. The two steps, consisting of the attachment of the amino acids to specific tRNA adaptors and the recognition of the mRNA codons by the anticodons of these adaptors, are the heart of the translation process. In effect, the tRNAs and the enzymes that attach amino acids to them (called aminoacyl-

mRNA nucleotide triplets, 5' to 3'

Second nucleotide

		U	C	A	G	
		Phe	Ser	Tyr	Cys	U
	U	Phe	Ser	Tyr	Cys	C
		Leu	Ser	Non	Non	A
		Leu	Ser	Non	Trp	G
		Leu	Pro	His	Arg	U
	C	Leu	Pro	His	Arg	C
First nucleotide		Leu	Pro	Gln	Arg	A
		Leu	Pro	Gln	Arg	G
		Ile	Thr	Asn	Ser	U
	A	Ile	Thr	Asn	Ser	C
		Ile	Thr	Lys	Arg	A
		Met	Thr	Lys	Arg	G
		Val	Ala	Asp	Gly	U
	G	Val	Ala	Asp	Gly	C
		Val	Ala	Glu	Gly	A
		Val	Ala	Glu	Gly	G

Third nucleotide

Fig. A2.6.   The genetic code. Each triplet is represented here by reading successively the left, top, and right-hand sides. The encoded amino acids are shown in 61 of the 64 positions. The nonsense codons ("Non") account for the remaining three triplets. The full names of the amino acids are given in the legend of figure 18.1.

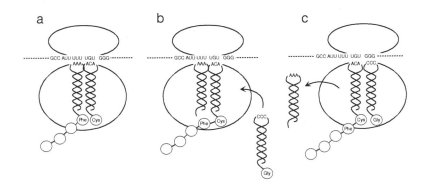

Fig. A2.7.   The ribosome cycle. The two-subunit ribosome associates with mRNA (the horizontal line shown with relevant codons). (*a*) Two positions on the ribosome are occupied with aminoacyl-tRNAs, the left one (peptidyl-tRNA) attached to a nascent polypeptide, the right one carrying a single amino acid (Cys). (*b*) The nascent polypeptide is transferred to the amino acid (Cys) carried in the right-hand position, thereby forming a peptide bond; Cys is thus added to the growing polypeptide. The ribosome moves one position to the right, evicting the free tRNA from the left-hand position as the peptidyl-tRNA moves into its position. The right-hand position on the ribosome now engages the appropriate aminoacyl-tRNA (carrying Gly), as selected by the GGG codon in the mRNA. (Recent research reveals that the movement of the ribosome has three positions [one an exit position] used in the course of these events. The figure should be taken only as a formal representation of the use of mRNA codons in the decoding process.)

tRNA synthetases) have an indispensable "dictionary" function in the translation process.

The chemical steps of peptide bond formation are complex (fig. A2.7) and take place in an orderly fashion on the surface of the ribosome. The mRNA is held in place and is threaded through the ribosome in a jerky process, three nucleotides at a time. At each step, the anticodon of a tRNA, with its attached amino acid, binds to an mRNA codon at one position on the ribosome. The ribosome carries a growing peptide chain, held in an adjacent position by the tRNA of the prior amino acid, which has just been added. The tRNA holding the peptide then transfers the peptide to the amino acid of the newly positioned tRNA-amino acid complex. The new amino acid has thus become the next amino acid of the peptide. Its tRNA and attached peptide then move, with the mRNA, to the position occupied by the previous tRNA, which is ejected from the ribosome in the process. (The ejected tRNA will be used again by attachment to another molecule of its particular amino acid.) This cycle is repeated until the ribosome encounters the *nonsense codon* at the end of the message. In eukaryotes, it departs the mRNA, and simultaneously the polypeptide is released into the cytoplasm. It completes folding at that time and may aggregate with other polypeptide(s) before it assumes its function in the cell.

The colinear sequence relationships and polarities of DNA, RNA, and polypeptide are shown in figure A2.8.

------------------ TRANSCRIBED REGION ------------------

↓ Start of transcription   End of transcription ↓

```
DNA 5' CTTCGTTTCAGAAGACACACTGTCAACAACCCCTCAAACCGACTCAAAATGTCTGGCCTCGGAGTTCCCGCCATTCCC.....AGAATGGGCGGAGGAGTTCTAAAGTGGTTCCAAGCGCCTG 3'
DNA 3' GAAGCAAAGTCTTCTGTGTGACAGTTGTTGGGGAGTTTGGGCTGAGTTTACAGACCCGAGCCTCAAGGGCGGTAAGGG......TCTTACCCGCTCCTCAAGATTTCACCAAGGTTCGCGGAC 5'
```

---------------- CODING REGION ----------------

↓ Start of translation   End of translation ↓

```
mRNA 5' CACUGUCAACAACCCCCUCAAACCGACUCAAAAUGUCUGGCCUCGGAGUUCCCGCCAUUCCC.....AGAAUGGGCGGAGGAGUUCUAAAGUGGUUCCAAGCGC

Polypeptide   M  S  G  L  G  V  P  A  I  P......R  M  G  E  E  F  *
```

Fig. A2.8. The colinear relationship of DNA, mRNA, and polypeptide. The 3' (lower) strand of the DNA is used as a template for synthesis of mRNA, polymerized in a 5' to 3' direction. Note that uridylic acid (U) is used, rather than thymidylic acid, as a complement for A nucleotides in DNA. The mRNA is then used for polypeptide synthesis, starting with the first AUG codon (in bold), which encodes methionine (M). The synthesis of the polypeptide is terminated when the ribosome encounters a nonsense codon (UAA, in bold) at the end of the coding region. The breaks in the sequences reflect the omission of the vast bulk (ca. 95%) of the information in the coding region and the polypeptide itself.

Appendix 3. Genetic Engineering

Some Enzymes

Restriction Enzymes

Most bacteria have DNA-cutting enzymes called *restriction enzymes*. These have evolved as a defense against foreign DNA, particularly the DNA of infecting bacteriophages. One large class of restriction enzyme recognizes specific nucleotide sequences (in most cases, four to six base-pairs long) called *restriction sites*. Almost all of these sites have a *palindromic*, or symmetrical sequence, in which the bases of the two chains are the same, read in opposite directions, e.g.,

$$5'- \ldots AGCGCT \ldots -3'$$
$$3'- \ldots TCGCGA \ldots -5'$$

Some restriction enzymes make a double-stranded cut in the middle of their cognate sequence, yielding fragments with "blunt ends." Others make two cuts, one in each nucleotide chain, but in positions several bases from one another, a *staggered cut*. These staggered cuts are a significant aid in genetic engineering because any DNA cut with a given restriction enzyme will have identical ends, in which there may be a four-base overhang in which one nucleotide chain is longer than the other. In the sequence above, a cut between the A and the G will give both pieces of DNA the same end:

$$5'- \ldots A \qquad GCGCT \ldots -3'$$
$$3'- \ldots TCGCG \qquad A \ldots -5'$$

The identical nature of the ends of fragments and the symmetry of the original restriction site allows transient complementary base-pairing between the single-stranded ends of any DNA resulting from a cut with a restriction enzyme that makes staggered cuts.

The restriction enzymes of a bacterium do not attack the DNA of the bacterium itself. This is because every restriction enzyme made by a bacterial cell is accompanied by a second enzyme activity, often carried by the same enzyme protein, that *methylates* a nucleotide of all restriction sites of the bacterium's DNA. Methylated sites cannot be cut by the restriction enzyme of the same bacterium, although they may be cut by enzymes of another species, many of which are "methyl-insensitive."

Most bacterial species have one or more restriction enzymes, but these differ in their restriction sites. Many restriction enzymes have been identified and purified, and they are now sold as reagents for genetic engineering work in universities and the biotechnology industry. They are named for the bacterium from which they were isolated: *Eco*RI is restriction enzyme I from *E. coli* and attacks the sequence GAATTC; *Hind*III is enzyme dIII from *Haemophilus influenzae* and attacks the sequence AAGCTT. A number of enzymes with the same restriction sites are known; some may be methyl-sensitive, others methyl-insensitive. Restriction sites vary in length (between 4 and 10 or more nucleotides), and some enzymes may not require specific nucleotides at all positions in the site.

DNA Ligase

DNA ligase attaches nucleotide chains to one another. If this enzyme is included in a solution containing fragments with complementary ends, it will catalyze the formation of the normal phosphate–sugar bonds between fragments, making the chains continuous, normal DNA. It will also attach the ends of the same fragment, making a circular DNA, or restore circularity to a plasmid cut once with such an enzyme. DNA ligase will also attach blunt ends of DNA molecules to one another, but less efficiently.

Cloning DNA

Restriction enzymes and DNA ligase (which is not specific for the ends it will join) permit investigators to cut DNA into fragments defined by the location of restriction sites and to rejoin different DNAs in virtually any combination. If a large preparation of a specific DNA molecule such as a plasmid or phage DNA of 30–200 kb is cut with one or two restriction enzymes, it will yield about 8–20 fragments of different sizes. After cutting the DNA, the preparation may be subjected to *agarose gel electrophoresis*, in which the DNA is put in a small indentation, or well, in an agar slab. This slab is held in an apparatus in which direct current can be applied from one end to the other. DNA is negatively charged and will move in the direction of the cathode (positive). However, the agarose matrix impedes the movement of the DNA, and the larger fragments move slowly and the small fragments move more rapidly. After an hour or two, the gel is stained with a dye (ethidium bromide) that binds to DNA, enabling one to visualize with ultraviolet light the position of the fragments in the gel.

The fragments may be removed by cutting out a piece of the gel and purifying them. If only one restriction enzyme is used, all of the molecules of the preparation will have the same sequence and the same ends. (If the starting molecule is linear, fragments carrying the original ends will, of course, be different.)

These tools (restriction enzymes, DNA ligase, and gel electrophoresis), are used in cloning DNA. Small plasmids replicate to high numbers (ca. 50–200) in a single cell of *E. coli* and can be isolated in large quantities from a population of bacteria carrying them. If a circular plasmid is cut with a restriction enzyme at one point, the plasmid becomes a linear piece of DNA. If DNA ligase is added, the free ends may be rejoined, restoring the plasmid to its circular form. However, if fragments of another organism's DNA are added before DNA ligase, some plasmids will incorporate a foreign fragment of DNA as they recircularize. Plasmids designed to carry foreign pieces of DNA are called *cloning vehicles* or *vectors*.

The larger circles, when introduced by artificial transformation methods into *E. coli,* replicate to high levels just like the original plasmids. If a single transformant is chosen and grown to a large population size, the fragment of DNA in the plasmid is said to be *cloned*. The specific fragment of foreign DNA can be isolated in pure form by cutting a large preparation of the plasmid with one or more restriction enzymes and disposing of the plasmid DNA.

DNA Libraries

The collection of clones that represents the entire genome is called a *DNA library*. In the early development of the field, methods for random breakage of DNA by physical or less discriminating enzymatic methods were designed so that at least some copies of all genes of a preparation remained intact, unbroken by a restriction enzyme. With larger genomes, gentle extraction and breakage of the DNA can yield much larger fragments suitable for cloning in vehicles derived from the *E. coli* F plasmid; these are called bacterial artificial chromosomes (BACs). The very long "inserts" in BACs are useful in establishing more securely the manner in which contiguous genes are linked in the original genome. Two methods of genomic sequencing differ in this respect. (i) In the "clone-by-clone" approach, the location of restriction sites of BAC inserts are characterized ("mapped"), using differrent restriction enzymes to cut the DNA. With several restriction enzymes, one can order the resulting sites and the fragments arising from cutting them fairly easily. These fragments are then isolated and subjected to nucleotide sequencing. (ii) In the "shotgun" technique, the entire genome is cut into small fragments and are later ordered according to the overlaps in their sequences by computer techniques.

Notes

Chapter 1

1. The historical material that follows is taken largely from Magner (1993).

2. Burian (1993). This is the last article of a special section, "The right organism for the job" in the journal cited, to which I will refer hereafter. Many of the points I make are well stated by Burian, but he does not include my stress on the central importance of genetic analysis in qualifying a species as a *model organism* as I define it later in this chapter.

3. Logan (2001). Logan traces the changes in strategy of embryological and behavioral work from observation of species differences to manipulation of standardized stocks of rats. A good example of the development of a biological system in the 20th century.

4. Mendel (1866). This paper summarizes two scientific presentations made in 1865.

5. Dunn (1965), pp. 21–24; Hartl and Orel (1992). Hartl and Orel analyze and refute a number of revisionist accounts that deny Mendel's claim as the founder of genetics.

6. Dunn (1965), p. 4.

7. Crow and Crow (2002).

8. Coleman (1965).

9. Morgan had been drawn out of his "naturalist" training, in which observation of normal embryogenesis in various species was the basis of generalization and speculation. Instead, he began to use the new experimental, interventionist, and empirical approaches (*Entwicklungsmechanik*) championed by Roux and Driesch in the late 1890s. His empirical mindset, which he retained thereafter, would delay his acceptance of Mendelian and Darwinian ideas and even the chromosomal basis of inheritance until he could point to hard evidence, free of speculation (Allen, 1978, pp. 72–79).

10. Allen (1978), pp. 106–121.

11. Kohler (1993).

12. Kohler (1993).

13. Morgan et al. (1915); Morgan (1926).

14. Benson (2001).

15. Morgan et al. (1915), p. ix.

16. Darden (1991).

17. National Research Council Committee on Models for Biomedical Research (1985).

18. Fink (1988); Ziman (2000), p. 149. Ziman applies the term "communal intellectual resource" to describe theoretical models, but it serves well here.

19. Rosenberg (1994), chaps. 1–4; Ziman (2000), pp. 321–327. The term "reductionism" is loaded, and the literature is turgid with conflicting definitions and philosophical argument on the matter. The fundamental issue is whether in theory or in practice biological phenomena are reducible to physics and chemistry. Rosenberg (1994) denies such reducibility, noting that natural selection acts on function, and that evolution offers different solutions to functional problems, solutions that are not distinguished by differences in fitness. Most of these solutions are complex aggregations of molecular attributes, all involving trade-offs, and therefore the lesser and more fundamental levels cannot even in theory be mapped on one another. Most modern commentators call attention to the historical and contingent aspects of biological systems that render them impossible to subsume into a world of laws drawn from the "ahistorical" physical sciences. In place of laws, biology can only offer generalizations, of which Mendel's laws are examples.

20. Keller (2000).

21. Keller (1992), p. 109; Kay (1996), pp. 87–100.

Chapter 2

1. Sapp (1987).

2. Sutton (1903).

3. Benson (2001).

4. Sturtevant (1965), p. 41.

5. Sturtevant (1965), p. 47.

6. Sturtevant (1913).

7. Morgan et al. (1915).

8. Morgan's work somewhat eclipsed the contributions of agricultural geneticists, who had established strong programs in Mendelian analysis of such plants as beans and corn well before Morgan embarked on his *Drosophila* work. These programs, of which R. A. Emerson's in corn became the most prominent, were understandably slanted to physiological aspects of genes and led to great improvement of crops in the United States thereafter. The development of hybrid corn is a model of such programs (Kimmelman, 1992).

9. Morgan (1934).

10. Morgan and Bridges (1919); Sturtevant (1920).

11. Burian et al. (1988); Sapp (1987).

12. Beadle (1974).

13. Horowitz (1996).

14. Burian et al. (1988).

15. Burian et al. (1988).

16. Scott-Moncrieff (1936).

17. Garrod (1909).

18. Sapp (1990); Bearn (1993).

19. Sapp (1990).
20. Beadle (1974).
21. Cited in Bearn (1993), pp. 114, 159.
22. Sapp (1990).
23. Sapp (1990).
24. Morange (1998), p. 29.
25. Much of the information about Muller comes from Carlson (1981). This work has done much to emphasize Muller's separate contributions to genetics, correcting some impressions imparted by others in the field.
26. Morgan et al. (1915).
27. Sturtevant (1965), p. 47.
28. Sturtevant (1965), p. 48–49; Carlson (1981), pp. 82–85.
29. Carlson (1981), p. 229.

Chapter 3

1. Shear and Dodge (1927).
2. Robbins (1962).
3. Dodge (1927).
4. The few *Neurospora* species described by Dodge include a "pseudohomothallic" species, *N. tetrasperma*. In this species, asci contain only four spores, but each of the spores includes two nuclei, each of a different mating type. The general nature of meiosis and segregation, however, are similar to the heterothallic species, in which the spores are homokaryotic, and of either the *A* or the *a* mating type.
5. Horowitz (1990).
6. Lindegren (1931); Lindegren (1973).
7. Lindegren's work is summarized in Lindegren (1942).
8. Lederberg (1990).
9. Beadle (1974).
10. Burian and Gayon (1999).
11. Beadle and Tatum (1941). See Carlson (1966), pp. 171–173, for further discussion of this point.
12. Brock (1990), pp. 36–37.
13. Letter from Beadle to Dodge, Feb. 27, 1941. Copy kindly provided by David Perkins.
14. Beadle and Tatum (1941).
15. Kay (1993).
16. Beadle and Tatum (1945).
17. Srb and Horowitz (1944).
18. Beadle and Coonradt (1944).
19. Tatum et al. (1949).
20. Woodward et al. (1954).
21. Horowitz (1991).
22. Beadle (1945), p. 19.
23. Perkins (1992).
24. Delbrück (1946). Discussion following paper by Bonner (1946), pp. 22–23.
25. Horowitz and Leupold (1951).
26. Horowitz (1950).
27. Pauling et al. (1949).

28. Ingram (1957).
29. Davis, R. H. (1960).
30. Mitchell and Lein (1948).
31. Horowitz (1956); Horowitz and Fling (1953).
32. McClintock (1945).
33. Barratt et al. (1954).
34. Ryan (1950).
35. Reviewed by Perkins and Barry (1977).
36. Mitchell (1955a,b).
37. R. H. Davis (2000).
38. Yanofsky (2001).

Chapter 4

1. Cohen (2000).
2. Pontecorvo and Gemmell (1944).
3. Beadle and Coonradt (1944).
4. Pontecorvo and Gemmell (1944), p. 514.
5. Pontecorvo (1946a).
6. Pontecorvo (1946b).
7. Pontecorvo (1994).
8. Pontecorvo (1994).
9. Pontecorvo (1948); Pontecorvo et al. (1949).
10. Pontecorvo et al. (1949).
11. Roper (1952).
12. Martinelli (1994a). The volume in which this article appears (Martinelli and Kinghorn, 1994) summarizes the major *A. nidulans* research areas up to 1994.
13. Lederberg and Tatum (1946).
14. Raffel and Muller (1940); summarized by Carlson (1966), pp. 144–157.
15. MacKendrick and Pontecorvo (1952).
16. Summarized in Lewis (1950).
17. Here we encounter a semantic problem: in the 1940s and early 1950s, "gene products" were spoken of both as proteins (e.g., enzymes) and products of their action (e.g., vitamins, amino acids). Genes, for all anyone knew, might be proteins, nucleoproteins, or structural templates for making proteins, with much confusion about whether protein synthesis took place in the nucleus, and the role of RNA in the process.
18. Roper (1950). This study is discussed in more detail in Pontecorvo (1954).
19. Pontecorvo (1956).
20. A more extravagant conception of gene action held that the chromosome as a whole was a large, integrated organelle, and that genes could not be localized to points along its length. Although mutations would have specific locations and effects, such damage did not imply that a wild-type function was similarly localized. The fact that certain rearrangements of chromosomes produced derangements of their function (another form of position effect) supported this idea. Richard Goldschmidt, who championed this view, offered an analogy: "If the A-string on a violin is stopped an inch from the end the tone of C is produced. Something has been done to a locus in the string, it has been changed in regard to its function. But nobody would conclude that there is a C-body at that point" (Goldschmidt 1951, p. 7).
21. Pontecorvo (1954), p. 229.

Chapter 5

1. Hall and Linder (1993). This volume is a rich source of background information that I have used extensively for this chapter.
2. von Borstel (1993).
3. Mortimer (1993a).
4. Mortimer (1993b).
5. Lindegren (1949), p. 27-1.
6. Pomper and Burkholder (1949).
7. Mortimer (1993b).
8. Lindegren and Lindegren (1946); Lindegren (1949).
9. Lindegren (1966).
10. These matters are discussed in more detail by Mortimer (1993b).
11. Roman et al. (1951).
12. Winge and Roberts (1950).
13. Hawthorne (1956).
14. Lindegren (1955).
15. Mitchell (1955a,b); Roman (1956).
16. Lindegren (1958).
17. Mortimer (1993b), p. 29.
18. Spiegelman et al. (1945).
19. Spiegelman et al. (1945), p. 100.
20. Lindegren (1949). Spiegelman became a prominent and highly respectable molecular biologist in the 1960s and 1970s. When asked about the experiments he had done with Lindegren, he reputed to have said, "Those data are no longer available."
21. Lindegren (1952).
22. Moor and Mühlethaler (1963); Matile et al. (1969).
23. Lindegren (1957).
24. Lindegren (1957), p. 439.
25. This section is based largely on Roman (1986), Hawthorne (1993), and Mortimer (1993c).
26. Lindegren (1949).
27. Roman (1986).
28. Roman (1956).
29. John Pringle, personal communication (2001).
30. Burian and Gayon (1999).
31. Sapp (1987).
32. Ephrussi (1953), p. 13.
33. Winge and Laustsen (1940).
34. Summarized in Ephrussi (1953).
35. Ephrussi and Hottinguer (1950).

Chapter 6

1. Blakeslee (1904).
2. Judson (1979), even in his early history of molecular biology, has no index entries for *Neurospora*, bread mold, *Aspergillus*, *Saccharomyces*, or yeast. Beadle's name appears four times, each referring to brief statements that he and Tatum discovered that genes specified enzymes. This failing, widespread in the literature, illustrates how

easy it is to separate cognitively the areas (and eras) of biochemical genetics and molecular biology.

Chapter 7

1. Cairns et al. (1966); Watson (1968).
2. Olby (1974); Judson (1979).
3. Kay (1993).
4. Keller (1995); Doyle (1997).
5. Sapp (1987, 1990). Rosalind Franklin, working in England on the structure of DNA at the same time as Watson and Crick, is perhaps the most conspicuously ill-used of the "also-ran" category, according to Sayre (1975). Sayre's account seeks to correct the impressions of Franklin conveyed by Watson's memoir (Watson, 1968). A more detached view of Franklin's contribution has been published recently (Maddox, 2002); Abir-Am (1999).
6. Penman (2001).
7. Brock (1990).
8. Lederberg and Tatum (1946).
9. Dubos (1945).
10. Avery et al. (1944).
11. Brock (1990).
12. Dubos (1945).
13. Luria and Delbrück (1943).
14. Avery et al. (1944).
15. Dubos (1945) includes an addendum by C. F. Robinow that summarizes the cytological observations of the latter.
16. Chapman and Hillier (1953); Porter and Blum (1953).
17. Stanier and van Niel (1962).
18. Dubos (1945), p. 183.
19. Luria (1984), p. 76. Brock (1990), however, calls attention to the fact that the paper describing the results indicates that the idea followed the initial results of tests for bacteriophage resistance.
20. Luria and Delbrück (1943).
21. Avery et al. (1944).
22. Dobzhansky (1941); Mayr (1942).
23. This account is based largely on memoirs of Lederberg (1986, 1987) and Zuckerman and Lederberg (1986). Additional information may be found in Brock (1990).
24. Tatum (1945), p. 215.
25. J. Lederberg, personal communication (19 July 2001).
26. Gray and Tatum (1944); Tatum (1945).
27. Tatum (1945), p. 215.
28. Brock (1990); J. Lederberg (1986, 1987).
29. J. Lederberg (1951).
30. J. Lederberg (1998).
31. Delbrück to Lederberg, 26 Feb. 1948: "I refuse to believe in your bacterial genetics (and therefore also refuse to take an interest in details of the type you mention) as long as the kinetics of the postulated mating reaction has not been worked out."

Made available courtesy of J. Lederberg [19 July 2001]; letter available for scholarly examination at the Lederberg archives at the National Library of Medicine.

32. Zuckerman and Lederberg (1986).

33. B. D. Davis (1950).

34. Zinder and Lederberg (1952).

35. Lederberg, J, personal communication (19 July 2001).

36. Hayes (1953); Cavalli et al. (1953). Much of the work described in this chapter is reviewed by the classic, early book of W. H. Hayes (1964). Brock (1990) is a second source of much of the material discussed here.

37. Watson and Crick (1953a).

38. Jacob and Wollman (1961).

39. Lederberg and Lederberg (1952).

40. Lederberg and Zinder (1948); B. D. Davis (1948).

41. Hayes (1964), p. 631.

42. Demerec and Latarget (1946).

43. Roberts et al. (1957).

Chapter 8

1. The early studies are summarized by Stent (1963) and Brock (1990), chapter 6. An interesting discussion of the matter among d'Herelle, Twort, Jules Bordet, and André Gratia, held at the annual meeting of the British Medical Association in 1922, is reprinted in Stent (1965), pp. 3–25.

2. Viruses had been discovered and characterized as "filtrable agents" in the 19th century by Pasteur, Beijerinck, and Klebs; the first viruses to be so characterized were tobacco mosaic virus and the agent of foot-and-mouth disease. The idea that they were replicating entities was prevalent up to about 1915, when bacteriophages were discovered. The notion that bacteriophages were in fact bacterial viruses was not accepted universally until about 1945 or later.

3. Muller (1922).

4. Muller (1927).

5. Carlson (1981), p. 188.

6. Timoféeff-Ressovsky et al. (1935).

7. Zimmer (1966).

8. Timoféeff-Ressovsky et al. (1935).

9. Schrödinger (1944).

10. See Watson (2000), in which a 1993 essay is reprinted. He recalls being inspired in 1945 by Schrödinger's book to study the gene.

11. Yoxen (1979).

12. The material on Delbrück's life and intellectual orientation are taken from Kay (1985).

13. Olby (1974).

14. Bawden et al. (1936); Fruton (1999).

15. Summers (1993).

16. Weaver (1970).

17. The source of my information on the choice of bacteriophage, its relation to d'Herelle's work, and the collaboration of Ellis and Delbrück comes largely from Summers (1993).

18. d'Herelle (1926).

19. Ellis (1966).

20. Ellis and Delbrück (1939).

21. Summers (1993).

22. Kay (1985).

23. Delbrück, quoted in Kay (1985), p. 231.

24. Luria (1984). I have relied on this book for much of Luria's biographical material.

25. Hartman (1988).

26. Luria (1984), p. 71–72.

27. Cohen (1984); Kozloff (1984); Stahl (2000); Anderson (1966).

28. Hartman (1988).

29. Luria and Delbrück (1943).

30. Luria and Anderson (1942).

31. Demerec and Fano (1945).

32. The Cold Spring Harbor course, taught by others in later years, ceased in 1970 (Susman, 1995).

33. Hershey (1946).

34. Delbrück and Bailey (1946).

35. Hershey and Rotman (1949).

36. Anderson (1952).

37. Herriott (1951); letter from Herriott to A. D. Hershey, 1951, quoted in Judson (1979), p. 57.

38. Hershey and Chase (1952).

39. Kay (1985), p. 244.

Chapter 9

1. The controversy and its resolution is nicely summarized by Brock (1990), pp. 163–179.

2. Brock (1990).

3. Burnet and McKie (1929).

4. Burnet and Lush (1936).

5. Wollman and Wollman (1939).

6. Lwoff and Gutmann (1950).

7. Brock (1990); Stent (1963, 1965).

8. Lwoff et al. (1950).

9. Zinder and Lederberg (1952).

10. This point is explicit in the discussion of the Zinder-Lederberg (1956) paper, in which the authors state that bacterial gene exchange involves cell fusion and meiosis and displays linkage of genes.

11. Lederberg et al. (1951).

12. Davis, B. D. (1950).

13. Zinder (1992) gives a first-hand account of the progress of these experiments.

14. Zinder and Lederberg (1952).

15. Hartman (1988).

16. J. Lederberg (1951); E. M. Lederberg and J. Lederberg (1953).

17. Lwoff et al. (1950).

18. A readable summary of the earlier days of phage λ biology can be found in Hayes (1980).

19. Lederberg and Lederberg (1952).

20. Wollman et al. (1956).

21. Wollman et al. (1956).

22. Morse et al. (1956).

23. Campbell (1961). The history of this conception and its confirmation is summarized by Campbell (1993).

24. Bertani (1951).

25. The autonomous nature of the P1 prophage represents the Lederbergs' initial picture of phage λ before they performed their *gal⁻* × *gal⁺* crosses.

26. Lennox (1955).

27. Zinder (1960).

28. Sanderson and MacLachlan (1987).

29. Demerec (1956).

Chapter 10

1. Watson and Crick (1953a).

2. Olby (1994); Pauling and Corey (1950).

3. Sanger (1959).

4. Ingram (1957).

5. Watson (1968).

6. Jacob (1988), pp. 270–271. Quoted with permission of author and Basic Books.

7. Chargaff (1975), p. 253. In his first use of this famous phrase, Chargaff omits the article: "without license."

8. Yanofsky (2001).

9. Yanofsky (1956).

10. Rachmeler and Yanofsky (1958).

11. Suskind et al. (1955).

12. Demerec (1956).

13. Crawford and Yanofsky (1958).

14. Helinski and Yanofsky (1962).

15. Watson and Crick (1953b).

16. Stent (1968, 1969).

Chapter 11

1. Stadler (1954).

2. Holmes (2000). This is an informative, detailed account of Benzer's development of the T4 system.

3. This cross is best visualized by considering different mutations carried by the two parental phages as a^+b^- and a^-b^+, with recombination reconstructing a normal phage, a^+b^+.

4. Benzer (1955).

5. Quoted in Benzer (1966), p. 161.

6. I thank Dr. Franklin Stahl for bringing this point to my attention.

7. Crick et al. (1961).

8. Gamow (1954).

9. Horowitz and Leupold (1951).

10. Stahl (1995).

11. Epstein et al. (1963).

12. Thomas (1963).

13. Brenner et al. (1961).

14. Jacob and Wollman (1961).

15. Cairns (1963).

16. Volff and Altenbuchner (2000).

17. Watson and Crick (1953b).

18. Meselson and Stahl (1958).

19. Roberts et al. (1963).

20. Roberts et al. (1963), pp. 436, 437.

21. Maaløe and Kjeldgaard (1966). This and later work is summarized in Bremer and Dennis (1987).

22. Cooper and Helmstetter (1968).

23. An early status report on this phenomenon may be found in Umbarger (1961).

24. Reviewed by Brock (1990).

25. Brock (1990), pp. 292–293.

26. Pardee et al. (1959).

27. Jacob and Monod (1961).

28. Jacob and Monod (1961).

29. Jacob (1988), p. 316–317.

30. Crick originally thought that the RNA was that of the ribosomes, the particles on which proteins are made. It soon became clear that the RNA delivering the genetic message was mRNA.

31. Stent (1968).

32. Meselson and Weigle (1961); Meselson (1964).

33. Stahl (1998).

34. Gilbert and Müller-Hill (1966); Ptashne (1967).

35. Ptashne (1986).

Chapter 12

1. Papers of the *Cold Spring Harbor Symposium of Quantitative Biology*, vol. 28 (1963), illustrate an early stage in our knowledge of protein synthesis, drawing heavily on in vitro experiments with mammalian liver and reticulocyte extracts.

2. The following paragraphs describe developments described in many texts in molecular biology and genetics, e.g., Snyder and Champness (1996).

3. Kornberg and Baker (1992).

4. McCann and Ames (1976).

5. Kowalczykowski et al. (1994).

6. Monod appears to have insisted on negative control as a universal mechanism, having seen it in the *lac* system, in repression of the λ prophage, and in repression of the enzymes of tryptophan synthesis. Jacob reports that he could not convince Monod of the legitimacy of positive control in 1960 (Jacob 1988, pp. 319–320).

7. Monod's opposition to positive control, which he had espoused in another context prior to the PaJaMo experiment, was based in part in his faith in the "unity in biochemistry" (see Maas 1991); Engelsberg et al. (1969).

8. Nomura (1997), p. 278. This article and references therein describe these experiments.

9. Shapiro (1983).

10. McClintock (1961).

11. Luria (1953).

12. Reviewed by Arber (1971).

13. Kelly and Smith (1970); Smith (1979).

14. Fox (2000).

15. Schaechter and the View from Here Group (2001).

16. Schaechter and the View from Here Group (2001), p. 124.

17. Ankeny (2001).

18. This transition can best be appreciated from some of the articles in Hall and Linder (1993).

19. Jacob (1998), p. 6. Quoted with permission of author and Harvard University Press.

20. Monod (1972), p. 98.

Chapter 13

Tracy Sonneborn often used this chapter's epigraph in conferences with research students.

1. Ephrussi in Sapp (1987), p.132.

2. Michaelis (1954), summarized by Sager (1972).

3. Sturtevant and Beadle (1939), pp. 332–333.

4. See Sapp (1987), pp. 92–93.

5. See Sapp (1987), p. 95.

6. Chatton and Lwoff (1935), cited in Lwoff (1950).

7. Much information on *Paramecium* in this section taken from Beale (1954).

8. Nanney (1986).

9. Allen and Nerad (1978).

10. Nanney and Simon (2000).

11. Sonneborn (1937).

12. Tsukii (1988).

13. Preer (2000).

14. Sonneborn (1937), p. 385.

15. See Beale (1954).

16. Sonneborn (1950).

17. Summarized in Beale (1954).

18. Beale (1954), pp. 74–75.

19. Quackenbush (1988).

20. Preer (1997). This is an aptly titled update of work on many of the interesting phenomena uncovered in the 1940s and 1950s with *Paramecium*.

21. These papers were published in 1949 by the Centre Nationale de la Recherche Scientifique (CNRS) under the title *Unités Biologique Douées de Continuité Génétique* in a series, *Colloques Internationaux du CNRS* (CNRS, 1949).

22. Lwoff (1949).

23. Lwoff (1950).

24. Quote in Lwoff (1950), p. 4, citing Darlington (1944), p. 164.

25. Sonneborn and Beale (1949).

26. Delbrück in Discussion following Sonneborn and Beale (1949), quoted in Thieffry (1996), p. 169. Beale acknowledged the legitimacy of the model in relation

to the antigen experiments during this discussion, though he favored the idea of particulate inheritance.

27. This picture resonates with Delbrück's picture of mutation: a quantum change in a metastable molecule (see chapter 8).

28. Beale and his student Ian Gibson struggled to maintain the hypothesis suggested by early results against increasingly unpromising attempts to extend it. This time, it was Sonneborn who kept his distance after contributing data on the phenomenon with Gibson, a Sonneborn postdoctoral fellow. In 1969, Sonneborn's student, Barbara Byrne (née McManamy), effectively did away with the metagon in her thesis work in Sonneborn's laboratory. The episode is reviewed by Hall (1998). On the decline of *Paramecium*, see Preer (2000).

29. Nanney (1988).

30. Preer (1997).

31. Hill (1972).

32. Nanney and Simon (2000).

33. Gall (1986).

34. Allen and Nanney (1958).

35. Yao (1996).

36. Blackburn (1986).

37. Collins (1996).

38. Cech et al. (1981).

39. Doherty and Doudna (2000).

Chapter 14

1. Ephrussi (1953).

2. Ephrussi (1953), pp. 5–6.

3. Ephrussi (1949).

4. Ephrussi (1949), p. 165 (translation by author).

5. Winge and Laustsen (1940).

6. Ephrussi (1953), pp. 13–47. The impairment of oxidative respiration forces yeast to use the less efficient process of fermentation to generate its energy from glucose. Thus normal yeast and petites have a similar colony size in anaerobic conditions.

7. Ephrussi (1953), pp. 118–119.

8. Interesting accounts of this era can be found in Burian and Gayon (1999), Burian et al. (1988), and Sapp (1987).

9. Guérin (1991).

10. Reviewed in Rabinowitz and Swift (1970) and Sager (1972).

11. Schatz et al. (1964).

12. Dujon (1981).

13. Wilkie (1970).

14. Pascher (1916, 1918).

15. Sapp (1990).

16. Sapp (1990), p. 56.

17. Sonneborn (1951).

18. Ryan (1955). Valid demonstrations of relative sexuality in many phycomycetous fungi were published in the 1930s (Raper, 1966), which may have given Moewus's and Hartmann's work on *Chlamydomonas* credence at the time.

19. Sapp (1990).

20. Harris (1989).
21. Sapp (1990).
22. These historical observations are drawn from Sapp (1990), pp. 171–174.
23. Sager (1972), pp. 49–50.
24. This spelling has been generally used in recent literature.
25. Gillham (1993).
26. Levine and Ebersold (1960).
27. Harris (1989).
28. Togasaki and Surzycki (1997), p. 15.
29. Levine (1968).
30. Sager (1954).
31. Sager and Ramanis (1963).
32. Gillham (1993).
33. Sager (1960, 1962).
34. Gillham and Levine (1962).
35. Sager and Ramanis (1965).
36. Gillham (1993).
37. Remacle and Matagne (1997).
38. Armbrust (1997).
39. Rochaix et al. (1997); Dentler and Witman (1995); Harris (2001).
40. Harris (1989).
41. Gillham (1993); Dujon (1981).
42. Dujon (1981). I thank Nicholas Gillham for bringing this to my attention.
43. Gillham (1993).
44. Williamson (2002).
45. Maleszka and Clark-Walker (1992).
46. Harris (2001); Maleszka and Clark-Walker (1992). The form of the chloroplast DNA (linear or circular) is not known at the time of printing.
47. Williamson (2002).
48. Reviewed by Dujon (1981).
49. Gillham (1993).
50. Margulis (1981).

Chapter 15

1. Fogel (1993), p. 139.
2. Carlson (1971) suggests that most prokaryotic molecular biologists came from nongenetically oriented fields, many of whom took the Three-Man Paper (Timoféeff-Ressovsky et al. 1935) or the Schrödinger book (1944) as their intellectual starting point.
3. von Borstel (1993).
4. von Borstel (1993).
5. Hartwell and McLaughlin (1968); Hartwell et al. (1974); reviewed recently in detail by Lew et al. (1997).
6. Botstein (1993).
7. Botstein (1993), p. 363.
8. Ratzkin and Carbon (1977).
9. The generic nomenclature bears explanation. *E. coli* genes are named with lowercase italics; gene symbols have three letters designating the phenotypic nature (here, leucine-requiring) followed by a capital letter designating the particular gene

of the pathway; hence *leuB*, with a minus superscript indicating the mutant allele. Yeast genes are also designated with three letters in italics, but they are followed by a number to designate the particular gene. Wild-type alleles are capitalized; mutant alleles are lowercase.

10. Hinnen et al. (1978).

11. Matile et al. (1969).

12. The first of these series was published by Cold Spring Harbor Laboratory Press: *The Molecular Biology of the Yeast Saccharomyces* in 1981 (vol. 1, *Life Cycle and Inheritance*) and 1982 (vol. 2, *Metabolism and Gene Expression*), edited by J. N. Strathern, E. W. Jones, and J. R. Broach. The second series, from the same press, was *The Molecular and Cellular Biology of the Yeast Saccharomyces* in 1991 (vol. 1, *Genome Dynamics, Protein Synthesis, and Energetics*, edited by J. R. Broach, E. W. Jones, and J. R. Pringle), 1992 (vol. 2, *Gene Expression*, edited by E. W. Jones, J. R. Pringle, and J. R. Broach), and 1997 (vol. 3, *Cell Cycle and Cell Biology*, edited by J. R. Pringle, E. W. Jones, and J. R. Broach). The subtitles of the volumes of the two series suggest the increasing scope of yeast studies. See the second series for further discussion of and references to material reviewed in this section.

13. Botstein and Fink (1988).

14. Szostak et al. (1983).

15. Summarized in Sherman and Stewart (1982).

16. Carsiotis and Lacy (1965).

17. Hinnebusch (1998).

18. Guarente (1992).

19. Hartwell (1993).

20. Su and Nurse (1997).

21. Herskowitz et al. (1992).

22. Novick et al. (1980).

23. Kaiser et al. (1997).

24. Johnson (1999).

25. Pruyne and Bretscher (2000).

26. Goffeau et al. (1996).

27. Keller (2000b).

28. Eisen et al. (1998).

29. Chu et al. (1998).

30. Winzeler et al. (1999).

31. Glaever et al (2002).

32. As one of many examples, see Jeong et al. (2001).

33. Yeadon et al. (2001).

34. Slayman (1970).

35. For review, see Rao and Slayman (1996).

36. For review, see R. H. Davis (1985).

37. Bowman and Bowman (1982).

38. Wiemken and Dürr (1974); Jones (1991); Klionsky et al. (1990); Wada and Anraku (1994).

39. This remark betrays a concern common to those who see their scientific activities superseded. In this case, the literature on yeast is vast and coherent. It would be naïve to expect the bulk of the yeast community to remain simultaneously aware of comparable studies in other organisms if they did not aid the yeast research agendas.

Chapter 16

1. Martinelli and Kinghorn (1994); R. H. Davis (2000).
2. Pontecorvo et al. (1953); Martinelli (1994b).
3. Meselson and Weigle (1961).
4. Perkins (1997).
5. Jinks (1964).
6. R. H. Davis (2000) and Martinelli and Kinghorn (1994) review this and later phases of work in both organisms.
7. R. H. Davis (1975).
8. Giles et al. (1989).
9. Ovádi (1991). This article states the case for the phenomenon and is followed by critical articles discussing it.
10. Srere (1987).
11. Fincham (1966).
12. R. H. Davis (2000).
13. Clutterbuck (1977); Champe et al. (1981).
14. Champe et al. (1994); Timberlake and Clutterbuck (1994).
15. Morris et al. (1982).
16. Osiewacz (2002).
17. Hinnen et al. (1978).
18. Bennett and Lasure (1985); Timberlake (1985).
19. I am indebted to Eric Selker for his account of this matter (personal communication, 9 January 2002). (I was a sabbatical visitor to the Yanofsky laboratory in the winter of 1985.)
20. Yanofsky (2001).
21. Davis and Perkins (2002).
22. Much of the work reported here is cited in the recent review by Perkins and Davis (2000).
23. Cogoni (2001).
24. Reviewed in R. H. Davis (2000).
25. Morris (2000).
26. Timberlake (1990); R. H. Davis (2000).
27. Morris et al. (1995).
28. See articles in Pringle et al. (1997).
29. *N. crassa* studies reviewed in R. H. Davis (2000); for *Aspergillus*, see Doonan (1994).
30. Feldman (1982).
31. Dunlap (1999).
32. Glass et al. (2000).
33. http://www/genome.wi.mit.edu/annotation/fungi/neurospora.
34. Braun et al. (2000).

Chapter 17

1. Kendrew (1967). This is a review of *Phage and the Origins of Molecular Biology* (Cairns et al., 1966), reprinted, together with Stent (1968) in the expanded edition of the original volume reissued in 1992 by Cold Spring Harbor Laboratory Press.

2. Hess (1970).
3. See Fruton (1999).
4. Hess (1970).
5. Abir-Am (1991, 1992, 1995); De Chadevarian (1996); Gilbert (1982).
6. Abir-Am (1992).
7. Gilbert (1982).
8. I am indebted to Norman Horowitz for emphasizing this point to me.
9. See R. H. Davis (2000), pp. 174–178, for an elaborate example.
10. Fruton (1999).
11. Ellis and Delbrück (1939).
12. See Fruton (1999), p. 422.
13. See Kay (2000).
14. Kornberg (1976), p. 248.
15. Kay (2000).
16. Thieffry (1997); Brachet (1989); see also Fruton (1999).
17. For a summary of the main advances in this field, see Siekevitz (1995).
18. Crick (1958).
19. Brenner et al. (1961). Crick (1988) describes the evening that Brenner and others recognized the significance of the Volkin-Astrachan work.
20. Rheinberger (1996).
21. Nirenberg and Matthaei (1961).

Chapter 18

1. Kay (2000), p. 175.
2. Lily Kay died in December 2000.
3. Ziman (2000).
4. Watson (2000).
5. The history of the HGP and the competing project by Celera Genomics can be appreciated in the articles and the accompanying commentary describing their sequencing efforts. The HGP project and commentary is published in *Nature*, 409:813–958 (2001). The Celera Genomics project and commentary is published in *Science*, 291:1177–1351 (2001). My brief treatment of the history of the subject relies heavily on both.
6. Woese (1977, 1987).
7. Zuckerkandl and Pauling (1965); Margoliash and Smith (1965).
8. Fitch and Margoliash (1967).
9. Graur and Li (2000).
10. Dacks and Doolittle (2001).
11. Riley and Serres (2000).
12. Riley and Serres (2000).
13. Harold (2001). Harold presents (without espousing) an alternative view for the apparent heterogeneous ancestry of genomes of bacteria; namely, an "annealing" of lines of descent as they formed from a widely shared set of genetic resources.
14. Ochman et al. (2000).
15. Detection methods for DNA independent of the organism depends on the powerful polymerase chain reaction (PCR), now used widely in forensics.
16. Pace (1997).
17. Lockhart and Winzeler (2000).

18. Pandey and Mann (2000); Ideker et al. (2001b).
19. Winzeler et al. (1999); Ross-Macdonald et al. (1999).
20. Kitano (2002).
21. Eisen et al. (1998); Hartwell et al. (1999).
22. Spellman et al. (1998); Chu et al. (1998); Roberts et al. (2000).
23. Goffeau (2000).
24. Ravasz et al. (2002).
25. Ideker et al. (2001b); Uetz et al. (2000).
26. Jeong et al. (2001); Gavin et al. (2002).
27. Ideker et al. (2001a).
28. Winzeler et al. (1999); Ross-Macdonald et al. (1999).
29. Rutherford (2000).
30. Hartman et al. (2001), Ideker et al. (2001a).
31. Keller (2000a); Harold (2001).
32. Keller (2000a).
33. Rutherford (2000).

Chapter 19

1. Nelkin (2001).
2. Dobzhansky (1941); Lewontin (1974).
3. Abir-Am (1999).
4. Olby (1994). The metaphor, for the few readers unfamiliar with *The Wizard of Oz*, refers to the route Dorothy took to the Emerald City, once she and her dog Toto had been transported from Kansas to Oz by a tornado.
5. Rosenberg (1994).
6. Abir-Am (1992, 1985); Keller (2000b); Kay (2000); Beurton et al. (2000); Ziman (2000). Of particular interest in Beurton et al. (2000) is the tension among authors who think the word "gene" might be defined in a unitary way, those who think that the word unwisely denies underlying conceptual ambiguities, and still others that recognize the usefulness of a fuzzy "consensus gene" concept. The book also illustrates the usefulness and dangers of reductionist thinking in the post-genomic era. See also Nelkin (2001).
7. Keller (1983, 1992).
8. Baldi (2001).

References

Abir-Am, P. (1985) Themes, genres and orders of legitimation in the consolidation of new scientific disciplines: deconstructing the historiography of molecular biology. Hist. Sci. 23:73–117.

———— (1991) Noblesse oblige: Lives of molecular biologists. Isis 82:326–343.

———— (1992) The politics of macromolecules. Molecular biologists, biochemists, and rhetoric. Osiris (2nd ser.) 7:164–191.

———— (1995) "New" trends in the history of molecular biology. Hist. Stud. Phys. Biol. Sci. 26:167–196.

———— (1999) The first American and French commemorations in molecular biology. From collective memory to comparative history. Osiris 14:324–370.

Allen, G. (1978) *Thomas Hunt Morgan. The Man and His Science*. Princeton University Press, Princeton, NJ.

Allen, S. L., and D. L. Nanney (1958) An analysis of nuclear differentiation in the selfers of Tetrahymena. Am. Nat. 92:139–160.

Allen, S. L., and T. A. Nerad (1978) Method for simultaneous establishment of many axenic cultures of *Paramecium*. J. Protozool. 25:134–139.

Anderson, T. F. (1952) Stereoscopic studies of cells and viruses in the electron microscope. Am. Nat. 86:91–100.

———— (1966) Electron microscopy of phages. In Cairns, J., G. S. Stent, and J. D. Watson (eds.), *Phage and the Origins of Molecular Biology*. Cold Spring Harbor Laboratory Press, Cold Spring Harbor, NY, pp. 63–78.

Ankeny, R. A. (2001) The natural history of *Caenorhabditis elegans* research. Nature Rev. Genetics 2:474–479.

Arber, W. (1971) Host controlled variation. In Hershey, A. D. (ed.), *The Bacteriophage Lambda*. Cold Spring Harbor Laboratory Press, Cold Spring Harbor, NY, pp. 83–96.

Armbrust, V. (1997) Uniparental inheritance of chloroplast genomes. In Rochaix, J.-D., M. Goldschmidt-Clermont, and S. Merchant (eds.), *The Molecular Biology of Chloroplasts and Mitochondria in Chlamydomonas*. Kluwer Academic Publishers, Boston, MA, pp. 93–113.

Avery, O. T., C. M. MacLeod, and M. McCarty (1944) Studies on the chemical nature of the substance inducing transformation of pneumococcal types. Induction of transformation by a desoxyribonucleic acid fraction isolated from pneumococcus type III. J. Exptl. Med. 79:137–159.

Baldi, P. (2001) *The Shattered Self.* MIT Press, Cambridge, MA.

Barratt, R. W., D. Newmeyer, D. D. Perkins, and L. Garnjobst (1954) Map construction in *Neurospora crassa.* Adv. Genet. 6:1–93.

Bawden, F. C., N. W. Pirie, J. D. Bernal, and I. Fankuchen (1936) Liquid crystalline substances from virus-infected plants. Nature 138:1051–1054.

Beadle, G. W. (1945) Biochemical genetics. Chem. Revs. 37:15–96.

——— (1974) Recollections. Annu. Rev. Biochem 43:1–13.

Beadle, G. W., and V. L. Coonradt (1944) Heterocaryosis in *Neurospora crassa.* Genetics 29:191–308.

Beadle, G. W., and E. L. Tatum (1941) Genetic control of biochemical reactions in Neurospora. Proc. Natl. Acad. Sci. USA 27:499–506.

——— (1945) Neurospora. II. Methods of producing and detecting mutations concerned with nutritional requirements. Am J. Bot. 32: 678–686.

Beale, G. H. (1954) *The Genetics of Paramecium aurelia.* Cambridge University Press, Cambridge.

Bearn, A. G. (1993) *Archibald Garrod and the Individuality of Man.* Clarendon Press, Oxford.

Bennett, J. W., and L. L. Lasure (eds.) (1985) *Gene Manipulations in Fungi.* Academic Press, Orlando, FL.

Benson, K. R. (2001) T. H. Morgan's resistance to the chromosome theory. Nature Rev. Genetics 2:469–474.

Benzer, S. (1955) Fine structure of a genetic region in bacteriophage. Proc. Natl. Acad. Sci. USA 41:344–354.

——— (1966) Adventures in the *rII* region. In Cairns, J., G. S. Stent, and J. D. Watson (eds.), *Phage and the Origins of Molecular Biology.* Cold Spring Harbor Laboratory Press, Cold Spring Harbor, NY, pp. 157–165.

Bertani, G. (1951) Studies on lysogenesis. I. The mode of phage liberation by lysogenic *Escherichia coli.* J. Bacteriol. 62:293–300.

Beurton, P., R. Falk, and H.-J. Rheinberger (eds.) (2000) *The Concept of the Gene in Development and Evolution.* Cambridge University Press, Cambridge.

Blackburn, E. H. (1986) Telomeres. In Gall, J. G. (ed.), *The Molecular Biology of Ciliated Protozoa.* Academic Press, New York.

Blakeslee, A. F. (1904) Sexual reproduction in the Mucorineae. Proc. Am. Acad. Arts Sci. 40:205–319.

Bonner, D. M. (1946) Biochemical mutations in *Neurospora.* Cold Spring Harbor Symp. Quant. Biol. 11:14–24.

Botstein, D. (1993) A phage geneticist turns to yeast. In Hall, M. N., and P. Linder (eds.), *The Early Days of Yeast Genetics.* Cold Spring Harbor Laboratory Press, Cold Spring Harbor, NY, pp. 361–673.

Botstein, D., and G. R. Fink (1988) Yeast: An experimental organism for modern biology. Science 240:1439–1443.

Bowman, E. J., and B. J. Bowman (1982) Identification and properties of an ATPase in vacuolar membranes of *Neurospora crassa.* J. Bacteriol. 151:1326–1337.

Brachet, J. (1989) Recollections on the origins of molecular biology. Biochim. Biophys. Acta 1000:1–5.

Braun, E. L., A. L. Halpern, M. A. Nelson, and D. O. Natvig (2000) Large-scale comparison of fungal sequence information: Mechanisms of innovation in *Neurospora crassa* and gene loss in *Saccharomyces cerevisiae*. Genome Res. 10:416–430.

Bremer, H., and P. P. Dennis (1987) Modulation of chemical composition and other parameters of the cell by growth rate. In Neidhart, F. C. (ed.), *Cellular and Molecular Biology*, vol. 2, *Escherichia coli and Salmonella typhimurium*. American Society of Microbiology, Washington, DC, pp. 1527–1542.

Brenner, S., F. Jacob, and M. Meselson (1961) An unstable intermediate carrying infomrration from genes to ribosomes for protein synthesis. Nature 190:576–581.

Broach, J. R., E. W. Jones, and J. R. Pringle (eds.) (1991) *The Molecular and Cellular Biology of the Yeast Saccharomyces*, vol. 1, *Genome Dynamics, Protein Synthesis, and Energetics*. Cold Spring Harbor Laboratory Press, Cold Spring Harbor, NY.

Brock, T. D. (1990) *The Emergence of Bacterial Genetics*. Cold Spring Harbor Laboratory Press, Cold Spring Harbor, NY.

Burian, R. M. (1993) How the choice of experimental organism matters: Epistemological reflections on an aspect of biological practice. J. Hist. Biol. 26:351–367.

Burian, R. M., and J. Gayon (1999) The French school of genetics: From physiological and population genetics to regulatory molecular genetics. Annu. Rev. Genet. 33:313–349.

Burian, R. M., J. Gayon, and D. Zallen (1988) The singular fate of Genetics in the history of French biology, 1900–1940. J. Hist. Biol. 21:357–402.

Burnet, F. M., and D. Lush (1936) Induced lysogeny and mutation of bacteriophage within lysogenic bacteria. Austral. J. Exptl. Biol. Med. Sci. 14:27–38.

Burnet, F. M., and M. McKie (1929) Observations on a permanently lysogenic strain of *B. enteritidis Gaertner*. Austral. J. Expt. Biol. Med. Sci. 6:277–284.

Cairns, J. (1963) The chromosome of *Escherichia coli*. Cold Spring Harbor Symp. Quant. Biol. 28:43–46.

Cairns, J., G. S. Stent, and J. D. Watson (eds.) (1966) *Phage and the Origins of Molecular Biology*. Cold Spring Harbor Laboratory Press, Cold Spring Harbor, NY.

Campbell, A. M. (1961) Episomes. Adv. Genet. 11:101–145.

——— (1993) Thirty years ago in GENETICS: Prophage insertion into bacterial chromosomes. Genetics 133:433–438.

Carlson, E. A. (1966) *The Gene: A Critical History*. W. B. Saunders, Philadelphia.

——— (1971) An unacknowledged founding of molecular biology: H. J. Muller's contribution to gene theory, 1910–1936. J. Hist. Biol. 4:149–170.

——— (1981) *Genes, Radiation, and Society. The Life and Work of H. J. Muller*. Cornell University Press, Ithaca, NY.

Carsiotis, M., and A. M. Lacy (1965) Increased activity of tryptophan biosynthetic enzymes in histidine mutants of *Neurospora crassa*. J. Bacteriol. 89:1472–1477.

Cavalli, L. L, J. Lederberg, and E. M. Lederberg (1953) An infective factor controlling sex compatibility in *Bacterium coli*. J. Gen. Microbiol. 8:89–103.

Cech, T. R. , A. J. Zaug, and P. J. Grabowski (1981) *In vitro* splicing of the ribosomal RNA precursor of Tetrahymena: Involvement of a guanosine nucleotide in the excision of the intervening sequence. Cell 27:487–496.

Champe, S. P., M. B. Kurtz, L. N. Yager, N. J. Butnick, and D. E. Axelrod (1981) Spore formation in *Aspergillus nidulans*: Competence and other developmental processes. In Turian, G., and H. Hohl (eds.), *The Fungal Spore: Morphogenetic Controls*. Academic Press, New York, pp. 255–276.

Champe, S. P., D. L. Nagle, and L. N. Yager (1994). Sexual sporulation. In Martinelli, S. D., and J. R. Kinghorn (eds.), *Aspergillus: 50 Years On*. Elsevier, London, pp. 429–454.

Chapman, G., and J. Hillier (1953) Electron microscopy of ultra-thin sections of bacteria. J. Bacteriol. 66:362–373.

Chargaff, E. (1975) Voices in the labyrinth: Dialogues around the study of nature. Perspect. Biol. Med. 18:251–285.

Chatton, E., and A. Lwoff (1935) Les ciliés apostomes. I. Aperçu historique et général; étude monographique des genres et des espèces. Arch. Zoll. Exp. Gén. 77:1–453.

Chu, S., J. DeRisi, M. Eisen, J. Mulholland, D. Botstein, P. O. Brown, and I. Herskowitz (1998) The transcriptional program of sporulation in budding yeast. Science 282: 699–705.

Clutterbuck, A. J. (1977) The genetics of conidiation in *Aspergillus nidulans*. In Pateman, J. A., and J. E. Smith (eds.), *The Genetics and Physiology of Aspergillus*. Academic Press, New York, pp. 305–317.

CNRS (1949) *Colloques Internationaux du Centre Nationale de la Recherche Scientifique: Unités Biologique Douées de Continuité Génétique*. CNRS, Paris.

Cogoni, C. (2001) Homology-dependent gene silencing mechanisms in fungi. Annu. Rev. Microbiol. 55:381–406.

Cohen, B. I. (2000) Guido Pontecorvo ("Ponte"), 1907–1999. Genetics 154:497–501.

Cohen, S. S. (1984) The biochemical origins of molecular biology. Trends Biochem. Sci. 9:334–336.

Coleman, W. (1965) Cell, nucleus, and inheritance: an historical study. Proc. Am. Phil. Soc. 109:124–158.

Collins, K. (1996) Structure and function of telomerase. Curr. Opin. Cell Biol. 8:374–380.

Cooper, S., and C. Helmstetter (1968) Chromosome replication and the division cycle of *Escherichia coli* B/r. J. Mol. Biol. 31:519–540.

Crawford, I. P., and C. Yanofsky (1958) On the separation of the tryptophan synthetase of *Escherichia coli* into two protein components. Proc. Natl. Acad. Sci. USA 44: 1161–1170.

Crick, F. H. C. (1958) On protein synthesis. Symp. Soc. Exptl. Biol. 12:138–163.

———— (1988) *What Mad Pursuit. A Personal View of Scientific Discovery*. Basic Books, New York.

Crick, F. H. C., L. Barnett, S. Brenner, and R. J. Watts-Tobin (1961) General nature of the genetic code for proteins. Nature 192:1227–1232.

Crow, E. W., and J. F. Crow (2002) 100 years ago: Walter Sutton and the chromosome theory of heredity. Genetics 160:1–4.

Dacks, J. B., and W. F. Doolittle (2001) Reconstructing/deconstructing the earliest eukaryotes: how comparative genomics can help. Cell 107:419–425.

Darden, L. (1991) *Theory Change in Science. Strategies from Mendelian Genetics*. Oxford University Press, New York.

Darlington, C. D. (1944) Heredity, development and infection. Nature 154:164–169.

Davis, B. D. (1948) Isolation of biochemically deficient mutants of bacteria by penicillin. J. Am. Chem. Soc. 70:4267.

———— (1950) Non-filtrability of the agents of genetic recombination in *Escherichia coli*. J. Bacteriol. 60:507–508.

Davis, R. H. (1960) An enzymatic difference among *pyr-3* mutants of *Neurospora crassa*. Proc. Natl. Acad. Sci. USA 46:677–682.

———— (1975) Compartmentation and regulation of fungal metabolism: genetic approaches. Annu. Rev. Genetics 9:39–65.

———— (1985) Compartmental and regulatory mechanisms in the arginine pathway of *Neurospora crassa* and *Saccharomyces cerevisiae*. Microbiol. Rev. 50:280–313.

——— (2000) *Neurospora. Contributions of a Model Organism.* Oxford University Press, New York.

Davis, R. H., and D. D. Perkins (2002) *Neurospora:* A model of model microbes. Nature Rev. Genet. 3:397–403.

De Chadevarian, S. (1996) Sequences, conformation, information: Biochemists and molecular biologists in the 1950s. J. Hist. Biol. 29:361–386.

Delbrück, M. (1946) Discussion. [Following Bonner, D., Biochemical mutations in Neurospora.] Cold Spring Harbor Symp. Quant. Biol. 11:14–24.

——— (1949) Discussion. [Following paper by Sonneborn and Beale] in *Unités Biologique Douées de Continuité Génétique.* CNRS, Paris, pp. 33–35.

Delbrück, M., and W. T. Bailey, Jr. (1946) Induced mutations in bacterial viruses. Cold Spring Harbor Symp. Quant. Biol. 11:33–37.

Demerec, M. (1956) A comparative study of certain gene loci in *Salmonella.* Cold Spring Harbor Symp. Quant. Biol. 21:113–121.

Demerec, M., and U. Fano (1945) Bacteriophage resistant mutants of *Escherichia coli.* Genetics 30:119–136.

Demerec, M., and R. Latarget (1946) Mutations in bacteria induced by radiations. Cold Spring Harbor Symp. Quant. Biol. 11:38–50.

Dentler, W., and G. Witman (eds.) (1995) *Methods in Cell Biology,* vol. 47. *Cilia and Flagella.* Academic Press, San Diego, CA.

d'Herelle, F. (1926) *The Bacteriophage and Its Behavior.* Williams and Wilkins, Baltimore, MD.

Dobzhansky, T. (1941) *Genetics and the Origin of Species,* 2nd ed. Columbia University Press, New York.

Dodge, B. O. (1927) Nuclear phenomena associated with heterothallism and homothallism in the ascomycete *Neurospora.* J. Agr. Res. 35:289–305.

Doherty, E. A., and J. A. Doudna (2000) Ribozyme structures and mechanisms. Annu. Rev. Biochem. 69:597–615.

Doonan, J. H. (1994) Control of cell growth. In Martinelli, S. D., and J. R. Kinghorn (eds.), *Aspergillus: 50 Years On.* Elsevier, London, pp. 455–478.

Doyle, R. (1997) *On Beyond Living: Rhetorics of Vitality and Post Vitality in Molecular Biology.* Stanford University Press, Stanford, CA.

Dubos, R. J. (1945) *The Bacterial Cell. Its Relation to Problems of Virulence, Immunity and Chemotherapy.* Harvard University Press, Cambridge, MA.

Dujon, B. (1981) Mitochondrial genetics and functions. In Strathern, J. N., E. W. Jones, and J. R. Broach (eds.), *The Molecular Biology of the Yeast Sacchromyces.* Cold Spring Harbor Press, Cold Spring Harbor, NY, pp. 505–635.

Dunlap, J. C. (1999) Molecular bases for circadian clocks. Cell 96:271–290.

Dunn, L. C. (1965) *A Short History of Genetics.* McGraw-Hill, Inc., New York.

Eisen, M. B., P. T. Spellman, P. O. Brown, and D. Botstein (1998) Cluster analysis and display of genome-wide expression patterns. Proc. Natl. Acad. Sci. USA 95: 14863–14868.

Ellis, E. L. (1966) Bacteriophage: One-step growth. In Cairns, J., G. S. Stent, and J. D. Watson (eds.), *Phage and the Origins of Molecular Biology.* Cold Spring Harbor Laboratory Press, Cold Spring Harbor, NY, pp. 53–62.

Ellis, E., and M. Delbrück (1939) The growth of bacteriophage. J. Gen. Physiol. 22:365–384.

Engelsberg, E., C. Squires, and F. Meronk (1969) The arabinose operon in *Escherichia coli* B/r: a genetic demonstration of two functional states of the product of a regulator gene. Proc. Natl. Acad. Sci. USA 62:1100–1107.

Ephrussi, B. (1949) Action de l'acriflavine sur les levures. In *Unités Biologique Douées de Continuité Génétique*. CNRS, Paris, pp. 165–180.

——— (1953) *Nucleo-cytoplasmic Relations in Micro-organisms*. Oxford University Press, Oxford.

——— (1958) The cytoplasm and somatic cell variation. J. Cell. Comp. Physiol. 52:35–53.

Ephrussi, B., and H. Hottinguer (1950) Direct demonstration of the mutagenic action of euflavine on baker's yeast. Nature 166:956.

Epstein, R. H., A. Bolle, C. M Steinberg, E. Kellenberger, E. Boy de la Tour, R. Chevalley, R. S. Edgar, M. Susman, G. H. Denhardt, and A. Lielausis (1963) Physiological studies of conditional lethal mutants of bacteriophage T4D. Cold Spring Harbor Symp. Quant. Biol. 28:375–394.

Feldman, J. R. (1982) Genetic approaches to circadian clocks. Annu. Rev. Plant Physiol. 33:583–608.

Fincham, J. R. S. (1966) *Genetic Complementation*. W. A. Benjamin, New York.

Fink, G. R. (1988) Notes of a bigamous biologist. Genetics 118:547–550.

Fitch, W. M., and E. Margoliash (1967) The construction of phylogenetic trees. A generally applicable method utilizing estimates of the mutation distance obtained from cytochrome *c* sequences. Science 155: 279–284.

Fogel, S. (1993) The salad days of yeast genetics and meiotic gene conversion. In Hall, M. N., and P. Linder (eds.), *The Early Days of Yeast Genetics*. Cold Spring Harbor Laboratory Press, Cold Spring Harbor, NY, pp. 131–144.

Fox, J. L. (2000) Microbial genomes poised for progress. ASM News 66:727–731.

Fruton, J. S. (1999) *Proteins, Enzymes, Genes. The Interplay of Chemistry and Biology*. Yale University Press, New Haven, CT.

Gall, J. G. (ed.) (1986) *The Molecular Biology of Ciliated Protozoa*. Academic Press, New York.

Gamow, G. (1954) Possible relation between deoxyribonucleic acid and protein structures. Nature 173:318.

Garrod, A. E. (1909) *Inborn Errors of Metabolism*. Frowde, Hodder, and Stoughton, London.

Gavin, A.-C., M. Bosche, R. Krause, et al. (2002) Functional organization of the yeast proteome by systematic analysis of protein complexes. Nature 415:141–147.

Gilbert, S. F. (1982) Intellectual traditions in the life sciences: Molecular biology and biochemistry. Perspect. Biol. Med. 26:151–161.

Gilbert, W., and Müller-Hill (1966) Isolation of the *lac* repressor. Proc. Natl. Acad. Sci. USA 56:1891–1899.

Giles, N. H., M. E. Case, J. Baum, R. Geever, L. Huiet, V. Patel, and B. Tyler (1989) Gene organization and regulation in the *qa* (quinic acid) gene cluster in *Neurospora crassa* Microbiol. Rev. 49:338–358.

Gillham, N. W. (1993) *Organelle Genes and Genomes*. Oxford University Press, New York.

Gillham, N. W., and R. P. Levine (1962) Studies on the origin of streptomycin resistant mutants in *Chlamydomonas reinhardi*. Genetics 47:1463–1474.

Glaever, G., A. M. Chu, L. Ni, et al. (2002) Functional profiling of the *Saccharomyces cerevisiae* genome. Nature 418:387–391.

Glass, N. L., D. J. Jacobson, and P. K. Shiu (2000) The genetics of hyphal fusion and vegetative incompatibility in filamentous ascomycete fungi. Annu. Rev. Genet. 34:165–186.

Goffeau, A. (2000) Four years of post-genomic life with 6000 genes. FEBS Lett. 480:37–41.

Goffeau, A., B. G. Barrell, H. Bussey, R. W. Davis, B. Dujon, H. Feldmann, F. Galibert, J. D. Hoheisel, C. Jacq, M. Johnston, E. J. Louis, H. W. Mewes, Y. Murakami, P. Phillippsen, H. Tettelin, and S. G. Oliver (1996) Life with 6000 genes. Science 274:546–567.

Goldschmidt R. B. (1951) Chromosomes and genes. Cold Spring Harbor Symp. Quant. Biol. 16:1–12.

Graur, D., and W.-H. Li (2000) *Fundamentals of Molecular Evolution*. Sinauer Associates, Sunderland, MA.

Gray, C. H., and E. L. Tatum (1944) X-ray induced growth factor requirements in bacteria. Proc. Natl. Acad. Sci. USA 30:404–410.

Guarente, L. (1992) Messenger RNA transcription and its control in *Saccharomyces cerevisiae*. In Jones, E. W., J. R. Pringle, and J. R. Broach (eds.), *The Molecular and Cellular Biology of the Yeast Saccharomyces*, vol. *2. Gene Expression*. Cold Spring Harbor Laboratory Press, Cold Spring Harbor, NY, pp. 49–98.

Guérin, B. (1991) Mitochondria. In A. H. Rose and J. S. Harrison (eds.), *The Yeasts*, vol. IV (2nd ed.). Academic Press, New York, pp. 541–600.

Hall, M. N., and P. Linder (eds.) (1993) *The Early Days of Yeast Genetics*. Cold Spring Harbor Laboratory Press, Cold Spring Harbor, NY.

Hall, N. (1998) Metagons in killer paramecia: Problems of reproducibility and alternative hypotheses. J. Hist. Biol. 31:423–446.

Harold, F. M. (2001) *The Way of the Cell*. Oxford University Press, New York.

Harris, E. H. (1989) *The Chlamydomonas Source Book*. Academic Press, San Diego, CA.

———— (2001) *Chlamydomonas* as a model organism. Annu. Rev. Plant Physiol. Plant Mol. Biol. 52:363–406.

Hartl, D. L., and V. Orel (1992) What did Gregor Mendel think he discovered? Genetics 131:245–253.

Hartman, J. L., IV, B. Garvik, and L. Hartwell (2001) Principles for the buffering of genetic variation. Science 291:1001–1004.

Hartman, P. E. (1988) Between Novembers: Demerec, Cold Spring Harbor and the gene. Genetics 120:615–619.

Hartwell, L., and C. McLaughlin, (1968) Temperature-sensitive mutants of yeast exhibiting a rapid inhibition of protein synthesis. J. Bacteriol. 96:1664–1671.

Hartwell, L. H. (1993) Getting started in the cell cycle. In Hall, M. N., and P. Linder (eds.), *The Early Days of Yeast Genetics*. Cold Spring Harbor Laboratory Press, Cold Spring Harbor, NY, pp. 307–314.

Hartwell, L. H., J. Culotti, J. R. Pringle, and B. J. Reid (1974) Genetic control of the cell division cycle in yeast. Science 183:46–51.

Hartwell, L. H., J. J. Hopfield, S. Leibler, and A. W. Murray (1999) From molecular to modular cell biology. Nature 402 (suppl.):C47–C52.

Hawthorne, D. C. (1956) The genetics of alpha-methyl-glucoside fermentation in *Saccharomyces*. Heredity 12:273–284.

———— (1986) *Saccharomyces* studies, 1950–1960. In Hall, M. N., and P. Linder (eds.), *The Early Days of Yeast Genetics*. Cold Spring Harbor Laboratory Press, Cold Spring Harbor, NY, pp. 109–124.

Hayes, W. (1953) Observations on a transmissible agent determining sexual differentiation in *Bacterium coli*. J. Gen. Microbiol. 8:72–88.

———— (1964) *The Genetics of Bacteria and their Viruses*. John Wiley & Sons, New York.

———— (1980) Portraits of viruses: bacteriophage lambda. Intervirology 13:133–153.

Helinski, D. R., and C. Yanofsky (1962) Correspondence between genetic data and the position of amino acid alterations in a protein. Proc. Natl. Acad. Sci. USA 48:173–183.

Herriot, R. (1951/1979) Letter to A. D. Hershey. In H. F. Judson, *The Eighth Day of Creation* Simon and Schuster, NY, p. 57.

———— (1951) Nucleic-acid-free T2 virus 'ghosts' with specific biological action. J. Bacteriol. 61:752–754.

Hershey, A. D. (1946) Mutations in bacterial viruses. Cold Spring Harbor Symp. Quant. Biol. 11:67–77.

Hershey, A. D., and M. Chase (1952) Independent functions of viral protein and nucleic acid in growth of bacteriophage. J. Gen. Physiol. 36:39–56.

Hershey, A. D., and R. Rotman (1949) Genetic recombination between host-range and plaque-type mutants of bacteriophage in single bacterial cells. Genetics 34:44–71.

Herskowitz, I, J. Rine, and J. Strathern (1992) Mating-type determination and mating-type interconversion in *Saccharomyces cerevisiae*. In Jones, E. W., J. R. Pringle, and J. R. Broach (eds.), *The Molecular and Cellular Biology of the Yeast Saccharomyces*, vol. 2. *Gene Expression*. Cold Spring Harbor Laboratory Press, Cold Spring Harbor, NY, pp. 583–656.

Hess, E. L. (1970) Origins of molecular biology. Science 168:664–669.

Hill, D. L. (1972) *The Biochemistry and Physiology of Tetrahymena*. Academic Press, New York.

Hinnebusch, A. G. (1998) Mechanisms of gene regulation in the general control of amino acid biosynthesis in *Saccharomyces cerevisiae*. Microbiol. Rev. 52:248–273.

Hinnen, A., J. B. Hicks, and G. R. Fink (1978) Transformation of yeast. Proc. Natl. Acad. Sci. USA 75:1929–1933.

Holmes, F. L. (2000) Seymour Benzer and the definition of the gene. In Beurton, P., R. Falk, and H.-J. Rheinberger (eds.), *The Concept of the Gene in Development and Evolution*. Cambridge University Press, Cambridge, pp. 115–155.

Horowitz, N. H. (1950) Biochemical genetics of *Neurospora*. Adv. Genet. 3:33–71.

———— (1956) Progress in developing chemical concepts of genetic phenomena. Fed. Proc. 15:818–822.

———— (1990) George Wells Beadle. Biog. Mem. Natl. Acad. Sci. 59:26–52.

———— (1991) Fifty years ago: The *Neurospora* revolution. Genetics 127:631–635.

———— (1996) The sixtieth anniversary of biochemical genetics. Genetics 143:1–4.

Horowitz, N. H., and M. Fling (1953) Genetic determination of tyrosinase thermostability in *Neurospora*. Genetics 38:360–374.

Horowitz, N. H., and U. Leupold (1951) Some recent studies bearing on the one gene-one enzyme hypothesis. Cold Spring Harbor Symp. Quant. Biol. 16:65–74.

Ideker, T. , T. Galitsky, and L. Hood (2001a) A new approach to decoding life: systems biology. Annu. Rev. Genomics Hum. Genet. 2:343–372.

Ideker, T., V. Thorsson, J. A. Ranish, R. Christmas, J. Buhler, J. K. Eng, R. Bumgartner, D. R. Goodlett, R. Aebersopld, and L. Hood (2001b) Integrated genomic and proteomic analyses of a systematically perturbed metabolic network. Science 292:929–934.

Ingram V. M. (1957) Gene mutations in human haemoglobin: the chemical difference between normal and sickle cell haemoglobin. Nature 180:326–328.

Jacob, F. (1988) *The Statue Within* [*La statue intérieure*]. Trans. P. Franklin. Basic Books, New York.

———— (1998) *Of Flies, Mice, and Men* [*La souris, la mouche et l'homme*]. Trans. G. Weiss. Harvard Univ. Press, Cambridge, MA.

Jacob, F., and J. Monod (1961) On the regulation of gene activity. Cold Spring Harbor Symp. Quant. Biol. 26:193–211.

Jacob, F., and E. L. Wollman (1961) *Sexuality and the Genetics of Bacteria.* Academic Press, New York.

Jeong, H., S. P. Mason, A.-L. Barabási, and Z. N. Oltvai (2001) Lethality and centrality in protein networks. Nature 411:41–42.

Jinks, J. L. (1964) *Extrachromosomal Inheritance.* Prentice-Hall, Englewood Cliffs, NJ.

Johnson, D. I. (1999) Cdc42p: an essential rho-type GTPase controlling eukaryotic cell polarity. Microbiol. Mol. Biol. Rev. 63:54–105.

Jones, E. W. (1991) Three proteolytic systems in the yeast *Saccharomyces cerevisiae.* J. Biol. Chem. 266:7963–7966.

Jones, E. W., J. R. Pringle, and J. R. Broach (eds.) (1992) *The Molecular and Cellular Biology of the Yeast Saccharomyces,* vol. 2. *Gene Expression.* Cold Spring Harbor Laboratory Press, Cold Spring Harbor, NY.

Judson, H. F. (1979) *The Eighth Day of Creation. Makers of the Revolution in Biology.* Simon and Schuster, New York.

Kaiser, C. A., R. E. Gimeno, and D. A. Shaywitz (1997) Protein secretion, membrane biogenesis, and endocytosis. In Pringle, J. R., J. R. Broach, and E. W. Jones (eds.), *The Molecular and Cellular Biology of the Yeast Saccharomyces,* vol. 3. *Cell Cycle and Cell Biology.* Cold Spring Harbor Laboratory Press, Cold Spring Harbor, NY, pp. 91–227.

Kay, L. E. (1985) Conceptual models and analytical tools: The biology of physicist Max Delbrück. J. Hist. Biol. 18:207–246.

——— (1993) *The Molecular Vision of Life. Caltech, The Rockefeller Foundation, and the Rise of the New Biology.* Oxford University Press, New York.

——— (1996) Life as technology: Representing, intervening, and molecularizing. In Sarkar, S. (ed.), *The Philosophy and History of Modern Biology: New Perspectives.* Kluwer Academic Publishers, Boston, MA, pp. 87–100.

——— (2000) *Who Wrote the Book of Life? A History of the Genetic Code.* Stanford University Press, Stanford, CA.

Keller, E. F. (1983) *A Feeling for the Organism: The Life and Work of Barbara McClintock.* W. H. Freeman, New York.

——— (1992) *Secrets of Life, Secrets of Death.* Routledge, New York.

——— (1995) *Refiguring Life. Metaphors of Twentieth-Century Biology.* Columbia University Press, New York.

——— (2000a) Decoding the genetic program. Or, some circular logic in the logic of circularity. In Beurton, P., R. Falk, and H.-J. Rheinberger (eds.), *The Concept of the Gene in Development and Evolution.* Cambridge University Press, Cambridge, pp. 159–177.

——— (2000b) *The Century of the Gene.* Harvard University Press, Cambridge, MA.

Kelly, T. J., Jr., and H. O. Smith (1970) A restriction enzyme from *Hemophilus influenzae.* II Base sequence of the recognition site. J. Mol. Biol. 51:393–409.

Kendrew, J. C. (1967) How molecular biology started. Sci. Am. 216:141–144.

Kimmelman (1992) Organisms and interests in scientific research: R. A. Emerson's claims for the unique contributions of agricultural genetics. In Clarke, A. E., and J. H. Fujimura (eds.), *The Right Tools for the Job. At Work in Twentieth-Century Life Sciences.* Princeton Univ. Press, Princeton, NJ, pp. 198–232.

Kitano, H. (2002) Systems biology: A brief overview. Science 295:1662–1664.

Klionsky, D. J., P. K. Herman, and S. D. Emr (1990) The fungal vacuole: composition, function, and biogenesis. Microbiol. Rev. 54:266–292.

Kohler, R. E. (1993) *Drosophila*: a life in the laboratory. J. Hist. Biol. 26:281–310.

Kornberg, A. (1976) For the love of enzymes. In Kornberg, A., B. L. Horecker, L. Cornudella, and J. Oro (eds), *Reflections on Biochemistry. In Honor of Severo Ochoa*. Pergamon Press, Oxford, pp. 243–251.

Kornberg, A., and T. Baker (1992) *DNA Replication* (2nd ed.) W. H. Freeman, San Francisco, CA.

Kowalczykowski, S. C., D. A. Dixon, A. K. Eggleston, S. D. Lauder, and W. M. Rehrauer (1994) Biochemistry of homologous recombination in *Escherichia coli*. Microbiol. Rev. 58:401–465.

Kozloff, L. M. (1984) Phage biochemistry and the origin of molecular biology. Trends Biochem. Sci. 9:422–423.

Lederberg, E. M. (1951) Lysogenicity in *E. coli* K-12 [Abstr.]. Genetics 36:560.

Lederberg, E. M., and J. Lederberg (1953) Genetic studies of lysogenicity in *Escherichia coli*. Genetics 38:51–64.

Lederberg, J. (1951) Genetic studies with bacteria. In Dunn, L. C. (ed.), *Genetics in the Twentieth Century*. Macmillan, New York, pp. 263–289.

——— (1986) Forty years of genetic recombination in bacteria. A fortieth anniversary reminiscence. Nature 324:626–628.

——— (1987) Genetic recombination in bacteria: A discovery account. Annu. Rev. Genet. 21:23–46.

——— (1990) Edward Lawrie Tatum. Biog. Mem. Natl. Acad. Sci. USA 59:356–386.

——— (1998) *Escherichia coli*. In R. Bud and D. J. Warner (eds.), *Instruments of Science. An Historical Encyclopedia*. Garland Publishing, New York, pp. 230–232.

Lederberg, J., and E. M. Lederberg (1952) Replica plating and indirect selection of bacterial mutants. J. Bacteriol. 63:399–406.

Lederberg, J., E. M. Lederberg, N. D. Zinder, and E. R. Lively (1951) Recombination analysis of bacterial heredity. Cold Spring Harbor Symp. Quant. Biol. 16:413–443.

Lederberg, J., and E. L. Tatum (1946) Novel genotypes in mixed cultures of biochemical mutants of bacteria. Cold Spring Harbor Symp. Quant. Biol. 11:113–114.

Lederberg, J., and N. D. Zinder (1948) Concentration of biochemical mutants of bacteria with penicillin. J. Am. Chem. Soc. 70:4267–4268.

Lennox, E. S. (1955) Transduction of linked genetic characters of the host by bacteriophage PI. Virology 1:190–208.

Levine, R. P. (1968) Genetic dissection of photosynthesis in *Chlamydomonas reinhardtii*. Science 178:768–771.

Levine, R. P., and W. T. Ebersold (1960) The genetics and cytology of *Chlamydomonas*. Annu. Rev. Microbiol. 14:197–216.

Lew, D. J., T. Weinert, and J. R. Pringle (1997) Cell cycle control in *Saccharomyces cerevisiae*. In Pringle, J. R., J. R. Broach, and E. W. Jones (eds.), *The Molecular Biology of the Yeast Saccharomyces*, vol. 3. *Cell Cycle and Cell Biology*. Cold Spring Harbor Laboratory Press, Cold Spring Harbor, NY, pp. 607–695.

Lewis, E. B. (1950) The phenomenon of position effect. Adv. Genet. 3:73–115.

Lewontin, R. C. (1974) *The Genetic Basis of Evolutionary Change*. Columbia Univ. Press, New York.

Lindegren, C. C. (1931) *Genetic Study of Sex and Cultural Characters in Neurospora crassa*. Ph.D. Thesis, California Institute of Technology.

——— (1942) The use of the fungi in modern genetical analysis. Iowa State Coll. J. Sci. 16:271–290.

——— (1949) *The Yeast Cell, Its Genetics and Cytology*. Educational Publishers, St. Louis, MO.

———— (1952) The structure of the yeast cell. Symp. Soc. Exptl. Biol. 6:277–289.

———— (1955) Non-Mendelian segregation in a single tetrad of *Saccharomyces* ascribed to gene conversion. Science 121:605–607.

———— (1957) The integrated cell. Cytologia 22:415–441.

———— (1958) Priority in gene conversion. Experientia 14:444–445.

———— (1966) *The Cold War in Biology*. Planarian Press, Ann Arbor, MI.

———— (1973) Reminiscences of B. O. Dodge and the beginnings of Neurospora genetics. Neurospora Newslett. 20:13–14.

Lindegren, C. C., and G. Lindegren (1946) The cytogene theory. Cold Spring Harbor Symp. Quant. Biol. 11:115–129.

Lockhart, D. J., and E. A. Winzeler (2000) Genomics, gene expression and DNA arrays. Nature 405:827–836.

Logan, C. (2001) "Are Norway Rats . . . Things?": Diversity versus generality in the use of albino rats in experiments on development and sexuality. J. Hist. Biol. 34:287–314.

Luria, S. E. (1953) Host-induced modification of viruses. Cold Spring Harbor Symp. Quant. Biol. 18:237–244.

———— (1984) *A Slot Machine, A Broken Test Tube*. Harper & Row, New York.

Luria, S. E., and T. F. Anderson (1942) The identification and characterization of bacteriophages with the electron microscope. Proc. Natl. Acad. Sci. USA 28:127–130.

Luria, S. E., and M. Delbrück (1943) Mutations of bacteria from virus sensitivity to virus resistance. Genetics 28:491–511.

Lwoff, A. (1949) Les organites doués de continuité génétique chez les Protistes. In *Unités Biologiques Douées de Continuité Génétique*. CNRS, Paris, pp. 7–23.

———— (1950) *Problems of Morphogenesis in Ciliates. The Kinetosomes in Development, Reproduction, and Evolution*. John Wiley & Sons, New York.

Lwoff, A., and A. Gutmann (1950) Recherches sur un *Bacillus megatherium* lysogène. Ann. Inst. Pasteur 78:711–739. [Translated with modifications and reprinted in Stent, G. S., *Papers on Bacterial Viruses* (2nd ed.), Little Brown, Boston, MA, 1965, pp. 316–335.]

Lwoff, A., L. Siminovitch, and N. Kjeldgaard (1950) Induction de la lyse bactériophagique de la totalité d'une population microbiènne lysogène. C. R. Acad. Sci. 231:190–191.

Maaløe, O., and N. O. Kjeldgaard (1966) *Control of Macromolecular Synthesis*. W. A. Benjamin, New York.

Maas, W. K. (1991) The regulation of arginine biosynthesis: Its contribution to understanding the control of gene expression. Genetics 128:489–494.

MacKendrick, M. E., and G. Pontecorvo (1952) Crossing over between alleles at the W locus in *Drosophila melanogaster*. Experientia 8:390.

Maddox, B. (2002) *Rosalind Franklin: The Dark Lady of DNA*. Harper Collins, New York.

Magner, L. N. (1993) *A History of the Life Sciences* (2nd ed.). Marcel Dekker, New York.

Maleszka, R., and G. D. Clark-Walker (1992) In vivo conformation of mitochondrial DNA in fungi and zoosporic moulds. Curr. Genet. 22:341–344.

Margoliash, E., and E. L. Smith (1965) Structural and functional aspects of cytochrome *c* in evolution. In Bryson, V., and H. J. Vogel (eds.), *Evolving Genes and Proteins*. Academic Press, New York, pp. 221–242.

Margulis, L. (1981) *Symbiosis in Cell Evolution*. W. H. Freeman, San Francisco, CA.

Martinelli, S. D. (1994a) *Aspergillus nidulans* as an experimental organism. In

Martinelli, S. D., and J. R. Kinghorn (eds.), *Aspergillus: 50 Years On.* Elsevier, London, pp. 33–58.

——— (1994b) Pedigree of the authors. In Martinelli, S. D., and J. R. Kinghorn (eds.), *Aspergillus: 50 Years On.* Elsevier, London, pp. 25–31.

Martinelli, S. D., and J. R. Kinghorn (eds.) (1994) *Aspergillus: 50 Years On.* Elsevier, London.

Matile, P., H. Moor, and C. F. Robinow (1969) Yeast cytology. In Rose, A. H., and J. S. Harrison (eds.), *The Yeasts*, vol. 1. Academic Press, London, pp. 219–302.

Mayr, E. (1942) *Systematics and the Origin of Species.* Columbia University Press, New York.

McCann, J., and B. N. Ames (1976) Detection of carcinogens as mutagens in the Salmonella/microsome test assay of 100 chemicals: Discussion. Proc. Natl. Acad. Sci. USA 73:950–954.

McClintock, B. (1945) *Neurospora.* I. Preliminary observations of the chromosomes of *Neurospora crassa.* Am. J. Bot. 32:671–678.

——— (1961) Some parallels between gene control systems in maize and bacteria. Am. Nat. 95:265–277.

Mendel, G. (1866) Versuche über Pflanzenhybriden. Verh. Naturforsch. Verein Brünn 4:3–47.

Meselson, M. (1964) On the mechanism of genetic recombination between DNA molecules. J. Mol. Biol. 9:734–745.

Meselson, M., and F. W. Stahl (1958) The replication of DNA in *Escherichia coli.* Proc. Natl. Acad. Sci. USA 44:671–682.

Meselson, M., and J. J. Weigle (1961) Chromosome breakage accompanying genetic recombination in bacteriophage λ. Proc. Natl. Acad. Sci. USA 47:857–868.

Michaelis, P. (1954) Cytoplasmic inheritance in *Epilobium* and its theoretical significance. Adv. Genetics 6:287–401.

Mitchell, H. K., and J. Lein (1948) A *Neurospora* mutant deficient in the enzymatic synthesis of tryptophan. J. Biol. Chem. 175:481–482.

Mitchell, M. B. (1955a) Aberrant recombination of pyridoxine mutants of *Neurospora.* Proc. Natl. Acad. Sci. USA 41:215–220.

——— (1955b) Further evidence of aberrant recombination in *Neurospora.* Proc. Natl. Acad Sci. USA 41:935–937.

Monod, J. (1972) *Chance and Necessity.* Vintage Books, New York.

Moor, H., and K. Mühlethaler (1963) Fine structure in frozen-etched yeast cells. J. Cell Biol. 17:609–628.

Morange, M. (1998) *History of Molecular Biology.* Harvard University Press, Cambridge, MA.

Morgan, T. H. (1926) *The Theory of the Gene.* Yale Univ. Press, New Haven, CT. [reprint of revised ed., Hafner Publishing, New York, 1964].

——— (1934) *Embryology and Genetics.* Columbia University Press, New York.

Morgan, T. H., and C. B. Bridges (1919) The origin of gyanandromorphs. Carnegie Inst. Wash. Pub. 278:1–22.

Morgan, T. H., A. H. Sturtevant, H. J. Muller, and C. B. Bridges (1915) *The Mechanism of Mendelian Heredity.* Constable and Company, London. [reprinted with an introduction by Garland Allen by Johnson Reprint Corporation, New York, 1972].

Morris, N. R. (2000) Nuclear migration: From fungi to the mammalian brain. J. Cell Biol. 148:1097–1101.

Morris, N. R., G. Sheir-Neiss, and B. R. Oakley (1982). The biochemical genetics of mitosis in *Aspergillus nidulans*. In: P. Cappuccinelli and N.R. Morris (eds.), *Microtubules in Microorganisms*. Marcel Dekker, New York, pp. 257–273.

Morris, N. R., X. Xiang, S. Osmani, A. Osmani, S. Beckwith, Y.-H.Chiu, C. Roghi, D. Willins, G. Goldman, M. Xin (1995) Analysis of nuclear migration in *Aspergillus nidulans*. Cold Spring Harbor Symp. Quant. Biol. 60:813–820.

Morse, M., E. M. Lederberg, and J. Lederberg (1956) Transduction in *Escherichia coli* K-12. Genetics 41:121–156.

Mortimer, R. K. (1993a) Øjvind Winge: founder of yeast genetics. In Hall, M. N., and P. Linder (eds.), *The Early Days of Yeast Genetics*. Cold Spring Harbor Laboratory Press, Cold Spring Harbor, NY, pp. 3–16.

———— (1993b) Carl C. Lindegren: iconoclastic father of *Neurospora* and yeast genetics. In Hall, M. N., and P. Linder (eds.), *The Early Days of Yeast Genetics*. Cold Spring Harbor Laboratory Press, Cold Spring Harbor, NY, pp. 17–38.

———— (1993c) Some recollections on forty years of research in yeast genetics. In Hall, M. N., and P. Linder (eds.), *The Early Days of Yeast Genetics*. Cold Spring Harbor Laboratory Press, Cold Spring Harbor, NY, pp. 173–185.

Muller, H. J. (1922) Variation due to change in the individual gene. Am. Nat. 56:32–50.

———— (1927) Artificial transmutation of the gene. Science 66:84–87.

Nanney, D. L. (1986) Introduction. In Gall, J. G. (ed.), *The Molecular Biology of Ciliated Protozoa*. Academic Press, New York, pp. 1–26.

Nanney, D. L., and E. M. Simon (2000) Laboratory and evolutionary history of *Tetrahymena thermophila*. Meth. Cell Biol. 62:3–25.

National Research Council Committee on Models for Biomedical Research (1985) *Models for Biomedical Research. A New Perspective*. National Academy Press, Washington, D.C.

Nelkin, D. (2001) Molecular metaphors: the gene in popular discourse. Nature Rev. Genet. 2:555–559.

Nirenberg, M. W., and H. Matthaei (1961) The dependence of cell-free protein synthesis in *E. coli* upon naturally occurring or synthetic polyribonucleotides. Proc. Natl. Acad. Sci. USA 47:1588–1602.

Nomura, M. (1997) Reflections on the days of ribosome reconstitution research. Trends Biochem. Sci. 22:275–279.

Novick, P., C. Field, and R. Schekman (1980) Identification of 23 complementation groups required for post-translational events in the yeast secretory pathway. Cell 21:205–215.

Ochman, H., J. G. Lawrence, and E. A. Groisman (2000) Lateral gene transfer and the nature of bacterial innovation. Nature 405:299–304.

Olby, R. (1974) *The Path to the Double Helix*. Macmillan, London.

———— (1994) *The Path to the Double Helix*, 2nd ed. Macmillan, London.

Osiewacz, H. D. (ed.) (2002) *Molecular Biology of Fungal Development*. Marcel Dekker, New York.

Ovádi, J. (1991) Physiological significance of metabolic channeling. J. Theoret. Biol. 152:1–22.

Pace, N. R. (1997) A molecular view of microbial diversity and the biosphere. Science 276:734–740.

Pandey, A., and M. Mann (2000) Proteomics to study genes and genomes. Nature 405:837–846.

Pardee, A. B., F. Jacob, and J. Monod (1959) The genetic control and cytoplasmic expression of "inducibility" in the synthesis of β-galactosidase by *E. coli*. J. Mol. Biol. 1:165–178.

Pascher, A. (1916) Über die Kreutzung einzelliger, haploider Organismen: *Chlamydomonas*. Ber. Deutch. Bot. Ges. 34:228–242.

——— (1918) Über die Beziehung der Reduktionssteilung zur Mendelschen Spaltung. Ber. Deutch. Bot. Ges. 36:163–168.

Pauling, L., and R. B. Corey (1950) Two hydrogen-bonded spiral configurations of the polypeptide chain. J. Am. Chem. Soc. 72:5349.

Pauling, L., H. Itano, S. J. Singer, and I. C. Wells (1949) Sickle cell anemia, a molecular disease. Science 110:543–548.

Penman, S. (2001) What are genes, anyway? Am. Sci. 80:66–67.

Perkins, D. D. (1992) *Neurospora*: the organism behind the molecular revolution. Genetics 130:687–701.

——— (1997) Chromosome rearrangements in *Neurospora* and other filamentous fungi. Adv. Genet. 36:239–398.

Perkins, D. D., and E. G. Barry (1977) The cytogenetics of *Neurospora*. Adv. Genet. 19:133–285.

Perkins, D. D., and R. H. Davis (2000) *Neurospora* at the millennium. Fung. Genet. Biol. 31:153–167.

Pomper, S., and P. R. Burkholder (1949) Studies on the biochemical genetics of yeast. Proc. Natl. Acad Sci. USA 35:456–464.

Pontecorvo, G. (1946a) Microbiology, biochemistry and the genetics of micro-organisms. Nature 157:95–96.

——— (1946b) Genetic systems based on heterokaryosis. Cold Spring Harbor Symp. Quant. Biol. 11:193–201.

——— (1948) Genetical technique for self-fertile (homothallic) microorganisms [abstr.]. Proc. Intl. Cong. Genet. 8:642–643.

——— (1954) Genetical analysis of cell organization. Symp. Soc. Exptl. Biol. 6:218–229.

——— (1956) Allelism. Cold Spring Harbor Symp. Quant. Biol. 21:171–174.

——— (1994) Foreward. In Martinelli, S. D., and J. R. Kinghorn (eds.), *Aspergillus:50 Years On*. Elsevier, London, pp. xxiii–xxv.

Pontecorvo, G., E. Forbes, and O. B. Adam (1949) Genetics of the homothallic ascomycete *Aspergillus nidulans* [abstr.]. Heredity 3:385.

Pontecorvo, G., and A. R. Gemmell. (1944) Genetic proof of heterokaryosis in *Penicillium notatum*. Nature 154:514–515.

Pontecorvo, G., J. A. Roper, L. M. Hemmons, K. D. Macdonald, and A. W. J. Bufton (1953) The genetics of *Aspergillus nidulans*. Adv. Genet. 5:141–238.

Porter, K. R., and J. Blum (1953) A study in microtomy for electron microscopy. Anat. Rec. 117:685–708.

Preer, J. R., Jr. (1997) Whatever happened to *Paramecium* genetics? Genetics 145:217–225.

——— (2000) Epigenetic mechnisms affecting macronuclear development in *Paramecium* and *Tetrahymena*. J. Eukaryot. Microbiol. 47:515–524.

Pringle, J. R., E. W. Jones, and J. R. Broach (eds.) (1997) *The Molecular and Cellular Biology of the Yeast Saccharomyces*, vol. 3. *Cell Cycle and Cell Biology*. Cold Spring Harbor Laboratory Press, Cold Spring Harbor, NY.

Pruyne, D., and A. Bretscher (2000) Polarization of cell growth in yeast. J. Cell Sci. 113:571–585.

Ptashne, M. (1967) Isolation of the λ phage repressor. Proc. Natl. Acad. Sci. USA 57:306–313.

——— (1986) *A Genetic Switch. Gene Control and Phage* λ. Blackwell Scientific, Cambridge, MA.

Quackenbush, R. L. (1988) Endosymbionts of killer paramecia. In Görtz, H.-D. (ed.), *Paramecium*. Springer-Verlag, Berlin, pp. 406–418.

Rabinowitz, M., and H. Swift (1970) Mitochondrial nucleic acids and their relation to the biogenesis of mitochondria. Physiol. Rev. 50:376–427.

Rachmeler, M., and C. Yanofsky (1958) The exclusion of free indole as an intermediate of the biosynthesis of tryptophan in *Neurospora crassa*. Biochim. Biophys. Acta 28:640–641.

Raffel, D., and H. J. Muller (1940) Position effect and gene divisibility considered in connection with three strikingly similar scute mutations. Genetics 25:541–583.

Rao, R., and C. W. Slayman (1996) Plasma membrane and related ATPases. In Brambl, R., and G. A. Marzluf (eds.), *The Mycota*, vol. 3. *Biochemistry and Molecular Biology*. Springer-Verlag, Berlin, pp. 29–56.

Raper, J. R. (1966) Life cycles, basic patterns of sexuality, and sexual mechanisms. In Ainsworth, G. C., and A. S. Sussman (eds.), *The Fungi. An Advanced Treatise*, vol. 2. *The Fungal Organism*. Academic Press, New York, pp. 473–511.

Ratzkin, B., and J. Carbon (1977) Functional expression of a cloned yeast DNA in *Escherichia coli*. Proc. Natl. Acad. Sci. USA 74:487–491.

Ravasz, E., A. L. Somera, D. A. Mongru, Z. N. Oltvai, and A.-L. Barabási (2002) Hierarchical organization in metabolic networks. Science 297:1551–1555.

Remacle, C., and R. Matagne (1997) Mitochondrial genetics. In Rochaix, J.-D., M. Goldschmidt-Clermont, and S. Merchant (eds.), *Molecular Biology of Chloropasts and Mitochondria in Chlamydomonas*. Kluwer Academic, Boston, MA, pp. 13–23.

Rheinberger, H.-J. (1996) Comparing experimental systems: Protein synthesis in microbes and in animal tissue at Cambridge (Ernest F. Gale) and the Massachusetts General Hospital (Paul C. Zamecnik), 1945–1960. J. Hist. Biol. 29:387–417.

Riley, M., and M. H. Serres (2000) Interim report on genomics of *Escherichia coli*. Annu. Rev. Microbiol. 54:341–411.

Robbins, W. J. (1962) Bernard Ogilvie Dodge. Biog. Mem. Natl. Acad. Sci. USA 33:31–35.

Roberts, C. J., B. Nelson, M. J. Marton, R. Stoughton, M. R. Meyer, H. A. Bennett, Y. D. He, H. Dai, W. L. Walker, T. R. Hughes, M. Tyers, C. Boone, and S. H. Friend (2000) Signal circuitry of multiple MAPK pathways revealed by a matrix of global gene expression profiles. Science 287:873–880.

Roberts, R. B., P. H. Abelson, D. B. Cowie, E. T. Bolton, and R. J. Britten (1963) *Studies of Biosynthesis in* Escherichia coli. Carnegie Institution of Washington Publication 607. Washington, DC.

Rochaix, J.-D., M. Goldschmidt-Clermont, and S. Merchant (eds.) (1997) *The Molecular Biology of Chloropasts and Mitochondria in Chlamydomonas*. Kluwer Academic, Boston, MA.

Roman, H. (1956) Studies of gene mutation in *Saccharomyces cerevisiae*. Cold Spring Harbor Symp. Quant. Biol. 21:175–185.

——— (1986) The early days of yeast genetics: a personal narrative. Annu. Rev. Genet. 20:1–12.

Roman, H., D. C. Hawthorne, and H. C. Douglas (1951) Polyploidy in yeast and its bearing on the occurrence of irregular genetic ratios. Proc. Natl. Acad. Sci. USA 37:79–84.

Roper, J. A. (1950) Search for linkage between genes determining a vitamin requirement. Nature 166:956–957.

——— (1952) Production of heterozygous diploids in filamentous fungi. Experientia 8:14–15.

Rosenberg, A. (1994) *Instrumental Biology or the Disunity of Science*. University Chicago Press, Chicago.

Ross-Macdonald, P., S. R. Coelho, T. Roemer, S. Agarwal, A. Kumar, R. Jansen, K.-H. Cheung, A. Sheehan, D. Symonaitis, L. Umnansky, M. Heldtman, F. K. Nelson, H. Iwasaki, K. Hager, M. Gerstein, P. Miller, G. S. Roeder, and M. Snyder (1999) Large-scale analysis of the yeast genome by transposon tagging and gene disruption. Nature 402:413–418.

Rutherford, S. (2000) From genotype to phenotype: buffering mechanisms and the storage of genetic information. BioEssays 22:1095–1105.

Ryan, F. J. (1950) Selected methods of *Neurospora* genetics. Meth. Med- Res. 3:51–75.

———— (1955) Attempt to reproduce some of Moewus' experiments on *Chlamydomonas* and *Polytoma*. Science 122:470.

Sager, R. (1954) Mendelian and non-Mendelian inheritance of streptomycin resistance in *Chlamydomonas reinhardi*. Proc. Natl. Acad. Sci. USA 40:356–363.

———— (1960) Genetic systems in *Chlamydomonas*. Science 132:1459–1465.

———— (1962) Streptomycin as a mutagen for nonchromosomal genes. Proc. Natl. Acad. Sci. USA 48:2018–2026.

———— (1972) *Cytoplasmic Genes and Organelles*. Academic Press, New York.

Sager, R., and Z. Ramanis (1963) The particulate nature of nonchromosomal genes in *Chlamydomonas*. Proc. Natl. Acad. Sci. USA 50:260–268.

———— (1965) Recombination of nonchromosomal genes in *Chlamydomonas*. Proc. Natl. Acad. Sci. USA 53:1053–1061.

Sanderson, K. E., and RP. R. MacLachlan (1987) F-mediated conjugation, F+ strains, and Hfr Strains of *Salmonella typhpimurium* and *Salmonella ebony*. In Neidhart, F. C. (ed.), *Cellular and Molecular Biology*, vol. 2. *Escherichia coli and Salmonella typhimurium*. American Society of Microbiology Press, Washington, DC, pp. 1138–1144.

Sanger, F. (1959) Chemistry of insulin: Determination of the structure of insulin opens the way to greater understanding of life processes. Science 129:1340–1344.

Sapp, J. (1987) *Beyond the Gene. Cytoplasmic Inheritance and the Struggle for Authority in Genetics*. Oxford University Press, New York.

———— (1990) *Where the Truth Lies: Franz Moewus and the Origins of Molecular Biology*. Cambridge University Press, New York.

Sayre, A. (1975) *Rosalind Franklin and DNA*. W. W. Norton, New York.

Schaechter, M., and the View from Here Group (2001) *Escherichia coli* and *Salmonella* 2000: the view from here. Microbiol. Mol. Biol. Rev. 65:119–130.

Schatz, G. S., E. Haslbrunner, and H. Tuppy (1964) Deoxyribonucleic acid associated with yeast mitochondria. Biochim. Biophys. Acta 15:127–132.

Schrödinger, E. (1944) *What Is Life?* [reprinted in Schrödinger, E., *What Is Life? and Other Scientific Essays*. Doubleday, Garden City, NY, 1956].

Scott-Moncrieff, R. (1936) A biochemical survey of some Mendelian factors for flower-color. J. Genet. 32:117–170.

Shapiro, J. A. (ed.) (1983) *Mobile Genetic Elements*. Academic Press, New York.

Shear, C. L., and B. O. Dodge (1927) Life histories and heterothallism of the red bread-mold fungi of the *Monilia sitophila* group. J. Agr. Res. 34:1019–1042.

Sherman, F., and J. W. Stewart (1982) Mutations altering initiation of translation of yeast iso-1-cytochrome *c*; contrasts between the eukaryotic and prokaryotic initiation process. In Strathern, J. N., E. W. Jones, and J. R. Broach (eds.), *The Molecular Biology of the Yeast* Saccharomyces. *Metabolism and Gene Expression*. Cold Spring Harbor Laboratory Press, Cold Spring Harbor, NY, pp. 301–360.

Siekevitz, P. (1995) Protein synthesis and the ribosome. In Ord, M. G., and L. A. Stocken (eds), *Quantum Leaps in Biochemistry*. JAI Press, Greenwich, CT, pp. 109–131.

Slayman, C. L. (1970) Movement of ions and electrogenesis in microorganisms. Am. Zool 10:377–392.

Smith, H. O. (1979) Nucleotide sequence specificity of restriction endonucleases. Science 205:455–462.

Snyder, L., and W. Champness (1996) *Molecular Genetics of Bacteria*. American Society for Microbiology Press, Washington, DC.

Sonneborn, T. (1937) Sex, sex inheritance, and sex determination in *Paramecium aurelia*. Proc. Natl. Acad. Sci. USA 23:378–395.

———— (1950) The cytoplasm in heredity. Heredity 4:11–36.

———— (1951) Some current problems of genetics in the light of investigations on *Chlamydomonas* and *Paramecium*. Cold Spring Harbor Symp. Quant. Biol. 16:483–503.

Sonneborn, T. M., and G. H. Beale (1949) Influence des gènes, des plasmagènes et du milieu dans le détermination de caractères antigéniques chez *Paramecium aurelia* (variété 4). In CNRS, *Unités Biologique Douées de Continuité Génétique*. CNRS, Paris, pp. 25–32.

Spellman, P. T., G. Sherlock, M. Q. Zhang, V. R. Iyer, K. Anders, M. B. Eisen, P. O. Brown, D. Botstein, and B. Futcher (1998) Comprehensive identification of cell cycle-regulated genes of the yeast *Saccharomyces cerevisiae* by microarray hybridization. Mol. Biol. Cell 9:3273–3297.

Spiegelman, S., C. C. Lindegren, and G. Lindegren (1945) Maintenance and increase of a genetic character by a substrate-cytoplasmic interaction in the absence of the specific gene. Proc. Natl. Acad. Sci. USA 31:95–102.

Srb, A. M., and N. H. Horowitz (1944) The ornithine cycle in Neurospora and its genetic control. J. Biol. Chem. 154:129–139.

Srere, P. A. (1987) Complexes of sequential metabolic enzymes. Annu. Rev. Biochem. 56:89–124.

Stadler, L. J. (1954) The gene. Science 120:811–819.

Stahl, F. W. 1995. The amber mutants of phage T4. Genetics 141:439–442.

———— (1998) Recombination in phage λ: one geneticist's historical perspective. Gene 223:95–102.

———— (ed.) (2000) *We Can Sleep Later. Alfred D. Hershey and the Origins of Molecular Biology*. Cold Spring Harbor Laboratory Press, Cold Spring Harbor, NY.

Stanier, R. Y., and C. B. van Niel (1962) The concept of a bacterium. Arch. Mikrobiol. 42:17–35.

Stent, G. S. (1963) *Molecular Biology of Bacterial Viruses*. W. H. Freeman, San Francisco, CA.

———— (1968) That was the molecular biology that was. Science 160:390–395.

———— (1969) *The Coming of the Golden Age: A View of the End of Progress*. Natural History Press, Garden City, NJ.

———— (ed.) (1965) *Papers on Bacterial Viruses* (2nd ed.). Little Brown, Boston, MA.

Strathern, J. N., E. W. Jones, and J. R. Broach (eds.) (1981) *The Molecular Biology of the Yeast Saccharomyces. Life Cycle and Inheritance*. Cold Spring Harbor Laboratory Press, Cold Spring Harbor, NY.

———— (eds.) (1982) *The Molecular Biology of the Yeast* Saccharomyces. *Metabolism and Gene Expression*. Cold Spring Harbor Laboratory Press, Cold Spring Harbor, NY.

Sturtevant, A. H. (1913) The linear arrangement of six sex-linked factors in Drosophila, as shown by their mode of association. J. Exptl. Zool. 14:43–59.

——— (1920) The vermilion gene and gynandromorphism. Proc. Soc. Exptl. Biol. Med. 17:70–71.

——— (1965) *A History of Genetics*. Harper and Row, NY.

Sturtevant, A. H., and G. W. Beadle (1939) *An Introduction to Genetics*. W. B. Saunders, NY [reprinted by Dover Publications, New York, 1962].

Su, S. S. Y., and P. Nurse (1997) Cell cycle control in fission yeast. In Pringle, J. R., J. R. Broach, and E. W. Jones (eds.), *The Molecular Biology of the Yeast* Saccharomyces, vol. 3. *Cell Cycle and Cell Biology*. Cold Spring Harbor Laboratory Press, Cold Spring Harbor, NY, pp. 697–763.

Summers, W. C. (1993) How bacteriophage came to be used by the phage group. J. Hist. Biol. 26:255–267.

Suskind, S. R., C. Yanofsky, and D. M. Bonner (1955) Allelic strains of *Neurospora* lacking tryptophan synthetase: a preliminary immunochemical characterization. Proc. Natl. Acad. Sci. USA 42:577–582.

Susman, M. (1995) The Cold Spring Harbor Phage Course (1945–1970): A 50th anniversary remembrance. Genetics 139:1101–1106.

Sutton, W. S. (1903) The chromosomes in heredity. Biol. Bull. 4:231–251.

Szostak, J. W., T. Orr-Weaver, R. Rothstein, and F. Stahl (1983) The double-strand-break repair model for recombination. Cell 33: 25–35.

Tatum, E. L. (1945) X-ray induced mutant strains of *Escherichia coli*. Proc. Natl. Acad. Sci. USA 31:215–219.

Tatum, E. L., R. W. Barratt, and V. M. Cutter (1949) Chemical induction of colonial paramorphs in *Neurospora* and *Syncephalastrum*. Science 109:509–511.

Thieffry, D. (1996) *Escherichia coli* as a model system with which to study cell differentiation. Hist. Phil. Life Sci. 18:163–193.

——— (1997) Contributions of the 'Rouge-Cloître Group' to the notion of 'messenger RNA.' Hist. Phil. Life Sci. 19:89–111.

Thomas, C. A. (1963) The arrangement of nucleotide sequences in T2 and T5 DNA molecules. Cold Spring Harbor Symp. Quant. Biol. 28:395–396.

Timberlake, W. E. (1990) Molecular genetics of *Aspergillus* development. Annu. Rev. Genet. 24:5–36.

Timberlake, W. E. (ed.) (1985) *Molecular Genetics of Filamentous Fungi*. Alan R. Liss, New York.

Timberlake, W. E., and A. J. Clutterbuck (1994). Genetic regulation of conidiation. In Martinelli, S. D., and J. R. Kinghorn (eds.), *Aspergillus: 50 Years On*. Elsevier, London, pp. 383–427.

Timoféeff-Ressovsky, N., K. G. Zimmer, M. Delbrück (1935) Über die Natur der Genmutation und der Genstruktur. Nachr. Ges. Wiss. Göttingen. Math.-phys. Kl. 6:190–245.

Togasaki, R. K., and S. J. Surzycki (1997) Perspectives on early research on photosynthesis in Chlamydomonas. In Rochaix, J.-D., M. Goldschmidt-Clermont, and S. Merchant (eds.), *The Molecular Biology of Chloropasts and Mitochondria in Chlamydomonas*. Kluwer Academic, Boston, MA, pp. 13–23.

Tsukii, Y. (1988) Mating type inheritance. In Görtz, H.-D. (ed.), *Paramecium*. Springer-Verlag, Berlin, pp. 69–79.

Uetz, P., L. Giot, G. Cagney, T. A. Mansfield, R. S. Judson, J. R. Knight, D. Lockshon, V. Narayan, M. Srinivasan, P. Pochart, A. Qureshi-Emili, Y. Li, B. Godwin, D. Conover, T. Kalbfleisch, G. Vijayadamodar, M. Yang, M. Johnston, S. Fields, and J. M. Rothberg (2000) A comprehensive analysis of protein-protein interactions in *Saccharomyces cerevisiae*. Nature 403:623–627.

Umbarger, H. E. (1961) Feedback control by endproduct inhibition. Cold Spring Harbor Symp. Quant. Biol. 26:301–312.

Volff, J.-N., and J. Altenbuchner. 2000. A new beginning with new ends: Linearisation of circular chromosomes during bacterial evolution. FEMS Microbiol. Lett. 186: 143–150.

von Borstel, R. C. (1993) The international yeast community. In Hall, M. N. and P. Linder (eds.), *The Early Days of Yeast Genetics*. Cold Spring Harbor Laboratory Press, Cold Spring Harbor, NY, pp 453–460.

Wada, Y., and Y. Anraku (1994) Chemiosmotic coupling of ion-transport in the yeast vacuole—its role in acidification inside organelles. J. Bioenerget. Biomemb. 26:631–637.

Watson, J. D. (1968) *The Double Helix*. Atheneum, New York.

———— (2000) *A Passion for DNA. Genes, Genomes, and Societies*. Cold Spring Harbor Laboratory Press, Cold Spring Harbor, NY.

Watson, J. D., and F. H. C. Crick (1953a) A structure for deoxyribose nucleic acid. Nature 171:737–738.

———— (1953b) Genetical implications of the structure of deoxyribonucleic acid. Nature 171:964–967.

Weaver, W. (1970) Molecular biology: Origin of the term. Science 170:582–583.

Wiemken, A., and M. Dürr (1974) Characterization of amino acid pools in the vacuolar compartment of *Saccharomyces cerevisiae*. Arch. Microbiol. 101:45–57.

Wilkie, D. (1970) Analysis of mitochondrial drug resistance in *Saccharomyces cerevisiae*. Symp. Soc. Exptl. Biol. 24:71–83.

Williams, L. J., G. R. Barnett, J. L. Ristow, J. Pitkin, M. Perriere, and R. H. Davis (1992) Ornithine decarbozylase gene of *Neurospora crassa*: Isolation, sequence, and polyamine-mediated regulation of its mRNA. Mol. Cell. Biol. 12:347–359.

Williamson, D. (2002) The curious history of yeast mitochondrial DNA. Nature Rev. Genet. 3:475–481.

Winge, Ø., and O. Laustsen (1940) On a cytoplasmic effect of inbreeding in homogeneous yeast. C. R. Acad. Carlsberg 23:17–40.

Winge, Ø., and C. Roberts (1950) Non-Mendelian segregation from heterozygotic yeast asci. Nature 165: 157–158.

Winzeler, E. A., D. D. Shoemaker, A. Astromoff, et al. (1999) Functional characterization of the *S. cerevisiae* genome by gene deletion and parallel analysis. Science 285:901–906.

Woese, C. R. (1977) Phylogenetic structure of the prokaryotic domain: the primary kingdoms. Proc. Natl. Acad. Sci. USA 74:5088–5090.

———— (1987) Bacterial evolution. Microbiol. Rev. 51:221–271.

Wollman, E., F. Jacob, and W. Hayes (1956) Conjugation and genetic recombination in *Escherichia coli* K-12. Cold Spring Harbor Symp. Quant. Biol. 21:141–162.

Wollman, E., and E. Wollman (1939) Les "phases" de la fonction lysogène. Action successive du lysozyme et de la trypsine sur un germe lysogène. C. R. Soc. Biol. 131:442–445.

Woodward, V. W., J. R. de Zeeuw, and A. M. Srb (1954) Selection of particular biochemical mutants in Neurospora from differentially germinated conidial filtrates. Proc. Natl. Acad Sci. USA 40:192–200.

Yanofsky, C. (1956) The enzymatic conversion of anthranilic acid to indole. J. Biol. Chem. 223:171–184.

———— (2001) Advancing our knowledge of biochemistry, genetics, and microbiology through studies on tryptophan metabolism. Annu. Rev. Biochem. 70:1–37.

Yao, M.-C. (1996) Programmed DNA deletions in *Tetrahymena*: Mechanisms and implications. Trends Genet. 12:26–30.

Yeadon, P. J., J. P. Rasmussen, and D. E. A. Catcheside (2001) Recombination events in *Neurospora crassa* may cross a translocation breakpoint by a template-switching mechanism. Genetics 159:571–579.

Yoxen, E. J. (1979) Where does Schrödinger's "What Is Life?" belong in the history of molecular biology? Hist. Sci. 17:17–52.

Ziman, J. (2000) *Real Science. What It Is, and What It Means.* Cambridge Univ. Press, Cambridge.

Zimmer, K. G. (1966) The target theory. In Cairns, J., G. S. Stent, and J. D. Watson (eds.), *Phage and the Origins of Molecular Biology.* Cold Spring Harbor Laboratory Press, Cold Spring Harbor, NY, pp. 33–42.

Zinder, N. D. (1960) Sexuality and mating in Salmonella. Science 131:924–926.

———— (1992) Forty years ago: The discovery of bacterial transduction. Genetics 132:291–294.

Zinder, N. D., and J. Lederberg (1952) Genetic exchange in *Salmonella.* J. Bacteriol. 64:679–699.

Zuckerkandl, E., and L. Pauling (1965) Evolutionary divergence and convergence in proteins. In Bryson, V., and H. J. Vogel (eds.), *Evolving Genes and Proteins.* Academic Press, New York, pp. 97–166.

Zuckerman, H., and J. Lederberg (1986) Forty years of genetic recombination in bacteria. Postmature scientific discovery? Nature 324:629–631.

Name Index

Subject Index

Owing to their ubiquity in the text, the following terms have no main entries: *Aspergillus nidulans, Chlamydomonas reinhardtii, Drosophila melanogaster, Escherichia coli, Neurospora crassa, Paramecium spp.*, phage λ, *Saccharomyces cerevisiae*, T phage, yeast.

The brief appendices have not been indexed. A number of technical terms and concepts may be found there if not entered below.

ghosts of, 110
growth of, 104–8, 110–11
use of by Ellis and Delbrück, 104
pheromones (*see also* hormones), 24,
    36, 189
photosynthesis, genetics of, 193
Phycomycetes, 73, 226
phylogenetic relationships, 6, 251
plaques, 99, 105, 106, 110, 114
plasmagenes, 173, 175, 177–9
plasmids, 159. *See also* F plasmid
    mitochondrial, 222
    use of in filamentous fungi, 223
    use of in yeast, 202–4
Plasmon, 166–7, 177
plastids. *See* chloroplast inheritance
pleomorphism, 82–3
Pneumococcus, 86, 90, 111
Pneumocystis carinii, 74
Podospora, 213, 218
point mutation. *See* mutation, point
polymorphism, molecular, 222, 240
    in natural populations, 241–2, 256
polypeptide. *See also* proteins
    colinearity of with DNA, 128
    definition of, 42
    synthesis of, 133–8, 235–6
polyploidy, yeast, 62, 70
population genetics, 12, 22, 199, 221–2,
    237, 252
positive control, 157, 206–7, 219
prokaryotes, definition of, 84
prophage, 113, 114–15, 119, 122,
    142, 148. *See also* phage,
    temperate
proteins. *See also* polypeptide
    development of knowledge of, 229–
    30, 231–2
    secretion and targeting of, 159–60,
    209–10
    structure of, 125, 128, 235
    synthesis of, 133–8, 206–7, 235–6
protoperithecium, 30–1
prototroph, definition of, 65
pseudoallelism, 56–60, 70, 124, 130

rII region
    cis-trans tests in, 131–2
    complementation tests in, 132–3
    in defining allelism, 131–3

in defining genetic code properties,
    133–8
mutants of, 131–3, 133–8
radiation inactivation, 70
rapid-lysis mutants. *See rII* region,
    mutants of
reading frame, 137, 211
recombination. *See also* chromosome;
    crossing over; gene conversion;
    linkage maps; pseudoallelism;
    tetrad analysis
    in *Aspergillus*, 50–51, 53, 124
    in *Chlamydomonas*, 193, 194–6
    in *Drosophila*, 11, 16, 19, 21, 24, 26,
    in *Escherichia coli*, 50, 92, 118, 122,
    128, 142–4
    at high resolution, 45, 53, 118, 128,
    131–3
    in λ phage, 151–2
    mitotic, 54–5
    molecular mechanisms of, 151–2, 218
    in *Neurospora*, 32–4, 39, 44–5
    in organelle genomes 188, 193, 194–
    6, 196–8
    in *Saccharomyces*, 65, 70–71, 188,
    196–8, 205, 218
    in T phage, 50, 110, 130, 151–2
reductionism, 47, 81, 162, 163, 225,
    229, 238, 249, 254, 256–8
    defined, 13, 258, 286 n. 19
redundancy, 162
regulation, enzyme. *See* feedback
    inhibition; induction, enzyme;
    operon model; repression, enzyme;
    repressor protein; transcription,
    control of
regulatory cascade, 210, 226
relative sexuality, 74, 188–9
replica plating, 61, 78, 95, 127, 192
replication, semiconservative, 144
repression, enzyme, 148–51. *See also*
    induction; operon model; repressor
    protein; *lac* repressor;
    transcription, control of
repressor protein
    action of, 153
    of *lac* system (*see lac* repressor)
    nature of, 152
    of phage lambda, 113, 114, 115, 119–
    20, 122, 149, 152–3